Springer Finance

Robert J. Elliott and P. Ekkehard Kopp

Mathematics of Financial Markets

Second edition

Robert J. Elliott
Haskayne School of Business
University of Calgary
Calgary, Alberta
Canada T2N 1N4
robert.elliott@haskayne.ucalgary.ca

P. Ekkehard Kopp
Department of Mathematics
University of Hull
Hull HU6 7RX
Yorkshire
United Kingdom
p.e.kopp@hull.ac.uk

With 7 figures.

ISBN 978-1-4419-1942-7   e-ISBN 978-0-387-22640-8

Elliott, Robert J. (Robert James), 1940–
    Mathematics of financial markets / Robert J. Elliott and P. Ekkehard Kopp.—2nd ed.
    p. cm. — (Springer finance)
    Includes bibliographical references and index.

    1. Investments—Mathematics.   2. Stochastic analysis.   3. Options
(Finance)—Mathematical models.   4. Securities—Prices—Mathematical models.
I. Kopp, P. E., 1944–   II. Title.   III. Series.
HG4515.3.E37   2004
332.6′01′51—dc22

Printed on acid-free paper.

(EB)

9 8 7 6 5 4 3 2 1

springeronline.com

Preface

This work is aimed at an audience with a sound mathematical background wishing to learn about the rapidly expanding field of mathematical finance. Its content is suitable particularly for graduate students in mathematics who have a background in measure theory and probability.

The emphasis throughout is on developing the mathematical concepts required for the theory within the context of their application. No attempt is made to cover the bewildering variety of novel (or 'exotic') financial instruments that now appear on the derivatives markets; the focus throughout remains on a rigorous development of the more basic options that lie at the heart of the remarkable range of current applications of martingale theory to financial markets.

The first five chapters present the theory in a discrete-time framework. Stochastic calculus is not required, and this material should be accessible to anyone familiar with elementary probability theory and linear algebra.

The basic idea of pricing by arbitrage (or, rather, by non-arbitrage) is presented in Chapter 1. The unique price for a European option in a single-period binomial model is given and then extended to multi-period binomial models. Chapter 2 introduces the idea of a martingale measure for price processes. Following a discussion of the use of self-financing trading strategies to hedge against trading risk, it is shown how options can be priced using an equivalent measure for which the discounted price process is a martingale. This is illustrated for the simple binomial Cox-Ross-Rubinstein pricing models, and the Black-Scholes formula is derived as the limit of the prices obtained for such models. Chapter 3 gives the 'fundamental theorem of asset pricing', which states that if the market does not contain arbitrage opportunities there is an equivalent martingale measure. Explicit constructions of such measures are given in the setting of finite market models. Completeness of markets is investigated in Chapter 4; in a complete market, every contingent claim can be generated by an admissible self-financing strategy (and the martingale measure is unique). Stopping times, martingale convergence results, and American options are discussed in a discrete-time framework in Chapter 5.

The second five chapters of the book give the theory in continuous time. This begins in Chapter 6 with a review of the stochastic calculus. Stopping times, Brownian motion, stochastic integrals, and the Itô differentiation rule are all defined and discussed, and properties of stochastic differential equations developed.

The continuous-time pricing of European options is developed in Chapter 7. Girsanov's theorem and martingale representation results are developed, and the Black-Scholes formula derived. Optimal stopping results are applied in Chapter 8 to a thorough study of the pricing of American options, particularly the American put option.

Chapter 9 considers selected results on term structure models, forward and future prices, and change of numéraire, while Chapter 10 presents the basic framework for the study of investment and consumption problems.

Acknowledgments   Sections of the book have been presented in courses at the Universities of Adelaide and Alberta. The text has consequently benefited from subsequent comments and criticism. Our particular thanks go to Monique Jeanblanc-Piqué, whose careful reading of the text and valuable comments led to many improvements. Many thanks are also due to Volker Wellmann for reading much of the text and for his patient work in producing consistent TeX files and the illustrations.

Finally, the authors wish to express their sincere thanks to the Social Sciences and Humanities Research Council of Canada for its financial support of this project.

Edmonton, Alberta, Canada   Robert J. Elliott
Hull, United Kingdom   P. Ekkehard Kopp

Preface to the Second Edition

This second, revised edition contains a significant number of changes and additions to the original text. We were guided in our choices by the comments of a number of readers and reviewers as well as instructors using the text with graduate classes, and we are grateful to them for their advice. Any errors that remain are of course entirely our responsibility.

In the five years since the book was first published, the subject has continued to grow at an astonishing rate. Graduate courses in mathematical finance have expanded from their business school origins to become standard fare in many mathematics departments in Europe and North America and are spreading rapidly elsewhere, attracting large numbers of students. Texts for this market have multiplied, as the rapid growth of the *Springer Finance* series testifies. In choosing new material, we have therefore focused on topics that aid the student's understanding of the fundamental concepts, while ensuring that the techniques and ideas presented remain up to date. We have given particular attention, in part through revisions to Chapters 5 and 6, to linking key ideas occurring in the two main sections (discrete- and continuous-time derivatives) more closely and explicitly.

Chapter 1 has been revised to include a discussion of risk and return in the one-step binomial model (which is given a new, extended presentation) and this is complemented by a similar treatment of the Black-Scholes model in Chapter 7. Discussion of elementary bounds for option prices in Chapter 1 is linked to sensitivity analysis of the Black-Scholes price (the 'Greeks') in Chapter 7, and call-put parity is utilised in various settings.

Chapter 2 includes new sections on superhedging and the use of extended trading strategies that include contingent claims, as well as a more elegant derivation of the Black-Scholes option price as a limit of binomial approximants.

Chapter 3 includes a substantial new section leading to a complete proof of the equivalence, for discrete-time models, of the no-arbitrage condition and the existence of equivalent martingale measures. The proof, while not original, is hopefully more accessible than others in the literature.

This material leads in Chapter 4 to a characterisation of the arbitrage interval for general market models and thus to a characterisation of complete models, showing in particular that complete models must be finitely generated.

The new edition ends with a new chapter on risk measures, a subject that has become a major area of research in the past five years. We include a brief introduction to Value at Risk and give reasons why the use of coherent risk measures (or their more recent variant, deviation measures) is to be preferred. Chapter 11 ends with an outline of the use of risk measures in recent work on partial hedging of contingent claims.

The changes we have made to the text have been informed by our continuing experience in teaching graduate courses at the universities of Adelaide, Calgary and Hull, and at the African Institute for Mathematical Sciences in Cape Town.

Acknowledgments   Particular thanks are due to Alet Roux (Hull) and Andrew Royal (Calgary) who provided invaluable assistance with the complexities of LaTeX typesetting and who read large sections of the text. Thanks are also due to the Social Sciences and Humanities Research Council of Canada for continuing financial support.

Calgary, Alberta, Canada   Robert J. Elliott
Hull, United Kingdom   P. Ekkehard Kopp
May 2004

Contents

Chapter 1

Pricing by Arbitrage

1.1   Introduction: Pricing and Hedging

The 'unreasonable effectiveness' of mathematics is evidenced by the frequency with which mathematical techniques that were developed without thought for practical applications find unexpected new domains of applicability in various spheres of life. This phenomenon has customarily been observed in the physical sciences; in the social sciences its impact has perhaps been less evident. One of the more remarkable examples of simultaneous revolutions in economic theory and market practice is provided by the opening of the world's first options exchange in Chicago in 1973, and the ground-breaking theoretical papers on preference-free option pricing by Black and Scholes [27] (quickly extended by Merton [222]) that appeared in the same year, thus providing a workable model for the 'rational' market pricing of traded options.

From these beginnings, financial derivatives markets worldwide have become one of the most remarkable growth industries and now constitute a major source of employment for graduates with high levels of mathematical expertise. The principal reason for this phenomenon has its origins in the simultaneous stimuli just described, and the explosive growth of these secondary markets (whose levels of activity now frequently exceed the underlying markets on which their products are based) continues unabated, with total trading volume now measured in trillions of dollars. The variety and complexity of new financial instruments is often bewildering, and much effort goes into the analysis of the (ever more complex) mathematical models on which their existence is predicated.

In this book,we present the necessary mathematics, within the context of this field of application, as simply as possible in an attempt to dispel some of the mystique that has come to surround these models and at the same time to exhibit the essential structure and robustness of the underlying theory. Since making choices and decisions under conditions

1

of *uncertainty* about their outcomes is inherent in all market trading, the area of mathematics that finds the most natural applications in finance theory is the modern theory of *probability and stochastic processes*, which has itself undergone spectacular growth in the past five decades. Given our current preoccupations, it seems entirely appropriate that the origins of probability, as well as much of its current motivation, lie in one of the earliest and most pervasive indicators of 'civilised' behaviour: *gambling*.

Contingent Claims

A *contingent claim* represents the potential liability inherent in a *derivative security*; that is, in an asset whose value is determined by the values of one or more underlying variables (usually securities themselves). The analysis of such claims, and their *pricing* in particular, forms a large part of the modern theory of finance. Decisions about the prices appropriate for such claims are made contingent on the price behaviour of these underlying securities (often simply referred to as the *underlying*), and the theory of derivatives markets is primarily concerned with these relationships rather than with the economic fundamentals that determine the prices of the underlying.

While the construction of mathematical models for this analysis often involves very sophisticated mathematical ideas, the *economic* insights that underlie the modelling are often remarkably simple and transparent. In order to highlight these insights we first develop rather simplistic mathematical models based on discrete time (and, frequently, finitely generated probability spaces) before showing how the analogous concepts can be used in the more widely known continuous models based on diffusions and *Itô processes*. For the same reason, we do not attempt to survey the range of contingent claims now traded in the financial markets but concentrate on the more basic *stock options* before attempting to discuss only a small sample of the multitude of more recent, and often highly complex, financial instruments that finance houses place on the markets in ever greater quantities.

Before commencing the mathematical analysis of market models and the options based upon them, we outline the principal features of the main types of financial instruments and the conditions under which they are currently traded in order to have a benchmark for the mathematical idealisations that characterise our modelling. We briefly consider the role of *forwards, futures, swaps,* and *options*.

Forward Contracts A *forward contract* is simply an agreement to buy or sell a specified asset S at a certain future time T for a price K that is specified now (which we take to be time 0). Such contracts are not normally traded on exchanges but are agreements reached between two sophisticated institutions, usually between a financial institution such as a bank and one of its corporate clients. The purpose is to share risk: one party assumes a *long position* by agreeing to buy the asset, and the other takes a *short position* by agreeing to sell the asset for the *delivery price* K at the *delivery date* T. Initially neither party incurs any costs in entering into the contract, and the *forward price* of the contract at time $t \in [0, T]$ is the delivery price that would give the contract zero value. Thus, at time 0, the forward price is K, but at later times movement in the market value of the underlying commodity will suggest different values. The *payoff* to the holder of the long position at time T is simply $S_T - K$, and for the short position it is $K - S_T$. Thus, since both parties are obliged to honour the contract, in general one will lose and the other gain the same amount.

Trading in forwards is not closely regulated, and the market participant bears the risk that the other party may default-the instruments are not traded on an exchange but 'over-the-counter' (OTC) worldwide, usually by electronic means. There are no price limits (as could be set by exchanges), and the object of the transaction is delivery; that is, the contracts are not usually 'sold on' to third parties. Thus the problem of determining a 'fair' or rational price, as determined by the collective judgement of the market makers or by theoretical modelling, appears complicated.

Intuitively, averaging over the possible future values of the asset may seem to offer a plausible approach. That this fails can be seen in a simple one-period example where the asset takes only two future values.

Example 1.1.1. Suppose that the current (time 0) value of the stock is $100 and the value at time 1 is $120 with probability $p = \frac{3}{4}$ and $80 with probability $1 - p = \frac{1}{4}$. Suppose the riskless interest rate is $r = 5\%$ over the time period. A contract price of $\frac{3}{4} \times \$120 + \frac{1}{4} \times \$80 = \$110$ produces a 10% return for the seller, which is greater than the riskless return, while $p = \frac{1}{2}$ would suggest a price of $100, yielding a riskless benefit for the buyer.

This suggests that we should look for a pricing mechanism that is independent of the probabilities that investors may attach to the different future values of the asset and indeed is independent of those values themselves.

The simple assumption that investors will always prefer having more to having less (this is what constitutes 'rational behaviour' in the markets) already allows us to price a forward contract that provides no dividends or other income. Let S_t be the *spot price* of the underlying asset S (i.e., its price at time $t \in [0, T]$); then the forward price $F(t, T)$ at that time is simply the value at the time T of a riskless investment of S_t made at time t whose value increases at a constant riskless interest rate $r > 0$. Under continuous compounding at this rate, an amount of money M_s in the bank will grow exponentially according to

$$\frac{dM_s}{M_s} = rds, s \in [t, T].$$

To repay the loan S_t taken out at t, we thus need $M_T = S_t e^{r(T-t)}$ by time T.

We therefore claim that

$$F(t,T) = S_t e^{r(T-t)} \text{ for } t \in [0,T].$$

To see this, consider the alternatives. If the forward price is higher, we can borrow S_t for the interval $[t,T]$ at rate r, buy the asset, and take a short position in the forward contract. At time T, we need $S_t e^{r(T-t)}$ to repay our loan but will realise the higher forward price from the forward contract and thus make a riskless profit. For $F(t,T) < S_t e^{r(T-t)}$, we can similarly make a sure gain by shorting the asset (i.e., 'borrowing' it from someone else's account, a service that brokers will provide subject to various market regulations) and taking a long position in the contract. Thus, simple 'arbitrage' considerations (in other words, that we cannot expect riskless profits, or a 'free lunch') lead to a definite forward price at each time t.

Forward contracts can be used for reducing risk (*hedging*). For example, large corporations regularly face the risk of currency fluctuations and may be willing to pay a price for greater certainty. A company facing the need to make a large fixed payment in a foreign currency at a fixed future date may choose to enter into a forward contract with a bank to fix the rate now in order to lock in the exchange rate. The bank, on the other hand, is acting as a *speculator* since it will benefit from an exchange rate fluctuation that leaves the foreign currency below the value fixed today. Equally, a company may speculate on the exchange rate going up more than the bank predicts and take a long position in a forward contract to lock in that potential advantage-while taking the risk of losses if this prediction fails. In essence, it is betting on future movements in the asset. The advantage over actual purchase of the currency now is that the forward contract involves no cost at time 0 and only potential cost if the gamble does not pay off. In practice, financial institutions will demand a small proportion of the funds as a deposit to guard against default risk; nonetheless, the *gearing* involved in this form of trading is considerable.

Both types of traders, *hedgers* and *speculators*, are thus required for forward markets to operate. A third group, *arbitrageurs*, typically enter two or more markets simultaneously, trying to exploit local or temporary disequilibria (i.e., *mispricing* of certain assets) in order to lock in riskless profits. The fundamental economic assumption that (ideal) markets operate in equilibrium makes this a hazardous undertaking requiring rapid judgements (and hence well-developed underlying mathematical models) for sustained success-their existence means that assets do not remain mispriced for long or by large amounts. Thus it is reasonable to build models and calculate derivative prices that are based on the assumption of the absence of arbitrage, and this is our general approach.

Futures Contracts Futures contracts involve the same agreement to trade an asset at a future time at a certain price, but the trading takes place on an exchange and is subject to regulation. The parties need not know each other, so the exchange needs to bear any default risk-hence the contract requires standardised features, such as daily settlement arrangements known as *marking to market*. The investor is required to pay an initial deposit, and this *initial margin* is adjusted daily to reflect gains and losses since the *futures price* is determined on the floor of the exchange by demand and supply considerations. The price is thus paid over the life of the contract in a series of instalments that enable the exchange to balance long and short positions and minimise its exposure to default risk. Futures contracts often involve commodities whose quality cannot be determined with certainty in advance, such as cotton, sugar, or coffee, and the delivery price thus has reference points that guarantee that the asset quality falls between agreed limits, as well as specifying contract size.

The largest commodity futures exchange is the Chicago Board of Trade, but there are many different exchanges trading in futures around the world; increasingly, financial futures have become a major feature of many such markets. Futures contracts are written on stock indices, on currencies, and especially on movements in interest rates. Treasury bills and Eurodollar futures are among the most common instruments.

Futures contracts are traded heavily, and only a small proportion are actually delivered before being sold on to other parties. Prices are known publicly and so the transactions conducted will be at the best price available at that time. We consider futures contracts in Chapter 9, but only in the context of interest rate models.

Swaps A more recent development, dating from 1981, is the exchange of future cash flows between two partners according to agreed prior criteria that depend on the values of certain underlying assets. Swaps can thus be thought of as portfolios of forward contracts, and the initial value as well as the final value of the swap is zero. The cash flows to be exchanged may depend on interest rates. In the simplest example (a *plain vanilla* interest rate swap), one party agrees to pay the other cash flows equal to interest at a fixed rate on a notional principal at each payment date. The other party agrees to pay interest on the same notional principal and in the same currency, but the cash flow is based on a floating interest rate. Thus the swap transforms a floating rate loan into a fixed rate one and vice versa. The floating rate used is often LIBOR (the London Interbank Offer Rate), which determines the interest rate used by banks on deposits from other banks in Eurocurrency markets; it is quoted on deposits of varying duration-one month, three months, and so on. LIBOR operates as a reference rate for international markets: three-month LIBOR is the rate underlying Eurodollar futures contracts, for example.

There is now a vast range of swap contracts available, with *currency swaps* (whereby the loan exchange uses fixed interest rate payments on loans in different currencies) among the most heavily traded. We do not study swaps in this book; see [232] or [305] for detailed discussions. The latter text focuses on options that have derivative securities, such as forwards, futures, or swaps, as their underlying assets; in general, such instruments are known as *exotics*.

Options An *option* on a stock is a contract giving the owner the right, but not the obligation, to trade a given number of shares of a common stock for a fixed price at a future date (the *expiry date T*). A *call* option gives the owner the right to buy stocks, and a *put* option confers the right to sell, at the fixed *strike price K*. The option is *European* if it can only be exercised at the fixed expiry date T. The option is *American* if the owner can exercise his right to trade at any time up to the expiry date. Options are the principal financial instruments discussed in this book.

In Figures 1.1 and 1.2, we draw the simple graphs that illustrate the *payoff* function of each of these options. In every transaction there are two parties, the *buyer* and the seller, more usually termed the *writer*, of the option. In the case of a European call option on a stock $(S_t)_{t \in \mathbb{T}}$ with strike price K at time T, the payoff equals $S_T - K$ if $S_T > K$ and 0 otherwise. The payoff for the writer of the option must balance this quantity; that is, it should equal $K - S_T$ if $S_T < K$ and 0 otherwise. The option writer must honour the contract if the buyer decides to exercise his option at time T.

Fair Prices and Hedge Portfolios

The problem of *option pricing* is to determine what value to assign to the option at a given time (e.g. at time 0). It is clear that a trader can make a riskless profit (at least in the absence of inflation) unless she has paid an 'entry fee' that allows her the chance of exercising the option favourably at the expiry date. On the other hand, if this 'fee' is too high, and the stock price seems likely to remain close to the strike price, then no sensible trader would buy the option for this fee. As we saw previously, operating on a set $\mathbb{T}$ of possible *trading dates* (which may typically be a finite set of natural numbers of the form $\{0, 1, \ldots, T\}$, or, alternatively, a finite interval $[0, T]$ on the real line), the buyer of a European call option on a stock with price process $(S_t)_{t \in \mathbb{T}}$ will have the opportunity of receiving a payoff at time T of $C(t) = \max\{S_T - K, 0\}$, since he will exercise the option if, and only if, the final price of the stock S_T is greater than the previously agreed strike price K.

With the call option price set at C_0, we can draw the graph of the gain (or loss) in the transaction for both the buyer and writer of the option. Initially we assume for simplicity that the riskless interest rate is 0 (the 'value of money' remains constant); in the next subsection we shall drop this assumption, and then account must be taken of the rate at which money held in a savings account would accumulate. For example, with continuous compounding over the interval $\mathbb{T} = [0, T]$, the price C_0 paid for the option at time 0 would be worth $C_0 e^{rT}$ by time T. With the rate $r = 0$,

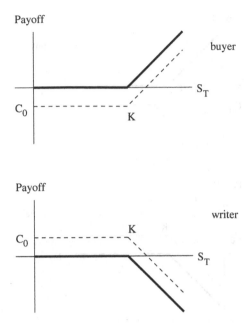

Figure 1.1: Payoff and gain for European call option

the buyer's gain from the call option will be $S_T - K - C_0$ if $S_T > K$ and $-C_0$ if $S_T \leq K$. The writer's gain is given by $K - S_T + C_0$ if $S_T > K$ and C_0 if $S_T \leq K$. Similar arguments hold for the buyer and writer of a European put option with strike K and option price P_0. The payoff and gain graphs are given in Figures 1.1 and 1.2.

Determining the option price entails an assessment of a price to which both parties would logically agree. One way of describing the *fair price* for the option is as the *current* value of a portfolio that will yield exactly the same return as does the option by time T. Strictly, this price is fair only for the writer of the option, who can calculate the fair price as the smallest initial investment that would allow him to *replicate* the value of the option throughout the time set $\mathbb{T}$ by means of a portfolio consisting of stock and a riskless *bond* (or savings account) alone. The buyer, on the other hand, will want to cover any potential losses by borrowing the amount required to buy the option (the buyer's option price) and to invest in the market in order to reduce this liability, so that at time T the option payoff at least covers the loan. In general, the buyer's and seller's option prices will not coincide-it is a feature of *complete market models*, which form the main topic of interest in this book, that they do coincide, so that it becomes possible to refer to *the* fair price of the option. Our first problem is to determine this price uniquely.

When option replication is possible, the replicating portfolio can be

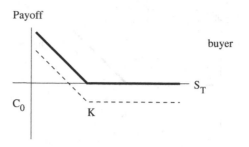

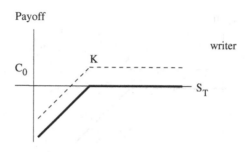

Figure 1.2: Payoff and gain for European put option

used to offset, or *hedge*, the risk inherent in writing the option; that is, the risk that the writer of the option may have to sell the share S_T for the fixed price K even though, with small probability, S_T may be much larger than K. Our second problem is therefore to construct such a *hedge portfolio*.

Call-Put Parity

Our basic market assumption enables us to concentrate our attention on call options alone. Once we have dealt with these, the solutions of the corresponding problems for the European put option can be read off at once from those for the call option. The crucial assumption that ensures this is that our market model rules out *arbitrage*; that is, no investor should be able to make riskless profits, in a sense that we will shortly make more precise. This assumption is basic to option pricing theory since there can be no market *equilibrium* otherwise. It can be argued that the very existence of 'arbitrageurs' in real markets justifies this assumption: their presence ensures that markets will quickly adjust prices so as to eliminate disequilibrium and hence will move to eliminate arbitrage.

So let C_t (resp. P_t) be the value at time t of the European call (resp. put) option on the stock $(S_t)_{t \in \mathbb{T}}$. Writing

$$x^+ = \begin{cases} x & \text{if } x > 0 \\ 0 & \text{if } x \leq 0 \end{cases},$$

we can write the payoff of the European call as $(S_T - K)^+$ and that of the corresponding put option as $(K - S_T)^+$.

It is obvious from these definitions that, at the expiry date T, we have

$$C_T - P_T = (S_T - K)^+ - (K - S_T)^+ = S_T - K. \qquad (1.1)$$

Assume now that a constant interest rate $r > 0$ applies throughout $\mathbb{T} = [0, T]$. With continuous compounding, a sum X deposited in the bank (or money-market account) at time $t < T$ accumulates to $Xe^{r(T-t)}$ by time T. Hence a cash sum of K, needed at time T, can be obtained by depositing $Ke^{-r(T-t)}$ at time t.

We claim that, in order to avoid arbitrage, the call and put prices on our stock S must satisfy (1.1) at all times $t < T$, with the appropriate discounting of the cash sum K; i.e.,

$$C_t - P_t = S_t - e^{-r(T-t)}K \text{ for all } t \in \mathbb{T}. \qquad (1.2)$$

To see this, compare the following 'portfolios':

(i) Buy a call and sell a put, each with strike K and horizon T. The fair price we should pay is $C_t - P_t$.

(ii) Buy one share at price S_t and borrow $e^{-r(T-t)}K$ from the bank. The net cost is $S_t - e^{-r(T-t)}K$.

The value of these portfolios at time T is the same since the first option yields $C_T - P_T = S_T - K$, while the net worth of the second portfolio at that time is also $S_T - K$. Hence, if these two portfolios did *not* have the same value at time t, we could make a riskless profit over the time interval $[t, T]$ by simultaneously taking a long position in one and a short position in the other. Equation (1.2) follows.

Exercise 1.1.2. Give an alternative proof of (1.2) by considering the possible outcomes at time T of the following trades made at time $t < T$: buy a call and write a put on S, each with strike K, and sell one share of the stock. Deposit the net proceeds in the bank account at constant riskless interest rate $r > 0$. Show that if (1.2) fails, these transactions will always provide a riskless profit for one of the trading partners.

More generally, the relation

$$C_t - P_t = S_t - \beta_{t,T}K \text{ for all } t \in \mathbb{T} \qquad (1.3)$$

holds, where $\beta_{t,T}$ represents the discount at the riskless rate over the interval $[t, T]$. In our examples, with r constant, we have $\beta_{t,T} = \beta^{T-t} = e^{-r(T-t)}$ in the continuous case and $\beta_{t,T} = \beta^{T-t} = (1 + r)^{-(T-t)}$ in the discrete case.

1.2   Single-Period Option Pricing Models

Risk-Neutral Probability Assignments

In our first examples, we restrict attention to markets with a single trading period, so that the time set $\mathbb{T}$ contains only the two trading dates 0 and T. The mathematical tools needed for contingent claim analysis are those of probability theory: in the absence of complete information about the time evolution of the risky asset $(S_t)_{t \in \mathbb{T}}$ it is natural to model its value at some future date T as a random variable defined on some probability space $(\Omega, \mathcal{F}, P)$. Similarly, any contingent claim H that can be expressed as a function of S_T or, more generally, a function of $(S_t)_{t \in \mathbb{T}}$, is a non-negative random variable on $(\Omega, \mathcal{F}, P)$.

The probabilistic formulation of option prices allows us to attack the problem of finding the fair price H_0 of the option in a different way: since we do not know in advance what value S_T will take, it seems logical to estimate H by $E(\beta H)$ using the discount factor β; that is, we estimate H by its average discounted value. (Here $E(\cdot) = E_P(\cdot)$ denotes expectation relative to the probability measure P.)

This averaging technique has been known for centuries and is termed the 'principle of equivalence' in actuarial theory; there it reflects the principle that, on average, the (uncertain) discounted future benefits should be equal in value to the present outlay. We are left, however, with a crucial decision: how do we determine the probability measure? At first sight it is not clear that there is a 'natural' choice at all; it seems that the probability measure (i.e., the assignment of probabilities to every possible event) must depend on investors' risk preferences.

However, in particular situations, one can obtain a 'preference-free' version of the option price: the theory that has grown out of the mathematical modelling initiated by the work of Black and Scholes [27] provides a framework in which there *is* a natural choice of measure, namely a measure under which the (discounted) price process is a *martingale*. Economically, this corresponds to a market in which the investors' probability assignments show them to be 'risk-neutral' in a sense made more precise later. Although this framework depends on some rather restrictive conditions, it provides a firm basis for mathematical modelling as well as being a test bed for more 'economically realistic' market models. To motivate the choice of the particular models currently employed in practice, we first consider a simple numerical example.

Example 1.2.1. We illustrate the connection between the 'fair price' of a claim and a replicating (or 'hedge') portfolio that mimics the value of the claim. For simplicity, we again set the discount factor $\beta \equiv 1$; that is, the riskless interest rate (or 'inflator') r is set at 0. The only trading dates are 0 and 1, so that any portfolio fixed at time 0 is held until time 1. Suppose a stock S has price 10 (dollars, say) at time 0, and takes one of only two possible values at time 1:

$$S_1 = \begin{cases} 20 & \text{with probability } p \\ 7.5 & \text{with probability } 1-p \end{cases}.$$

Consider a European call option $H = (S_1 - K)^+$ with strike price $K = 15$ written on the stock. At time 1, the option H yields a profit of \$5 if $S_1 = 20$ and \$0 otherwise. The probability assignment is $(p, 1-p)$, which, in general, depends on the investor's attitude toward risk: an inaccurate choice could mean that the investor pays more for the option than is necessary. We look for a 'risk-neutral' probability assignment $(q, 1-q)$; that is, one under which the stock price S is constant on average. Thus, if Q denotes the probability measure given by $(q, 1-q)$, then the expected value of S under Q should be constant (i.e., $E_Q(S_1) = S_0$), which we can also write as $E_Q(\Delta S) = 0$, where $\Delta S = S_1 - S_0$. (This makes S into a 'one-step martingale'.) In our example, we obtain

$$10 = 20q + 7.5(1-q),$$

so that $q = 0.2$. With the probability assignment $(0.2, 0.8)$, we then obtain the option price $\pi(H) = 5q = 1$.

To see why this price is the *unique* 'rational' one, consider the hedge portfolio approach to pricing: we attempt to replicate the final value of the option by means of a portfolio (η, θ) of cash and stock alone and determine what initial capital is needed for this portfolio to have the same time 1 value as H in all contingencies. The portfolio (η, θ) can then be used by the option writer to insure, or *hedge*, perfectly against all the risk inherent in the option.

Recall that the discount rate is 0, so that the bank account remains constant. The *value* of our portfolio is

$$V_t = \eta + \theta S_t \text{ for } t = 0, 1.$$

Here we use \$1 as our unit of cash, so that the value of cash held is simply η, while θ represents the *number* of shares of stock held during the period. Changes in the value of the portfolio are due solely to changes in the value of the stock. Hence the *gain* from trade is simply given by $G = \theta \Delta S$, and $V_1 = V_0 + G$. By the choice of the measure Q, we also have

$$V_0 = E_Q(V_0) = E_Q(V_1 - G) = E_Q(V_1) \tag{1.4}$$

since $E_Q(\theta \Delta S) = \theta E_Q(\Delta S) = 0$. To find a hedge (η, θ) replicating the option, we must solve the following equations at time 1:

$$5 = \eta + 20\theta, \qquad\qquad 0 = \eta + 7.5\theta.$$

These have the solution $\eta = -3$ and $\theta = 0.4$. Substituting into $V_0 = \eta + \theta S_0$ gives $V_0 = -3 + 0.4(10) = 1$.

The *hedging strategy* implied by the preceding situation is as follows. At time 0, sell the option in order to obtain capital of $1, and borrow $3 in order to invest the sum of $4 in shares. This buys 0.4 shares of stock. At time 1, there are two possible outcomes:

1. If $S_1 = 20$, then the option is exercised at a cost of $5; we repay the loan (cost $3) and sell the shares (gain $0.4 \times \$20 = \$8$).

   Net balance of trade: 0.

2. If $S_1 = 7.5$, then the option is not exercised (cost $0); we repay the loan (cost $3) and sell the shares, gaining $0.4 \times \$7.5 = \$3$.

   Net balance of trade: 0.

Thus, selling the option and holding the hedge portfolio exactly balance out in each case, *provided* the initial price of the option is set at $\pi(H) = 1$. It is clear that no other initial price has this property: if $\pi(H) > 1$ we can make a riskless profit by selling the option in favour of the portfolio (η, θ) and gain $(\pi(H) - 1)$, while if $\pi(H) < 1$ we simply exchange roles with the buyer in the same transaction! Moreover, since $\pi(H) = 5q = 1$, the natural (risk-neutral) probability is given by $q = 0.2$ as before.

Remark 1.2.2. This example shows that the risk-neutral valuation of the option is the unique one that prevents arbitrage profits, so that the price $\pi(H)$ will be fixed by the market in order to maintain market equilibrium. The preceding simple calculation depends crucially on the assumption that S_1 can take only *two* values at time 1: even with a three-splitting it is no longer possible, in general, to find a hedge portfolio (see Exercise 1.4.6). The underlying idea can, however, be adapted to deal with more general situations and to identify the *intrinsic risk* inherent in the particular market commodities. We illustrate this first by indicating briefly how one might construct a more general single-period model, where the investor has access to external funds and/or consumption.

1.3 A General Single-Period Model

We now generalise the hedge portfolio approach to option pricing by examining the *cost function* associated with various trading strategies and minimising its mean-square variation. Suppose that our stock price takes the (known) value S_0 at time 0 and the random value S_1 at time 1. (These are again the only trading dates in the model.) In order to express all values in terms of time-0 prices, we introduce a discount factor $\beta < 1$ and use the notation $\overline{X} = \beta X$ for any random variable X. So write $\overline{S}_1 = \beta S_1$ for the discounted value of the stock price.

The stock price S and a quite general contingent claim H are both taken to be random variables on some probability space $(\Omega, \mathcal{F}, P)$, and we wish to hedge against the obligation to honour the claim; that is, to pay out

$H(\omega)$ at time 1. (Here we are assuming that an underlying probability P is known in advance.) To this end, we build a portfolio at time 0 consisting of θ shares of stock and η_0 units of cash. The initial value of this portfolio is $V_0 = \eta_0 + \theta S_0$. We place the cash in the savings account, where it increases by a factor β^{-1} by time 1. We wish this portfolio to have value $V_1 = H$ at time 1; in discounted terms, $\overline{V}_1 = \overline{H}$.

Assuming that we have access to external funds, this can be achieved very simply by adjusting the savings account from η_0 to the value $\eta_1 = H - \theta S_1$ since this gives the portfolio value $V_1 = \theta S_1 + \eta_1 = \theta S_1 + H - \theta S_1 = H$. As H is given, it simply remains to choose the constants θ and V_0 to determine our *hedging strategy* (η, θ) completely. The *cost* of doing this can be described by the process (C_0, C_1), where $C_0 = V_0$ is the initial investment required, and $\Delta C = C_1 - C_0 = \eta_1 - \eta_0$ since the only change at time 1 was to adjust η_0 to η_1. Finally, write $\Delta \overline{X} = \beta X_1 - X_0$ for any 'process' $X = (X_0, X_1)$, in order to keep all quantities in discounted terms. From the preceding definitions, we obtain

$$\Delta \overline{C} = \beta C_1 - C_0 = \beta \eta_1 - \eta_0$$
$$= \beta(V_1 - \theta S_1) - (V_0 - \theta S_0)$$
$$= \overline{H} - (V_0 + \theta \Delta \overline{S}). \tag{1.5}$$

Equation (1.5) exhibits the discounted cost increment $\Delta \overline{C}$ simply as the difference between the discounted claim $\overline{H}$ and its approximation by *linear estimates* based on the discounted price increment $\Delta \overline{S}$. A rather natural choice of the parameters θ and V_0 is thus given by *linear regression*: the parameter values θ and V_0 that minimise the *risk function*

$$R = E\left((\Delta \overline{C})^2\right) = E\left((\overline{H} - (V_0 + \theta \Delta \overline{S}))^2\right)$$

are given by the *regression estimates*

$$\theta = \frac{\operatorname{cov}\left(\overline{H}, \Delta \overline{S}\right)}{\operatorname{var}\left(\Delta \overline{S}\right)}, \qquad V_0 = E\left(\overline{H}\right) - \theta E\left(\Delta \overline{S}\right).$$

In particular, $E\left(\Delta \overline{C}\right) = 0$, so that the *average* discounted cost remains constant at V_0. The minimal risk obtained when using this choice of the parameters is

$$R_{\min} = \operatorname{var}\left(\overline{H}\right) - \theta^2 \operatorname{var}\left(\Delta \overline{S}\right) = \operatorname{var}\left(\overline{H}\right)\left(1 - \rho^2\right),$$

where $\rho = \rho\left(\overline{H}, \overline{S}_1\right)$ is the correlation coefficient. Thus, the *intrinsic risk* of the claim H cannot be completely eliminated unless $|\rho| = 1$.

In general pricing models, therefore, we cannot expect all contingent claims to be *attainable* by some hedging strategy that eliminates all the risk-where this *is* possible, we call the model *complete*. The essential feature that distinguishes complete models is a *martingale representation property*:

it turns out that in these cases the (discounted) price process is a *basis* for a certain vector space of martingales.

The preceding discussion is of course much simplified by the fact that we have dealt with a single-period model. In the general case, this rather sophisticated approach to option pricing (due to [136]; see [134] and [268] for its further development, which we do not pursue here) can only be carried through at the expense of using quite powerful mathematical machinery. In this chapter we consider in more detail only the much simpler situation where the probabilities arise from a binomial splitting.

1.4   A Single-Period Binomial Model

We look for pricing models in which we can take $\eta_1 = \eta_0 = \eta$, that is, where there is no recourse to external funds. Recall that in the general single-period model the initial holding is

$$V_0 = \eta + \theta S_0,$$

which becomes

$$V_1 = \eta + \theta S_1 = V_0 + \theta \Delta S$$

at time 1.

Pricing

The simplest complete model has the binomial splitting of ΔS that we exploited in Example 1.2.1. We assume that the random variable S_1 takes just two values, denoted by $S_b = (1+b)S_0$ and $S_a = (1+a)S_0$, respectively, where a, b are real numbers. For any contingent claim H, we find θ and V_0 such that, at time 1, the discounted value of βH coincides with the discounted value βV_1 of its replicating portfolio (η, θ), where $\eta = V_0 - \theta S_0$. Writing H_b and H_a for the two possible time 1 values of H, we require V_0 and θ to satisfy the equations

$$\beta H_b = V_0 + \theta(\beta S_b - S_0), \qquad \beta H_a = V_0 + \theta(\beta S_a - S_0).$$

Their unique solution for (V_0, θ) is given by

$$\theta = \frac{H_b - H_a}{S_b - S_a} \tag{1.6}$$

and

$$V_0 = \beta H_a - \frac{H_b - H_a}{S_b - S_a}(\beta S_a - S_0) = \beta \left(H_b \frac{\beta^{-1}S_0 - S_a}{S_b - S_a} + H_a \frac{S_b - \beta^{-1}S_0}{S_b - S_a} \right).$$

Hence we also have

$$\eta = V_0 - \theta S_0 = \beta \frac{S_b H_a - S_a H_b}{S_b - S_a} = \beta \frac{(1+b)H_a - (1+a)H_b}{b - a}. \tag{1.7}$$

Since $V_1 = H$ for these choices of θ and V_0,

$$\theta = \frac{V_b - V_a}{S_b - S_a} = \frac{\delta V}{\delta S}$$

represents the rate of change in the value of the portfolio (or that of the contingent claim it replicates) per unit change in the underlying stock price. We shall meet this parameter again in more general pricing models (where it is known as the delta of the contingent claim and is usually denoted by Δ).

Setting

$$q = \frac{\beta^{-1} S_0 - S_a}{S_b - S_a},$$

it follows that

$$V_0 = \beta(q H_b + (1 - q) H_a)$$

since $1 - q = \frac{S_b - \beta^{-1} S_0}{S_b - S_a}$. In the special case where the discount rate β is $(1+r)^{-1}$ for some fixed $r > 0$, we see that $q \in (0,1)$ if and only if $r \in (a,b)$ (i.e., the riskless interest rate must lie between the two rates of increase in the stock price). This condition is therefore necessary and sufficient for the one-step binomial model to have a risk-neutral probability assignment $Q = (q, 1 - q)$ under which the fair price of the claim H is given as the expectation of its discounted final value, namely

$$\pi(H) = V_0 = E_Q(\beta V_T) = E_Q(\beta H). \tag{1.8}$$

These choices of θ and V_0 provide a linear estimator with perfect fit for H. The fair price V_0 for H therefore does not need to be adjusted by any risk premium in this model, and it is uniquely determined, irrespective of any initial probability assignment (i.e., it does not depend on the investor's attitude toward risk). The binomial model constructed here therefore allows preference-free or *arbitrage* pricing of the claim H. Since the cost function C has constant value V_0, we say that the replicating strategy (η, θ) is *self-financing* in this special case. No new funds have to be introduced at time 1 (recall that $\eta = V_0 - \theta S_0$ by definition).

In the general single-period model, it is not possible to ensure that C is constant. However, the pricing approach based on cost-minimisation leads to an optimal strategy for which the cost function is constant on average. Hence we call such a strategy *mean-self-financing* (see [141]).

The pricing formula (1.8) is valid for any contingent claim in the one-period binomial model. The following example shows how this simplifies for a European call option when the riskless interest rate is constant and the strike price lies between the two future stock price values.

Example 1.4.1. Assume that $H = (S_1 - K)^+$, $\beta = (1 + r)^{-1}$, and

$$(1 + a)S_0 < K \le (1 + b)S_0.$$

Then we have

$$H_b = (1 + b)S_0 - K, \qquad\qquad H_a = 0,$$

so that

$$\theta = \frac{H_b - H_a}{S_b - S_a} = \frac{S_0(1 + b) - K}{S_0(b - a)}.$$

The call option price is therefore

$$H_0 = V_0 = \frac{1}{1 + r} \frac{r - a}{b - a} \left(S_0(1 + b) - K\right).$$

Note that differentiation with respect to b and a, respectively, shows that, under the above assumptions, the call option price is an increasing function of b and a decreasing function of a, in accord with our intuition.

Risk and Return

We can measure the 'variability' of the stock S by means of the variance of the random variable $\frac{S_1}{S_0}$, which is the same as the variance of the return on the stock, $R_S = \frac{S_1 - S_0}{S_0}$. This is a Bernoulli random variable taking values b and a with probability p and $1 - p$, respectively. Hence its mean μ_S and variance σ_S^2 are given by

$$\mu_S = \frac{pS_b + (1 - p)S_a}{S_0} - 1 = a + p(b - a) \tag{1.9}$$

and

$$\sigma_S^2 = p(1 - p) \left(\frac{S_b - S_a}{S_0}\right)^2 = p(1 - p)(b - a)^2,$$

respectively.

We take the standard deviation $\sigma_S = \sqrt{p(1 - p)}(b - a)$ as the measure of risk inherent in the stock price S. We call it the *volatility* of the stock. Thus, with a given initial probability assignment $(p, 1 - p)$, the risk is proportional to $(b - a)$ and hence increases with increasing 'spread' of the values a, b, as expected. However, contrary to a frequently repeated assertion, the call option price H_0 does *not* necessarily increase with increasing σ_S, as is shown in the following simple example due to Marek Capinski (oral communication).

Example 1.4.2. Take $r = 0$ and let the call option begin at the money (i.e., let $K = S_0 = 1$). Then $(1 + b)S_0 - K = b$, so that the option price computed via (1.8) reduces to $V_0 = \frac{-ab}{b - a}$. The choice of $b = -a = 0.05$ yields $V_0 = 0.025$, while $\sigma_S = 0.1\sqrt{p(1 - p)}$. On the other hand, $b = 0.01$, $a = -0.19$ gives $V_0 = 0.0095$, and $\sigma_S = 0.2\sqrt{p(1 - p)}$.

Nonetheless, under any fixed initial probability assignment $P = (p, 1 - p)$, we can usefully compare the risk and return associated with holding the stock S with those for the option (or any contingent claim H). The treatment given here is a variant of that given in [69] and provides a foretaste of the sensitivity analysis undertaken for continuous-time models in Chapter 7.

In the single-period binomial model, the calculations reduce to consideration of the mean and standard deviation of Bernoulli random variables since the mean and variance of the claim H under P are given analogously by

$$\mu_H = \frac{pH_b + (1-p)H_a}{H_0} - 1, \qquad \sigma_H^2 = p(1-p)\left(\frac{H_b - H_a}{H_0}\right)^2. \qquad (1.10)$$

Define the *elasticity* (also known as the *beta* of the claim) as the covariance of the returns R_S and R_H normalised by the variance of R_S. Since both are Bernoulli random variables, it is easy to see that

$$E_H = \frac{p(1-p)\frac{H_b - H_a}{H_0}\frac{S_b - S_a}{S_0}}{p(1-p)\left(\frac{S_b - S_a}{S_0}\right)^2} = \frac{H_b - H_a}{H_0} \div \frac{S_b - S_a}{S_0}. \qquad (1.11)$$

Noting that $\theta = \frac{H_b - H_a}{S_b - S_a}$, we obtain $E_H = \frac{S_0}{H_0}\theta$, and therefore $\sigma_H = E_H\sigma_S$, so that the volatility of the claim H is proportional to that of the underlying stock S, with E_H as the constant of proportionality.

What about their rates of return? We shall consider the case of a constant riskless rate $r > 0$ and compare the *excess mean returns* $\mu_H - r$ and $\mu_S - r$. Recall that the replicating portfolio (η, θ) computed for H in (1.6) and (1.7) satisfies

$$\eta(1 + r) + \theta S_b = H_b, \qquad \eta(1 + r) + \theta S_a = H_a,$$

while also determining the option price $H_0 = \eta + \theta S_0$. Thus, with this portfolio we obtain

$$\theta S_b - H_b = (1 + r)(\theta S_0 - H_0) = \theta S_a - H_a.$$

Hence, for any $p \in (0, 1)$, we have

$$p(\theta S_b - H_b) + (1 - p)(\theta S_a - H_a) = (1 + r)(\theta S_0 - H_0);$$

i.e.,

$$\theta(pS_b + (1 - p)S_a) - (pH_b + (1 - p)H_a) = (1 + r)(\theta S_0 - H_0),$$

so that

$$\theta S_0 \left(\frac{pS_b + (1-p)S_0}{S_0} - 1\right) - H_0\left(\frac{pH_b + (1-p)H_a}{H_0} - 1\right) = r(\theta S_0 - H_0).$$

Using the definitions of μ_S and μ_H given by (1.9) and (1.10), we have

$$\theta S_0 \mu_S - H_0 \mu_H = r\theta S_0 - r H_0.$$

Rearranging terms, and recalling that $E_H = \frac{S_0}{H_0}\theta$, we have therefore shown that

$$\mu_H - r = E_H(\mu_S - r).$$

These relations are valid for any contingent claim H in the single-period binomial model and any fixed probability assignment $P = (p, 1-p)$. Now recall that the *risk-neutral* probabilities

$$(q, 1-q) = \left(\frac{(1+r)S_0 - S_a}{S_b - S_a}, \frac{S_b - (1+r)S_0}{S_b - S_a} \right) \tag{1.12}$$

provide the price of the claim H as the discounted expectation of its final values: $H_0 = (\frac{1}{1+r})(qH_b + (1-q)H_a)$. This leads to the identity

$$(1+r)\left(S_0(H_b - H_a) - H_0(S_b - S_a)\right) + (S_b H_a - S_a H_b) = 0. \tag{1.13}$$

Exercise 1.4.3. Show that (1.13) indeed holds true.

In particular, if $H = (S_1 - K)^+$ is a European call, then

$$S_b H_a - S_a H_b \le 0, \tag{1.14}$$

irrespective of the relationship between the values of K, H_b, and H_a.

Exercise 1.4.4. Verify that (1.14) holds true in all three cases.

Hence, for a European call option, the elasticity satisfies $E_H \ge 1$. This shows that holding the option is intrinsically riskier than holding the stock but also leads to a greater mean excess rate of return over the riskless interest rate.

Note further that for the risk-neutral probability $Q = (q, 1-q)$, the mean excess return is zero, as $E_Q\left(\frac{1}{1+r}H_1\right) = H_0$; i.e.,

$$E_Q(R_H) = \frac{qH_b + (1-q)H_a}{H_0} - 1 = r.$$

It is easy to verify that, for any given $P = (p, 1-p)$, we have

$$E_P(R_H) - E_Q(R_H) = (p-q)\frac{H_b - H_a}{H_0},$$

so that for any P with positive excess mean return (i.e., $E_P(R_H) \ge r$), we can express the mean return as

$$E_P(R_H) - r = E_P(R_H) - E_Q(R_H) = |p-q|\frac{H_b - H_a}{H_0} = |p-q|\frac{\sigma_H}{\sqrt{p(1-p)}}.$$

This justifies the terminology used to describe Q: the excess mean return under any probability assignment P is directly proportional to the standard deviation σ_H of the return R_H calculated under P. However, the mean return under Q is just the riskless rate r, and this holds irrespective of the 'riskiness' of H calculated under any other measure. The investor using the probability q to calculate the likelihood that the stock will move to $(1+b)S_0$ is therefore *risk-neutral*.

Thus, by choosing the risk-neutral measure Q, we can justify the long-standing actuarial practice of averaging the value of the discounted claim, at least for the case of our single-period binomial model. Moreover, we have shown that in this model *every* contingent claim can be *priced by arbitrage*; that is, there exists a (unique) self-financing strategy (η, θ) that replicates the value of H, so that the pricing model is *complete*. In a complete model, the optimal choice of strategy completely eliminates the risk in trading H, and the fair price of H is uniquely determined as the initial value V_0 of the optimal strategy, which can be computed explicitly as the expectation of H relative to the risk-neutral measure Q.

Before leaving single-period models, we review some of the preceding concepts in a modification of Example 1.2.1.

Example 1.4.5. Suppose that the stock price S_1 defined in Example 1.2.1 can take three values, namely 20, 15, and 7.5. In this case, there are an infinite number of risk-neutral probability measures for this stock. Since $\beta = 1$ in this example, the risk-neutral probability assignment requires $E_Q(S_1) = S_0$. This leads to the equations

$$20q_1 + 15q_2 + 7.5q_3 = 10, \qquad q_1 + q_2 + q_3 = 1,$$

with solutions $\left(\lambda, \frac{1}{3}(1 - 5\lambda), \frac{1}{3}(2 + 2\lambda)\right)$ for arbitrary λ. For nondegenerate probability assignments, we need $q_i \in (0,1)$ for $i - 1, 2, 3$; hence we require $\lambda \in \left(0, \frac{1}{5}\right)$. For each such λ, we obtain a different risk-neutral probability measure Q_λ.

Let $X = (X_1, X_2, X_3)$ be a contingent claim based on the stock S. We show that there exists a replicating portfolio for X if and only if

$$3X_1 - 5X_2 + 2X_3 = 0. \tag{1.15}$$

Indeed, recall that a hedge portfolio (η, θ) for X needs to satisfy $V_1 = \eta + \theta S_1 = X$ in all outcomes, so that

$$\eta + 20\theta = X_1, \qquad \eta + 15\theta = X_2, \qquad \eta + 7.5\theta = X_3.$$

This leads to

$$\theta = \frac{X_1 - X_3}{12.5} = \frac{X_2 - X_3}{7.5},$$

which, in turn, leads to (1.15). Thus, a contingent claim in this model is attainable if and only if equation (1.15) holds.

Finally, we verify that the value of an attainable claim X is the same under every risk-neutral measure: we have

$$E_{Q_\lambda}(X) = \lambda X_1 + \frac{1}{3}(1 - 5\lambda)X_2 + \frac{1}{3}(2 + 2\lambda)X_3$$

$$= \frac{1}{3}\left(\lambda(3X_1 - 5X_2 + 2X_3) + X_2 + 2X_3\right).$$

This quantity is independent of λ precisely when the attainability criterion (1.15) holds.

If the claim is not attainable, we cannot determine the price uniquely. Its possible values lie in the interval $(\inf_\lambda E_{Q_\lambda}(X), \sup_\lambda E_{Q_\lambda}(X))$, where $\lambda \in \left(0, \frac{1}{5}\right)$. For example, if $X = (S_1 - K)^+$ is a European call with strike 12, then we obtain $E_{Q_\lambda}(X) = \frac{1}{3}(\lambda(24 - 15) + 3) = 1 + 3\lambda$. Hence, the possible option values lie in the range $(1, 1.6)$. The choice of the 'optimal' value now depends on the optimality criterion employed. One such criterion was described in Section 1.3, but there are many others. The study of optimal pricing in incomplete models remains a major topic of current research and is largely beyond the scope of this book.

Exercise 1.4.6. Extend the market defined in the previous example by adding a second stock S' with $S_0' = 5$ and $S_1' = 6$, 6, or 4, so that the vector of stock prices (S, S') reads

$$(S_0, S_0') = (10, 5), \qquad (S_1, S_1') = \begin{cases} (20, 6) & \text{with probability } p_1 \\ (15, 6) & \text{with probability } p_2 \\ (7.5, 4) & \text{with probability } p_3 \end{cases}.$$

Verify that in this case there is *no* risk-neutral probability measure for the market-recall that we would need $p_i > 0$ for $i = 1, 2, 3$. We say that this market is not *viable*. Show that it is possible to construct arbitrage opportunities in this situation.

Exercise 1.4.7. Suppose the one-period market has riskless rate $r > 0$ and that the risky stock S has $S_0 = 4$ while S_1 can take the three values 2.5, 5, and 3. Find all the risk-neutral probabilities $Q = (q_1, q_2, q_3)$ in this model in terms of r. Show that there is no risk-neutral probability assignment for this model when $r = 0.25$. With this riskless rate, find an explicit strategy for making a profit with no net investment. When $r < 0.25$, find a sufficient condition (in terms of r) for a claim $X = (X_1, X_2, X_3)$ to be attainable.

1.5   Multi-period Binomial Models

Consider a binomial pricing model with trading dates $0, 1, 2, \ldots, T$ for some fixed positive integer T. By this we mean that the price of the stock takes values $S_0, S_1, S_2, \ldots, S_T$, and, for each $t \leq T$,

$$S_t = \begin{cases} (1 + b)S_{t-1} & \text{with probability } p \\ (1 + a)S_{t-1} & \text{with probability } 1 - p \end{cases}.$$

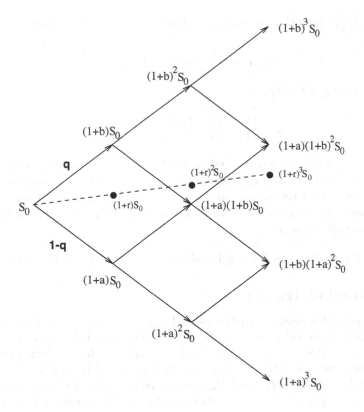

Figure 1.3: Event tree for the CRR model

As before, $r > 0$ is the riskless interest rate (so that $\beta = (1 + r)^{-1}$) and $r \in (a, b)$.

The event tree that describes the behaviour of stock prices in this model is depicted in Figure 1.3. Each arrow points 'up' with probability q and 'down' with probability $1 - q$.

A One-Step Risk-Neutral Measure

Assume that H is a contingent claim to be exercised at time T. Consider the current value of H at time $T - 1$, that is, one period before expiry. We can consider this as the initial value of a claim in the single-period model discussed previously, and so there is a hedging strategy (η, θ) that replicates the value of H on the time set $\{T - 1, T\}$ and a risk-neutral measure Q; we can therefore compute the current value of βH as its expectation under Q.

To be specific, assume that $H = (S_T - K)^+$ is a European call option with strike price K and expiry date T. Writing H_b for the value of H if $S_T = (1 + b)S_{T-1}$ and H_a similarly, the current value of H is given by

$E_Q\left(\frac{H}{1+r}\right)$, where the measure Q is given by $(q, 1-q)$ as defined in (1.12). Hence

$$V_{T-1} = \frac{1}{1+r}(qH_b + (1-q)H_a) \qquad (1.16)$$

with (writing S for S_{T-1})

$$q = \frac{(1+r)S - (1+a)S}{(1+b)S - (1+a)S} = \frac{r-a}{b-a}.$$

This again illustrates why we called Q the 'risk-neutral' measure since a risk-neutral investor is one who is indifferent between an investment with a certain rate of return and another whose uncertain rate of return has the same expected value. Under Q, the expectation of S_T, given that $S_{T-1} = S$, is given by

$$E_Q(S_T \mid S_{T-1} = S) = q(1+b)S + (1-q)(1+a)S = (1+r)S.$$

Two-Period Trading

Now apply this analysis to the value V_{T-2} of the call H at time $T-2$: the stock, whose value S_{T-2} is now written as S, can take one of the three values $(1+b)^2S$, $(1+a)(1+b)S$, and $(1+a)^2S$ at time T; hence the call H must have one of three values at that time (see Figure 1.3). We write these values as H_{bb}, H_{ab}, and H_{aa}, respectively. From (1.8), and using the definition of q in (1.12), we can read off the possible values of V_{T-1} as

$$V_b = \beta(qH_{bb} + (1-q)H_{ab}), \qquad V_a = \beta(qH_{ab} + (1-q)H_{aa}),$$

respectively. For each of these cases, we have now found the value of the option at time $T-1$ and can therefore select a hedging portfolio as before. The value of the parameters θ and η is determined at each stage exactly as in the single-period model. We obtain

$$\begin{aligned}
V_{T-2} &= \beta(qV_b + (1-q)V_a) \\
&= \beta\left\{q\beta(qH_{bb} + (1-q)H_{ab}) + (1-q)\beta(qH_{ab} + (1-q)H_{aa})\right\} \\
&= \beta^2\left\{q^2\left[(1+b)^2S - K\right]^+ + 2q(1-q)\left[(1+a)(1+b)S - K\right]^+ \right. \\
&\quad \left. + (1-q)^2\left[(1+a)^2S - K\right]^+\right\}.
\end{aligned}$$

Hence the current value of the claim is completely determined by quantities that are known to the investor at time $T-2$.

The CRR Formula

We can continue this backward recursion to calculate the value process $(V_t)_{t\in\mathbb{T}}$. In particular, with $\beta = (1+r)^{-1}$, the initial investment needed to replicate the European call option H is

$$V_0 = \beta^T \sum_{t=0}^{T} \binom{T}{t} q^t (1-q)^{T-t} \left((1+b)^t (1+a)^{T-t} S_0 - K \right)^+$$

$$= S_0 \sum_{t=A}^{T} \binom{T}{t} q^t (1-q)^{T-t} \frac{(1+b)^t (1+a)^{T-t}}{(1+r)^T}$$

$$- K(1+r)^{-T} \sum_{t=A}^{T} \binom{T}{t} q^t (1-q)^{T-t}, \qquad (1.17)$$

where A is the smallest integer k for which $S_0(1+b)^k(1+a)^{T-k} > K$.

Using

$$q = \frac{r-a}{b-a}, \qquad\qquad q' = q\frac{1+b}{1+r},$$

we obtain $q' \in (0,1)$ and $1-q' = (1-q)\frac{1+a}{1+r}$. We can finally write the fair price for the European call option in (1.17) in this multi-period binomial pricing model as

$$V_0 = S_0 \Psi(A;T,q') - K(1+r)^{-T} \Psi(A;T,q), \qquad (1.18)$$

where Ψ is the complementary binomial distribution function; that is,

$$\Psi(m;n,p) = \sum_{j=m}^{n} \binom{n}{j} p^j (1-p)^{n-j}.$$

Formula (1.18) is known as the *Cox-Ross-Rubinstein* (or CRR, see [59]) binomial option pricing formula for the European call. We shall shortly give an alternative derivation of this formula by computing the expectation of H under the risk-neutral measure Q directly, utilising the martingale property of the discounted stock price under this measure.

Recall the event tree in Figure 1.3. At each node there are only two branches, that is, one more than the number of stocks available. It is this simple splitting property that ensures that the model is complete since it allows us to 'cover' the two random outcomes at each stage by adjusting the quantities θ and η.

The Hedge Portfolio

More generally, it is clear that the value V_t of the option at time $t \leq T$ is given by the formula

$$V_t = S_t \Psi(A_t; T-t, q') - K(1+r)^{-T-t} \Psi(A_t; T-t, q), \qquad (1.19)$$

where A_t is the smallest integer k for which $S_t(1+b)^k(1+a)^{T-t-k} > K$. An analysis similar to that outlined in Section 1.4 provides the components of the trading strategy (η, θ): the portfolio $(\eta_{t-1}, \theta_{t-1})$ is held over the time interval $[t-1, t]$ and is required to replicate V_t; i.e.,

$$\theta_{t-1} S_t + \eta_{t-1}(1+r) = V_t.$$

Thus V_t is determined by S_{t-1} and the price movement in the time interval $[t-1, t]$, so that it takes two possible values, depending on whether $S_t = (1+b)S_{t-1}$ or $S_t = (1+a)S_{t-1}$. Writing V_t^b and V_t^a, respectively, for the resulting values, we need to solve the equations

$$\theta_{t-1}(1+b)S_{t-1} + \eta_{t-1}(1+r) = V_t^b, \quad \theta_{t-1}(1+a)S_{t-1} + \eta_{t-1}(1+r) = V_t^a.$$

Again we obtain

$$\theta_{t-1} = \frac{V_t^b - V_t^a}{(b-a)S_{t-1}}, \qquad \eta_{t-1} = \frac{(1+b)V_t^a - (1+a)V_t^b}{(1+r)(b-a)}. \tag{1.20}$$

This leads to the explicit formulas

$$\left.\begin{array}{c} \theta_t = \displaystyle\sum_{s=A_t}^{T-t} \binom{T-t}{s}(q')^s(1-q')^{T-t-s} \\[3mm] \eta_t = -K(1+r)^{-(T-t)} \displaystyle\sum_{s=A_t}^{T-t} \binom{T-t}{s}q^s(1-q)^{T-t-s} \end{array}\right\} \tag{1.21}$$

for θ_t and η_t.

Exercise 1.5.1. Verify the formulas in (1.21) by writing down binomial expressions for V_t^b and V_t^a analogously with (1.16).

1.6 Bounds on Option Prices

We conclude this chapter with a few simple observations concerning bounds on option prices. We restrict attention to call options, though similar arbitrage considerations provide bounds for put options. The bounds described here are quite crude but are independent of the model used, relying solely on the assumption of 'no arbitrage'. In this section, we denote the call price by C_0 and the put price by P_0.

It should be obvious that American options are, in general, more valuable than their European counterparts since the holder has greater flexibility in exercising them. We can illustrate this by constructing a simple arbitrage. For example, if the price $C_0(E)$ of a European call with strike K and exercise date T were greater than the price $C_0(A)$ of an American option with the same K and T, then we would make a riskless profit by writing the European option and buying the American one, while pocketing the difference $C_0(E) - C_0(A)$. We keep this riskless profit by holding the American option until time T when both options have the same value. Thus, in the absence of arbitrage, the relations

$$0 \leq C_0(E) \leq C_0(A) \tag{1.22}$$

will always hold.

Both option prices must lie below the current value S_0 of the underlying share (and will in practice be much less): if $C_0(A)$ were greater than S_0, we could buy a share at S_0 and write the option. The profit made is secure since the option liability is covered by the share. By (1.22), both option values are therefore less than S_0.

Call-put parity for European options (see (1.3)) demands that

$$C_0(E) - P_0(E) = S_0 - \beta^T K.$$

As $P_0(E) \geq 0$, it follows that $C_0(E) \geq S_0 - \beta^T K$. We conclude that the European call option price lies in the interval $\left[\min\left\{ 0, S_0 - \beta^T K \right\}, S_0 \right]$. While this remains a crude estimate, it holds in all option pricing models.

These bounds provide a simple, but initially surprising, relationship between European and American call option prices for shares that (as here) pay no dividends. Note first that

$$C_0(A) \geq C_0(E) \geq S_0 - \beta^T K \geq S_0 - K \tag{1.23}$$

since the discount factor β is less than or equal to 1. This means that the option price is, in either case, at least equal to the gain achieved by immediate exercise of the option. Hence (as long as our investor prefers more to less) the option will not be exercised immediately. But the same argument applies at any starting time $t < T$, so that the European option's value $C_t(E)$ at time t (which must be the same as that of an option written at t with strike K and exercise date T) satisfies $C_t(E) \geq S_t - \beta^{T-t} K$, and, as previously, $C_t(A) \geq S_t - K$, which is *independent* of the time to expiry $T - t$. Consequently, *an American call option on a stock that pays no dividends will not be exercised before expiry*, so that in this case $C_0(E) = C_0(A)$.

Exercise 1.6.1. Derive the following bounds for the European put option price $P_0(E)$ by arbitrage arguments:

$$\max\left\{ 0, \beta^T K - S_0 \right\} \leq P_0(E) \leq \beta^T K.$$

Call-put parity allows a calculation of the riskless interest rate from European put and call prices since we can write

$$e^{-r(T-t)} K = S_t - C_t(E) + P_t(E) \text{ for } t < T,$$

so that

$$r = \frac{1}{T-t} \left[\log K - \log(S_t + P_t(E) - C_t(E)) \right]. \tag{1.24}$$

However, as European options are much less frequently traded than their American counterparts, it is more useful to have an estimate of r in terms of the latter. This follows at once: as we have just seen for the case $t = 0$, we must have $C_t(A) = C_t(E)$ for all $t < T$, while $P_t(A) \geq P_t(E)$ by the same argument as was established in (1.22) for calls. Hence, for American options during whose lifetime the underlying stock pays no dividends, we have

$$r \geq \frac{1}{T-t} \left[\log K - \log(S_t + P_t(A) - C_t(A)) \right]. \tag{1.25}$$

In practice, this inequality is used to check put and call prices against the prevailing riskless rate (e.g. , LIBOR rate); where it fails, market prices offer (usually short-lived) arbitrage opportunities. It can also serve to provide estimates of r for use in the simulation of the evolution of the stock price from options on the stock (see, e.g. , [210].

Chapter 2

Martingale Measures

2.1 A General Discrete-Time Market Model

Information Structure

Fix a time set $\mathbb{T} = \{0, 1, \ldots, T\}$, where the *trading horizon* T is treated as the terminal date of the economic activity being modelled, and the points of $\mathbb{T}$ are the admissible *trading dates*. We assume as given a fixed probability space $(\Omega, \mathcal{F}, P)$ to model all 'possible states of the market'.

In most of the simple models discussed in Chapter 1, Ω is a *finite* probability space (i.e., has a finite number of points ω each with $P(\{\omega\}) > 0$). In this situation, the σ-field $\mathcal{F}$ is the power set of Ω, so that every subset of Ω is $\mathcal{F}$-measurable.

Note, however, that the finite models can equally well be treated by assuming that, on a general sample space Ω, the σ-field $\mathcal{F}$ in question is finitely generated. In other words, there is a finite partition $\mathcal{P}$ of Ω into mutually disjoint sets $A_1, A_2, \ldots, A_n$ whose union is Ω and that generates $\mathcal{F}$ so that $\mathcal{F}$ also contains only finitely many events and consists precisely of those events that can be expressed in terms of $\mathcal{P}$. In this case, we further demand that the probability measure P on $\mathcal{F}$ satisfies $P(A_i) > 0$ for all i.

In both cases, the only role of P is to identify the events that investors agree are *possible*; they may disagree in their assignment of probabilities to these events. We refer to models in which either of the preceding additional assumptions applies as *finite market models*. Although most of our examples are of this type, the following definitions apply to general market models. Real-life markets are, of course, always finite; thus the additional 'generality' gained by considering arbitrary sample spaces and σ-fields is a question of mathematical convenience rather than wider applicability!

The *information structure* available to the investors is given by an increasing (finite) sequence of *sub-σ-fields* of $\mathcal{F}$: we assume that $\mathcal{F}_0$ is trivial; that is, it contains only sets of P-measure 0 or 1. We assume that $(\Omega, \mathcal{F}_0)$ is complete (so that any subset of a null set is itself null and $\mathcal{F}_0$ contains all

P-null sets) and that $\mathcal{F}_0 \subset \mathcal{F}_1 \subset \mathcal{F}_2 \subset \cdots \subset \mathcal{F}_T = \mathcal{F}$. An increasing family of σ-fields is called a *filtration* $\mathbb{F} = (\mathcal{F}_t)_{t \in \mathbb{T}}$ on $(\Omega, \mathcal{F}, P)$. We can think of $\mathcal{F}_t$ as containing the information available to our investors at time t: investors learn without forgetting, but we assume that they are not prescient-insider trading is not possible. Moreover, our investors think of themselves as 'small investors' in that their actions will not change the probabilities they assign to events in the market. Again, note that in a finite market model each σ-field $\mathcal{F}_t$ is generated by a minimal finite partition $\mathcal{P}_t$ of Ω and that $\mathcal{P}_0 = \{\Omega\} \subset \mathcal{P}_1 \subset \mathcal{P}_2 \subset \cdots \subset \mathcal{P}_T = \mathcal{P}$. At time t, all our investors know which cell of $\mathcal{P}_t$ contains the 'true state of the market', but none of them knows more.

Market Model and Numéraire

The definitions developed in this chapter will apply to general discrete market models, where the sample space need not be finite. Fix a probability space $(\Omega, \mathcal{F}, P)$, a natural number d, the *dimension* of the market model, and assume as given a $(d + 1)$-dimensional stochastic process $S = \left\{ S_t^i : t \in \mathbb{T}, i = 0, 1, \ldots, d \right\}$ to represent the time evolution of the *securities price process*. The security labelled 0 is taken as a riskless (non-random) *bond* (or *bank account*) with price process S^0, while the d risky (random) stocks labelled $1, 2, \ldots, d$ have price processes $S^1, S^2, \ldots, S^d$. The process S is assumed to be *adapted* to the filtration $\mathbb{F}$, so that for each $i \leq d$, S_t^i is $\mathcal{F}_t$-measurable; that is, the prices of the securities at all times up to t are known at time t. Most frequently, we in fact take the filtration $\mathbb{F}$ as that generated by the price process $S = \left(S^1, S^2, \ldots, S^d \right)$. Then $\mathcal{F}_t = \sigma \left(S_u : u \leq t \right)$ is the smallest σ-field such that all the $\mathbb{R}^{d+1}$-valued random variables $\left\{ S_u = \left(S_u^0, S_u^1, \ldots, S_u^d \right), u \leq t \right\}$ are $\mathcal{F}_t$-measurable. In other words, at time t, the investors know the values of the price vectors $(S_u : u \leq t)$, but they have no information about later values of S.

The tuple $(\Omega, \mathcal{F}, P, \mathbb{T}, \mathbb{F}, S)$ is the *securities market model*. We require at least one of the price processes to be strictly positive throughout; that is, to act as a benchmark, known as the *numéraire*, in the model. As is customary, we generally assign this role to the bond price S^0, although in principle any strictly positive S^i could be used for this purpose.

Note on Terminology. The term 'bond' is the one traditionally used to describe the riskless security that we use here as numéraire, although 'bank account' and 'money market account' are popular alternatives. We continue to use 'bond' in this sense until Chapter 9, where we discuss models for the evolution of interest rates; in that context, the term 'bond' refers to a certain type of risky asset, as is made clear.

2.2 Trading Strategies

Value Processes

Throughout this section, we fix a securities market model $(\Omega, \mathcal{F}, P, \mathbb{T}, \mathbb{F}, S)$. We take S^0 as a strictly positive *bond* or riskless security, and without loss of generality we assume that $S^0(0) = 1$, so that the initial value of the bond S^0 yields the units relative to which all other quantities are expressed. The *discount factor* $\beta_t = \frac{1}{S^0_t}$ is then the sum of money we need to invest in bonds at time 0 in order to have 1 unit at time t. Note that we allow the discount *rate* - that is, the increments in β_t - to vary with t; this includes the case of a constant interest rate $r > 0$, where $\beta_t = (1 + r)^{-t}$.

The securities $S^0, S^1, S^2, \ldots, S^d$ are traded at times $t \in \mathbb{T}$: an investor's *portfolio* at time $t \geq 1$ is given by the $\mathbb{R}^{d+1}$-valued random variable $\theta_t = (\theta^i_t)_{0 \leq i \leq d}$ with *value process* $V_t(\theta)$ given by

$$V_0(\theta) = \theta_1 \cdot S_0, \qquad V_t(\theta) = \theta_t \cdot S_t = \sum_{i=0}^{d} \theta^i_t S^i_t \text{ for } t \in \mathbb{T}, \ t \geq 1.$$

The value $V_0(\theta)$ is the investor's initial endowment. The investors select their time t portfolio once the stock prices at time $t - 1$ are known, and they hold this portfolio during the time interval $(t - 1, t]$. At time t the investors can adjust their portfolios, taking into account their knowledge of the prices S^i_t for $i = 0, 1, \ldots, d$. They then hold the new portfolio θ_{t+1} throughout the time interval $(t, t + 1]$.

Market Assumptions

We require that the *trading strategy* $\theta = \{\theta_t : t = 1, 2, \ldots, T\}$ consisting of these portfolios be a *predictable* vector-valued stochastic process: for each $t < T$, θ_{t+1} should be $\mathcal{F}_t$-measurable, so θ_1 is $\mathcal{F}_0$-measurable and hence constant, as $\mathcal{F}_0$ is assumed to be trivial. We also assume throughout that we are dealing with a 'frictionless' market; that is, there are no transaction costs, unlimited short sales and borrowing are allowed (the random variables θ^i_t can take any real values), and the securities are perfectly divisible (the S^i_t can take any positive real values).

Self-Financing Strategies

We call the trading strategy θ *self-financing* if any changes in the value $V_t(\theta)$ result entirely from net gains (or losses) realised on the investments; the value of the portfolio after trading has occurred at time t and before stock prices at time $t + 1$ are known is given by $\theta_{t+1} \cdot S_t$. If the total value of the portfolio has been used for these adjustments (i.e., there are no withdrawals and no new funds are invested), then this means that

$$\theta_{t+1} \cdot S_t = \theta_t \cdot S_t \text{ for all } t = 1, 2, \ldots, T - 1. \tag{2.1}$$

Writing $\Delta X_t = X_t - X_{t-1}$ for any function X on $\mathbb{T}$, we can rewrite (2.1) at once as

$$\Delta V_t(\theta) = \theta_t \cdot S_t - \theta_{t-1} \cdot S_{t-1} = \theta_t \cdot S_t - \theta_t \cdot S_{t-1} = \theta_t \cdot \Delta S_t; \qquad (2.2)$$

that is, the gain in value of the portfolio in the time interval $(t-1, t]$ is the scalar product in $\mathbb{R}^d$ of the new portfolio vector θ_t with the vector ΔS_t of price increments. Thus, defining the *gains process* associated with θ by setting

$$G_0(\theta) = 0, \qquad G_t(\theta) = \theta_1 \cdot \Delta S_1 + \theta_2 \cdot \Delta S_2 + \cdots + \theta_t \cdot \Delta S_t,$$

we see at once that θ is self-financing if and only if

$$V_t(\theta) = V_0(\theta) + G_t(\theta) \text{ for all } t \in \mathbb{T}. \qquad (2.3)$$

This means that θ is self-financing if and only if the value $V_t(\theta)$ arises solely as the sum of the initial *endowment* $V_0(\theta)$ and the gains process $G_t(\theta)$ associated with the strategy θ.

We can write this relationship in yet another useful form: since $V_t(\theta) = \theta_t \cdot S_t$ for any $t \in \mathbb{T}$ and *any* strategy θ, it follows that we can write

$$\begin{aligned}
\Delta V_t &= V_t - V_{t-1} \\
&= \theta_t \cdot S_t - \theta_{t-1} \cdot S_{t-1} \\
&= \theta_t \cdot (S_t - S_{t-1}) + (\theta_t - \theta_{t-1}) \cdot S_{t-1} \\
&= \theta_t \cdot \Delta S_t + (\Delta \theta_t) \cdot S_{t-1}.
\end{aligned} \qquad (2.4)$$

Thus, the strategy θ is self-financing if and only if

$$(\Delta \theta_t) \cdot S_{t-1} = 0. \qquad (2.5)$$

This means that, for a self-financing strategy, the vector of changes in the portfolio θ is orthogonal in $\mathbb{R}^{d+1}$ to the *prior* price vector S_{t-1}. This property is sometimes easier to verify than (2.1). It also serves to justify the terminology: the cumulative effect of the time t variations in the investor's holdings (which are made *before* the time t prices are known) should be to balance each other. For example, if $d = 1$, we need to balance $\Delta \theta_t^0 S_{t-1}^0$ against $\Delta \theta_t^1 S_{t-1}^1$ since by (2.5) their sum must be zero.

Numéraire Invariance

Trivially, (2.1) and (2.3) each have an equivalent 'discounted' form. In fact, given any numéraire (i.e., any process (Z_t) with $Z_t > 0$ for all $t \in \mathbb{T}$), it follows that a trading strategy θ is self-financing relative to S if and only if it is self-financing relative to ZS since

$$(\Delta \theta_t) \cdot S_{t-1} = 0 \text{ if and only if } (\Delta \theta_t) \cdot Z_{t-1} S_{t-1} = 0 \text{ for } t \in \mathbb{T} \setminus \{0\}.$$

Thus, changing the choice of 'benchmark' security will not alter the class of trading strategies under consideration and thus will not affect market behaviour. This simple fact is sometimes called the 'numéraire invariance theorem'; in continuous-time models it is not completely obvious (see Chapter 9 and [102]). We will also examine the numéraire invariance of other market entities. While the use of different discounting conventions has only limited mathematical significance, economically it amounts to understanding the way in which these entities are affected by a change of currency.

Writing $\overline{X}_t = \beta_t X_t$ for the discounted form of the vector X_t in $\mathbb{R}^{d+1}$, it follows (using $Z = \beta$ in the preceding equation) that θ is self-financing if and only if $(\Delta\theta_t) \cdot \overline{S}_{t-1} = 0$, that is, if and only if

$$\theta_{t+1} \cdot \overline{S}_t = \theta_t \cdot \overline{S}_t \text{ for all } t = 1, 2, \dots, T-1, \qquad (2.6)$$

or, equivalently, if and only if

$$\overline{V}_t(\theta) = V_0(\theta) + \overline{G}_t(\theta) \text{ for all } t \in \mathbb{T}. \qquad (2.7)$$

To see the last equivalence, note first that (2.4) holds for any θ with $\overline{S}$ instead of S, so that for self-financing θ we have $\Delta\overline{V}_t = \theta_t \cdot \Delta\overline{S}_t$; hence (2.7) holds. Conversely, (2.7) implies that $\Delta\overline{V}_t = \theta_t \cdot \Delta\overline{S}_t$, so that $(\Delta\theta_t) \cdot \overline{S}_{t-1} = 0$ and so θ is self-financing.

We observe that the definition of $\overline{G}(\theta)$ does not involve the amount θ_t^0 held in bonds (i.e., in the security S^0) at time t. Hence, if θ is self-financing, the initial investment $V_0(\theta)$ and the predictable real-valued processes θ^i $(i = 1, 2, \dots, d)$ completely determine θ^0, just as we have seen in the one-period model in Section 1.4.

Lemma 2.2.1. *Given an $\mathcal{F}_0$-measurable function V_0 and predictable real-valued processes $\theta^1, \theta^2, \dots, \theta^d$, the unique predictable process θ^0 that turns*

$$\theta = \left(\theta^0, \theta^1, \theta^2, \cdots, \theta^d\right)$$

into a self-financing strategy with initial value $V_0(\theta) = V_0$ is given by

$$\theta_t^0 = V_0 + \sum_{u=1}^{t-1} \left(\theta_u^1 \Delta\overline{S}_u^1 + \cdots + \theta_u^d \Delta\overline{S}_u^d\right) - \left(\theta_t^1 \overline{S}_{t-1}^1 + \cdots + \theta_t^d \overline{S}_{t-1}^d\right). \quad (2.8)$$

Proof. The process θ^0 so defined is clearly predictable. To see that it produces a self-financing strategy, recall by (2.7) that we only need to observe that this value of θ^0 is the unique predictable solution of the equation

$$\overline{V}_t(\theta) = \theta_t^0 + \theta_t^1 \overline{S}_t^1 + \theta_t^2 \overline{S}_t^2 + \cdots + \theta_t^d \overline{S}_t^d$$

$$= V_0 + \sum_{u=1}^{t} \left(\theta_u^1 \Delta\overline{S}_u^1 + \theta_u^2 \overline{S}_u^2 + \cdots + \theta_u^d \Delta\overline{S}_u^d\right).$$

$\square$

Admissible Strategies

Let Θ be the class of all self-financing strategies. So far, we have not insisted that a self-financing strategy must at all times yield non-negative total wealth; that is, that $V_t(\theta) \geq 0$ for all $t \in \mathbb{T}$. From now on, when we impose this additional restriction, we call such self-financing strategies *admissible*; they define the class Θ_a.

Economically, this requirement has the effect of restricting certain types of short sales: although we can still borrow certain of our assets (i.e., have $\theta_t^i < 0$ for some values of i and t), the overall value process must remain non-negative for each t. But the additional restriction has little impact on the mathematical modelling, as we show shortly.

We use the class Θ_a to define our concept of 'free lunch'.

Definition 2.2.2. An *arbitrage opportunity* is an admissible strategy θ such that

$$V_0(\theta) = 0, \qquad V_t(\theta) \geq 0 \text{ for all } t \in \mathbb{T}, \qquad E\left(V_T(\theta)\right) > 0.$$

In other words, we require $\theta \in \Theta_a$ with initial value 0 but final value strictly positive with positive probability. Note, however, that the probability measure P enters into this definition only through its null sets: the condition $E\left(V_T(\theta)\right) > 0$ is equivalent to $P(V_T(\theta)) > 0) > 0$, justifying the following definition.

Definition 2.2.3. The market model is *viable* if it does not contain any arbitrage opportunities; that is, if $\theta \in \Theta_a$ has $V_0(\theta) = 0$, then $V_T(\theta) = 0$ a.s..

'Weak Arbitrage Implies Arbitrage'

To justify the assertion that restricting attention to admissible claims has little effect on the modelling, we call a self-financing strategy $\theta \in \Theta$ a *weak arbitrage* if

$$V_0(\theta) = 0, \qquad V_T(\theta) \geq 0, \qquad E\left(V_T(\theta)\right) > 0.$$

The following calculation shows that if a weak arbitrage exists then it can be adjusted to yield an admissible strategy - that is, an arbitrage as defined in Definition 2.2.2.

Note. If the price process is a martingale under some equivalent measure-as will be seen shortly-then any hedging strategy with zero initial value and positive final expectation will automatically yield a positive expectation at all intermediate times by the martingale property.

Suppose that θ is a weak arbitrage and that $V_t(\theta)$ is *not* non-negative a.s. for all t. Then there exists $t < T$, and $A \in \mathcal{F}_t$ with $P(A) > 0$ such that

$$(\theta_t \cdot S_t)(\omega) < 0 \text{ for } \omega \in A, \theta_u \cdot S_u \geq 0 \text{ a.s. for } u > t.$$

We amend θ to a new strategy ϕ by setting $\phi_u(\omega) = 0$ for all $u \in \mathbb{T}$ and $\omega \in \Omega \setminus A$, while on A we set $\phi_u(\omega) = 0$ if $u \le t$, and for $u > t$ we define

$$\phi_u^0(\omega) = \theta_u^0(\omega) - \frac{\theta_t \cdot S_t}{S_t^0(\omega)}, \; \phi_u^i(\omega) = \theta_u^i(\omega) \text{ for } i = 1, 2, \ldots, d.$$

This strategy is obviously predictable. It is also self-financing: on $\Omega \setminus A$ we clearly have $V_u(\phi) \equiv 0$ for all $u \in \mathbb{T}$, while on A we need only check that $(\Delta\phi_{t+1}) \cdot S_t = 0$ by the preceding construction (in which $\Delta\theta_u$ and $\Delta\phi_u$ differ only when $u = t + 1$) and (2.5). We observe that $\phi_t^i = 0$ on A^c for $i \ge 0$ and that, on A,

$$\Delta\phi_{t+1}^0 = \phi_{t+1}^0 = \theta_{t+1}^0 - \frac{\theta_t \cdot S_t}{S_t^0}, \; \Delta\phi_{t+1}^i = \theta_{t+1}^i \text{ for } i = 1, 2, \ldots, d.$$

Hence

$$(\Delta\phi_{t+1}) \cdot S_t = 1_A(\theta_{t+1} \cdot S_t - \theta_t \cdot S_t) = 1_A(\theta_t \cdot S_t - \theta_t \cdot S_t) = 0$$

since θ is self-financing.

We show that $V_u(\phi) \ge 0$ for all $u \in \mathbb{T}$, and $P(V_T(\phi) > 0) > 0$. First note that $V_u(\phi) = 0$ on $\Omega \setminus A$ for all $u \in \mathbb{T}$. On A we also have $V_u(\phi) = 0$ when $u \le t$, but for $u > t$ we obtain

$$V_u(\phi) = \phi_u \cdot S_u = \theta_u^0 S_u^0 - \frac{(\theta_t \cdot S_t) S_u^0}{S_t^0} + \sum_{i=1}^{d} \theta_u^i S_u^i = \theta_u \cdot S_u - (\theta_t \cdot S_t)\left(\frac{S_u^0}{S_t^0}\right).$$

Since, by our choice of t, $\theta_u \cdot S_u \ge 0$ for $u > t$, and $(\theta_t \cdot S_t) < 0$ while $S^0 \ge 0$, it follows that $V_u(\phi) \ge 0$ for all $u \in \mathbb{T}$. Moreover, since $S_t^0 > 0$, we also see that $V_T(\phi) > 0$ on A.

This construction shows that the existence of what we have called weak arbitrage immediately implies the existence of an arbitrage opportunity. This fact is useful in the fine structure analysis for finite market models we give in the next chapter.

Remark 2.2.4. Strictly speaking, we should deal separately with the possibility that the investor's initial capital is negative. This is of course ruled out if we demand that all trading strategies are admissible. We can relax this condition and consider a one-period model, where a trading strategy is just a portfolio θ, chosen at the outset with knowledge of time 0 prices and held throughout the period. In that case, an arbitrage is a portfolio that leads from a non-positive initial outlay to a non-negative value at time 1. Thus here we have two possible types of arbitrage since the portfolio θ leads to one of two conclusions:

a) $V_0(\theta) < 0$ and $V_1(\theta) \ge 0$ or

b) $V_0(\theta) = 0$ and $V_1(\theta) \ge 0$ and $P(V_1(\theta) > 0) > 0$.

In this setting, the assumption that there are no arbitrage opportunities leads to two conditions on the prices:

(i) $V_1(\theta) = 0$ implies $V_0(\theta) = 0$ or

(ii) $V_1(\theta) \geq 0$ and $P(V_1(\theta)) > 0$ implies $V_0(\theta) \geq 0$.

The reader will easily construct arbitrages if either of these conditions fails. In our treatment of multi-period models, we consistently use admissible strategies, so that Definition 2.2.3 is sufficient to define the viability of pricing models.

Uniqueness of the Arbitrage Price

Fix H as a contingent claim with maturity T so H is a non-negative $\mathcal{F}_T$-measurable random variable on $(\Omega, \mathcal{F}_T, P)$. The claim is said to be *attainable* if there is an admissible strategy θ that *generates* (or *replicates*) it, that is, such that
$$V_T(\theta) = H.$$

We should expect the value process associated with a generating strategy to be given *uniquely*: the existence of two admissible strategies θ and θ' with $V_t(\theta) \neq V_t(\theta')$ would violate the *Law of One Price*, and the market would therefore allow riskless profits and not be viable. A full discussion of these economic arguments is given in [241].

The next lemma shows, conversely, that in a viable market the *arbitrage price* of a contingent claim is indeed unique.

Lemma 2.2.5. *Suppose H is an attainable contingent claim in a viable market model. Then the value processes of all generating strategies for H are the same.*

Proof. If θ and ϕ are admissible strategies with

$$V_T(\theta) = H = V_T(\phi)$$

but $V(\theta) \neq V(\phi)$, then there exists $t < T$ such that

$$V_u(\theta) = V_u(\phi) \text{ for all } u < t, \qquad\qquad V_t(\theta) \neq V_t(\phi).$$

The set $A = \{V_t(\theta) > V_t(\phi)\}$ is in $\mathcal{F}_t$ and we can assume $P(A) > 0$ without loss of generality. The random variable $X = V_t(\theta) - V_t(\phi)$ is $\mathcal{F}_t$-measurable and defines a self-financing strategy ψ as by letting

$$\psi_u(\omega) = \theta_u(\omega) - \phi_u(\omega) \text{ for } u \leq t \text{ on } A, \text{ for } u \in \mathbb{T}, \text{ on } A^c,$$
$$\psi_u^0 = \beta_t X \text{ and } \psi_u^i = 0 \text{ for } i = 1, 2, \ldots, d \text{ for } u > t, \text{ on } A.$$

It is clear that ψ is predictable. Since both θ and ϕ are self-financing, it follows that (2.1) also holds with ψ for $u < t$, while if $u > t$, $\psi_{u+1} \cdot S_u =$

$\psi_u \cdot S_u$ on A^c similarly. On A, we have $\psi_{u+1} = \psi_u$. Thus we only need to compare $\psi_t \cdot S_t = V_t(\theta) - V_t(\phi)$ and $\psi_{t+1} \cdot S_t = 1_{A^c}(\theta_{t+1} - \phi_{t+1}) \cdot S_t + 1_A \beta_t X S_t^0$. Now note that $S_t^0 = \beta_t^{-1}$ and that $X = V_t(\theta) - V_t(\phi)$, while on A^c the first term becomes $(\theta_t - \phi_t) \cdot S_t = V_t(\theta) - V_t(\phi)$ and the latter vanishes. Thus $\psi_{t+1} \cdot S_t = V_t(\theta) - V_t(\phi) = \psi_t \cdot S_t$.

Since $V_0(\theta) = V_0(\phi)$, ψ is self-financing with initial value 0. But $V_T(\psi) = 1_A(\beta_t X S_t^0) = 1_A \beta_t \beta_T^{-1} X$ is non-negative a.s. and is strictly positive on A, which has positive probability. Hence ψ is a weak arbitrage, and by the previous section the market cannot be viable. $\qquad\square$

We have shown that in a viable market it is possible to associate a unique time t value (or *arbitrage price*) to any attainable contingent claim H. However, it is not yet clear how the generating strategy, and hence the price, are to be found in particular examples. In the next section, we characterise viable market models without having to construct explicit strategies and derive a general formula for the arbitrage price instead.

2.3 Martingales and Risk-Neutral Pricing

Martingales and Their Transforms

We wish to characterise viable market models in terms of the behaviour of the *increments* of the discounted price process $\overline{S}$. To set the scene, we first need to recall some simple properties of martingales. Only the most basic results needed for our purposes are described here; for more details consult, for example, [109], [199], [236], [299].

For these results, we take a general probability space $(\Omega, \mathcal{F}, P)$ together with any filtration $\mathbb{F} = (\mathcal{F}_t)_{t \in \mathbb{T}}$, where, as before, $\mathbb{T} = \{0, 1, \ldots, T\}$. Consider stochastic processes defined on this *filtered probability space* (also called *stochastic basis*) $(\Omega, \mathcal{F}, P, \mathbb{F}, \mathbb{T})$. Recall that a stochastic process $X = (X_t)$ is *adapted* to $\mathbb{F}$ if X_t is $\mathcal{F}_t$-measurable for each $t \in \mathbb{T}$.

Definition 2.3.1. An $\mathbb{F}$-adapted process $M = (M_t)_{t \in \mathbb{T}}$ is an $(\mathbb{F}, P)$-*martingale* if $E(|M_t|) < \infty$ for all $t \in \mathbb{T}$ and

$$E(M_{t+1} | \mathcal{F}_t) = M_t \text{ for all } t \in \mathbb{T} \setminus \{T\}. \tag{2.9}$$

If the equality in (2.9) is replaced by $\leq$ ($\geq$), we say that M is a *super-martingale* (*submartingale*).

Note that M is a martingale if and only if

$$E(\Delta M_{t+1} | \mathcal{F}_t) = 0 \text{ for all } t \in \mathbb{T} \setminus \{T\}.$$

Thus, in particular, $E(\Delta M_{t+1}) = 0$. Hence

$$E(M_{t+1}) = E(M_t) \text{ for all } t \in \mathbb{T} \setminus \{T\},$$

so that a martingale is 'constant on average'. Similarly, a submartingale increases, and a supermartingale decreases, on average. Thinking of M_t as representing the current capital of a gambler, a martingale therefore models a 'fair' game, while sub- and supermartingales model 'favourable' and 'unfavourable' games, respectively (as seen from the perspective of the gambler, of course!).

The linearity of the conditional expectation operator shows trivially that any linear combination of martingales is a martingale, and the tower property shows that M is a martingale if and only if

$$E\left(M_{s+t}\,|\mathcal{F}_s\right) = M_s \text{ for } t = 1, 2, \ldots, T - s.$$

Moreover, (M_t) is a martingale if and only if $(M_t - M_0)$ is a martingale, so we can assume $M_0 = 0$ without loss whenever convenient.

Many familiar stochastic processes are martingales. The simplest example is given by the successive conditional expectations of a single integrable random variable X. Set $M_t = E\left(X\,|\mathcal{F}_t\right)$ for $t \in \mathbb{T}$. By the tower property,

$$E\left(M_{t+1}\,|\mathcal{F}_t\right) = E\left(E\left(X\,|\mathcal{F}_{t+1}\right)|\mathcal{F}_t\right) = E\left(X\,|\mathcal{F}_t\right) = M_t.$$

The values of the martingale M_t are successive best mean-square estimates of X, as our 'knowledge' of X, represented by the σ-fields $\mathcal{F}_t$, increases with t.

More generally, if we model the price process of a stock by a martingale M, the conditional expectation (i.e., our best mean-square estimate at time s of the future value M_t of the stock) is given by its current value M_s. This generalises a well-known fact about processes with independent increments: if the zero-mean process W is adapted to the filtration $\mathbb{F}$ and $(W_{t+1} - W_t)$ is independent of $\mathcal{F}_t$, then $E\left(W_{t+1} - W_t\,|\mathcal{F}_t\right) = E\left(W_{t+1} - W_t\right) = 0$. Hence W is a martingale.

Exercise 2.3.2. Suppose that the centred (i.e., zero-mean) integrable random variables $(Y_t)_{t\in\mathbb{T}}$ are independent, and let $X_t = \sum_{u \leq t} Y_u$ for each $t \in \mathbb{T}$. Show that X is a martingale for the filtration it generates. What can we say when the Y_t have positive means?

Exercise 2.3.3. Let $(Z_n)_{n\geq1}$ be independent identically distributed random variables, adapted to a given filtration $(\mathcal{F}_n)_{n\geq0}$. Suppose further that each Z_n is non-negative and has mean 1. Show that $X(0) = 1$ and that $X_n = Z_1 Z_2 \cdots Z_n$ $(n \geq 1)$ defines a martingale for $(\mathcal{F}_n)$, provided all the products are integrable random variables, which holds, for example, if all $Z_n \in \mathcal{L}^\infty(\Omega, \mathcal{F}, P)$.

Note also that any *predictable* martingale is almost surely constant: if M_{t+1} is $\mathcal{F}_t$-measurable, we have $E\left(M_{t+1}\,|\mathcal{F}_t\right) = M_{t+1}$ and hence M_t and M_{t+1} are a.s. equal for all $t \in \mathbb{T}$. This is no surprise: if at time t we know the value of M_{t+1}, then our best estimate of that value will be perfect.

The construction of the gains process associated with a trading strategy now suggests the following further definition.

Definition 2.3.4. Let $M = (M_t)$ be a martingale and $\phi = (\phi_t)_{t \in \mathbb{T}}$ a predictable process defined on $(\Omega, \mathcal{F}, P, \mathbb{F}, \mathbb{T})$. The process $X = \phi \cdot M$ given for $t \geq 1$ by

$$X_t = \phi_1 \Delta M_1 + \phi_2 \Delta M_2 + \cdots + \phi_t \Delta M_t \tag{2.10}$$

and

$$X_0 = 0$$

is the *martingale transform* of M by ϕ.

Martingale transforms are the discrete analogues of the stochastic integrals in which the martingale M is used as the 'integrator'. The Itô calculus based upon this integration theory forms the mathematical backdrop to martingale pricing in continuous time, which comprises the bulk of this book. An understanding of the technically much simpler martingale transforms provides valuable insight into the essentials of stochastic calculus and its many applications in finance theory.

The Stability Property

If $\phi = (\phi_t)_{t \in \mathbb{T}}$ is bounded and predictable, then ϕ_{t+1} is $\mathcal{F}_t$-measurable and $\phi_{t+1} \Delta M_{t+1}$ remains integrable. Hence, for each $t \in \mathbb{T} \setminus \{T\}$, we have

$$E\left(\Delta X_{t+1} \,|\, \mathcal{F}_t\right) = E\left(\phi_{t+1} \Delta M_{t+1} \,|\, \mathcal{F}_t\right) = \phi_{t+1} E\left(\Delta M_{t+1} \,|\, \mathcal{F}_t\right) = 0.$$

Therefore $X = \phi \cdot M$ is a martingale with $X_0 = 0$. Similarly, if ϕ is also nonnegative and Y is a supermartingale, then $\phi \cdot Y$ is again a supermartingale.

This stability under transforms provides a simple, yet extremely useful, characterisation of martingales.

Theorem 2.3.5. *An adapted real-valued process M is a martingale if and only if*

$$E\left((\phi \cdot M)_t\right) = E\left(\sum_{u=1}^{t} \phi_u \Delta M_u\right) = 0 \text{ for } t \in \mathbb{T} \setminus \{0\} \tag{2.11}$$

for each bounded predictable process ϕ.

Proof. If M is a martingale, then so is the transform $X = \phi \cdot M$, and $X_0 = 0$. Hence $E\left((\phi \cdot M)_t\right) = 0$ for all $t \geq 1$ in $\mathbb{T}$.

Conversely, if (2.11) holds for M and every predictable ϕ, take $s > 0$, let $A \in \mathcal{F}_s$ be given, and define a predictable process ϕ by setting $\phi_{s+1} = 1_A$, and $\phi_t = 0$ for all other $t \in \mathbb{T}$. Then, for $t > s$, we have

$$0 = E\left((\phi \cdot M)_t\right) = E(1_A(M_{s+1} - M_s)).$$

Since this holds for all $A \in \mathcal{F}_s$, it follows that $E\left(\Delta M_{s+1} \,|\, \mathcal{F}_s\right) = 0$, so M is a martingale. $\qquad\square$

2.4   Arbitrage Pricing: Martingale Measures

Equivalent Martingale Measures

We now return to our study of viable securities market models. Recall that we assume as given an arbitrary complete measurable space $(\Omega, \mathcal{F})$ on which we consider various probability measures. We also consider a filtration $\mathbb{F} = (\mathcal{F}_t)_{t \in \mathbb{T}}$ such that $(\Omega, \mathcal{F}_0)$ is complete, and $\mathcal{F}_T = \mathcal{F}$. Finally, we are given a $(d+1)$-dimensional stochastic process $S = \{S_t^i : t \in \mathbb{T}, 0 \le i \le d\}$ with $S_0^0 = 1$ and S^0 interpreted as a riskless bond providing a discount factor $\beta_t = \frac{1}{S_t^0}$ and with S^i $(i = 1, 2, \ldots, d)$ interpreted as risky stocks. Recall that we are working in a general securities market model: we do *not* assume that the resulting market model is finite or that the filtration $\mathbb{F}$ is generated by S.

Suppose that the discounted vector price process $\bar{S}$ happens to be a martingale under some probability measure Q; that is,

$$E_Q\left(\Delta \bar{S}_t^i \,|\, \mathcal{F}_{t-1}\right) = 0 \text{ for } t \in \mathbb{T} \setminus \{0\} \text{ and } i = 1, 2, \ldots, d.$$

Note that, in particular, this assumes that the discounted prices are integrable with respect to Q. Suppose that $\theta = \left\{\theta_t^i : i \le d, t = 1, 2, \ldots, T\right\} \in \Theta_a$ is an admissible strategy whose discounted value process is also Q-integrable for each t. Recall from (2.7) that the discounted value process of θ has the form

$$\bar{V}_t(\theta) = V_0(\theta) + \overline{G}_t(\theta)$$

$$= \theta_1 \cdot S_0 + \sum_{u=1}^{t} \theta_u \cdot \Delta \bar{S}_u$$

$$= \sum_{i=1}^{d} \left(\theta_1^i S_0^i + \sum_{u=1}^{t} \theta_u \Delta \overline{S}_u^i\right).$$

Thus the discounted value process $\overline{V}(\theta)$ is a constant plus a finite sum of martingale transforms; and therefore it is a martingale with initial (constant) value $V_0(\theta)$. Hence we have $E\left(\overline{V}_t(\theta)\right) = E\left(V_0(\theta)\right) = V_0(\theta)$.

We want to show that this precludes the existence of arbitrage opportunities. If we know in advance that the value process of every admissible strategy is integrable with respect to Q, this is easy: if $V_0(\theta) = 0$ and $V_T(\theta) \ge 0$ a.s. (Q), but $E_Q\left(\overline{V}_t(\theta)\right) = 0$, it follows that $V_T(\theta) = 0$ a.s. (Q). This remains true a.s. (P), provided that the probability measure Q has the same null sets as P (we say that Q and P are *equivalent measures* and write $Q \sim P$). If such a measure can be found, then no self-financing strategy θ can lead to arbitrage; that is, the market is viable. This leads to an important definition.

Definition 2.4.1. A probability measure $Q \sim P$ is an *equivalent martingale measure* (EMM) for S if the discounted price process $\overline{S}$ is a (vector) martingale under Q for the filtration $\mathbb{F}$. That is, for each $i \leq d$ the discounted price process $\overline{S}^i$ is an $(\mathbb{F}, Q)$-martingale (recall that $\overline{S}^0 \equiv 1$).

To complete the argument, we need to justify the assumption that the value processes we have considered are Q-integrable. This follows from the following remarkable proposition (see also [132]).

Proposition 2.4.2. *Given a viable model $(\Omega, \mathcal{F}, P, \mathbb{T}, \mathbb{F}, S)$, suppose that Q is an equivalent martingale measure for S. Let H be an attainable claim. Then $\beta_T H$ is Q-integrable and the discounted value process for any generating strategy θ satisfies*

$$\overline{V}_t(\theta) = E_Q\left(\beta_T H \,|\, \mathcal{F}_t\right) \ \ a.s.\,(P) \ \ for \ all \ t \in \mathcal{F}. \tag{2.12}$$

Thus $\overline{V}(\theta)$ is a non-negative Q-martingale.

Proof. Choose a generating strategy θ for H and let $\overline{V} = \overline{V}(\theta)$ be its discounted value process. We show by backward induction that $\overline{V}_t \geq 0$ a.s.(P) for each t. This is clearly true for $t = T$ since $\overline{V}_T = \beta_T H \geq 0$ by definition. Hence suppose that $\overline{V}_t \geq 0$. If θ_t is unbounded, replace it by the bounded random vectors $\theta_t^n = \theta_t 1_{A_n}$, where $A_n = \{|\theta_t| \leq n\}$, so that $\overline{V}_{t-1}(\theta^n) = \overline{V}_{t-1}(\theta)1_{A_n}$ is $\mathcal{F}_{t-1}$-measurable and Q-integrable. Then we can write

$$\overline{V}_{t-1}(\theta^n) = \overline{V}_t(\theta^n) - \sum_{i=1}^{d}\theta_t^{n,i}\Delta\overline{S}_t^i \geq -\sum_{i=1}^{d}\theta_t^{n,i}\Delta\overline{S}_t^i,$$

so that

$$\overline{V}_{t-1}(\theta)1_{A_n} - \overline{V}_{t-1}(\theta^n)$$
$$= E_Q\left(\overline{V}_{t-1}(\theta^n)\,|\,\mathcal{F}_{t-1}\right)$$
$$\geq -\sum_{i=1}^{d}\theta_t^{n,i}E_Q\left(\Delta\overline{S}_t^i\,|\,\mathcal{F}_{t-1}\right)$$
$$= 0.$$

Letting n increase to ∞, we see that $\overline{V}_{t-1}(\theta) \geq 0$.

Thus we have a.s.(P) on each A_n that

$$E_Q\left(\overline{V}_t(\theta)\,|\,\mathcal{F}_{t-1}\right) - \overline{V}_{t-1}(\theta) = E_Q\left(\sum_{i=1}^{d}\theta_t^{n,i}\Delta\overline{S}_t^i\,|\,\mathcal{F}_{t-1}\right)$$
$$= \sum_{i=1}^{d}\theta_t^{n,i}E_Q\left(\Delta\overline{S}_t^i\,|\,\mathcal{F}_{t-1}\right)$$
$$= 0.$$

Again letting n increase to ∞, we have the identity

$$E_Q\left(\overline{V}_t(\theta)\,|\mathcal{F}_{t-1}\right) = \overline{V}_{t-1}(\theta)\ \text{a.s.}\,(P).\tag{2.13}$$

Finally, as $V_0 = \theta_1 \cdot S_0$ is a non-negative constant, it follows that $E_Q\left(\overline{V}_1\right) = V_0$. But by the first part of the proof $\overline{V}_1 \geq 0$ a.s. (P) and hence a.s. (Q), so $\overline{V}_1 \in L^1(Q)$. We can therefore begin an induction, using (2.13) at the inductive step, to conclude that $\overline{V}_t \in L^1(Q)$ and $E_Q\left(\overline{V}_t(\theta)\right) = V_0$ for all $t \in \mathbb{T}$. Thus $\overline{V}(\theta)$ is a non-negative Q-martingale, and since its final value is $\beta_T H$, it follows that $\overline{V}_t(\theta) = E_Q\left(\beta_T H\,|\mathcal{F}_t\right)$ a.s. (P) for each $t \in \mathbb{T}$. $\qquad\square$

Remark 2.4.3. The identity (2.12) not only provides an alternative proof of Lemma 2.2.5 by showing that the price of any attainable European claim is independent of the particular generating strategy, since the right-hand side does not depend on θ, but also provides a means of calculating that price without having to construct such a strategy. Moreover, the price does not depend on the choice of any particular equivalent martingale measure: the left-hand side does not depend on Q.

Exercise 2.4.4. Use Proposition 2.4.2 to show that if θ is a self-financing strategy whose final discounted value is bounded below a.s. $(P)$by a constant, then for any EMM Q the expected final value of θ is simply its initial value. What conclusion do you draw for trading only with strategies that have bounded risk?

We have proved that the existence of an equivalent martingale measure for S is *sufficient* for viability of the securities market model. In the next chapter, we discuss the *necessity* of this condition. Mathematically, the search for equivalent measures under which the given process $\overline{S}$ is a martingale is often much more convenient than having to show that no arbitrage opportunities exist for $\overline{S}$.

Economically, we can interpret the role of the martingale measure as follows. The probability assignments that investors make for various events do not enter into the derivation of the arbitrage price; the only criterion is that agents prefer more to less and would therefore become arbitrageurs if the market allowed arbitrage. The price we derive for the contingent claim H must thus be the same for all risk preferences (probability assignments) of the agents as long as they preclude arbitrage. In particular, an economy of risk-neutral agents will also produce the arbitrage price we derived previously. The equivalent measure Q, under which the discounted price process is a martingale represents the probability assignment made in this risk-neutral economy, and the price that this economy assigns to the claim will simply be the average (i.e., expectation under Q) discounted value of the payoff H.

Thus the existence of an equivalent martingale measure provides a general method for pricing contingent claims, which we now also formulate in terms of undiscounted value processes.

Martingale Pricing

We summarise the role played by martingale measures in pricing claims. Assume that we are given a viable market model $(\Omega, \mathcal{F}, P, \mathbb{F}, S)$ and some equivalent martingale measure Q. Recall that a *contingent claim* in this model is a non-negative ($\mathcal{F}$-measurable) random variable H representing a contract that pays out $H(\omega)$ dollars at time T if $\omega \in \Omega$ occurs. Its time 0 value or (current) *price* $\pi(H)$ is then the value that the parties to the contract would deem a 'fair price' for entering into this contract.

In a viable model, an investor could hope to evaluate $\pi(H)$ by constructing an admissible trading strategy $\theta \in \Theta_a$ that exactly replicates the returns (cash flow) yielded by H at time T. For such a strategy θ, the initial investment $V_0(\theta)$ would represent the price $\pi(H)$ of H. Recall that H is an *attainable claim* in the model if there exists a *generating strategy* $\theta \in \Theta_a$ such that $V_T(\theta) = H$, or, equivalently, $\overline{V}_t(\theta) = \beta_T H$. But as Q is a martingale measure for S, $\overline{V}(\theta)$ is, up to a constant, a martingale transform, and hence a martingale, under Q, it follows that for all $t \in \mathbb{T}$,

$$\overline{V}_t(\theta) = E_Q\left(\beta_T H \,|\, \mathcal{F}_t\right),$$

and thus

$$V_t(\theta) = \beta_t^{-1} E_Q\left(\beta_T H \,|\, \mathcal{F}_t\right) \tag{2.14}$$

for any $\theta \in \Theta_a$. In particular,

$$\pi(H) = \overline{V}_0(\theta) = E_Q\left(\beta_T H \,|\, \mathcal{F}_0\right) = E_Q\left(\beta_T H\right). \tag{2.15}$$

Market models in which all European contingent claims are attainable are called *complete*. These models provide the simplest class in terms of option pricing since any contingent claim can be priced simply by calculating its (discounted) expectation relative to an equivalent martingale measure for the model.

Uniqueness of the EMM

We have shown in Proposition 2.4.2 that for an attainable European claim H the identity $\overline{V}_0(\theta) = E_Q\left(\beta_T H\right)$ holds for *every* EMM Q in the model and for every replicating strategy θ.

This immediately implies that in a complete model the EMM must be unique. For if Q and R are EMMs in a complete pricing model, then any European claim is attainable. It follows that $E_Q\left(\beta_T H\right) = E_R\left(\beta_T H\right)$ and hence also

$$E_Q\left(H\right) = E_R\left(H\right), \tag{2.16}$$

upon multiplying both sides by β_T, which is non-random. In particular, equation (2.16) holds when the claim is the indicator function of an arbitrary set $F \in \mathcal{F}_T = \mathcal{F}$. This means that $Q = R$; hence the EMM is unique. Moreover, our argument again verifies that the *Law of One Price* (see Lemma 2.2.5) must hold in a viable model; that is, we cannot have two admissible trading strategies θ, θ' that satisfy $V_T(\theta) = V_T(\theta')$ but $V_0(\theta) \neq V_0(\theta')$. Our modelling assumptions are thus sufficient to guarantee consistent pricing mechanisms (in fact, this consistency criterion is strictly weaker than viability; see [241] for simple examples).

The Law of One Price permits valuation of an attainable claim H through the initial value of a self-financing strategy that generates H; the valuation technique using risk-neutral expectations gives the price $\pi(H)$ *without* prior determination of such a generating strategy. In particular, consider a single-period model and a claim H (an Arrow-Debreu security) defined by

$$H(\omega) = \begin{cases} 1 & \text{if } \omega = \omega' \\ 0 & \text{otherwise,} \end{cases}$$

where $\omega' \in \Omega$ is some specified state. If H is attainable, then

$$\pi(H) = E_Q\left(\beta_T H\right) = \frac{1}{\beta_T} Q(\{\omega'\}).$$

This holds even when β is random. The ratio $\frac{Q(\{\omega'\})}{\beta_T(\omega')}$ is known as the *state price* of ω'. In a finite market model, we can similarly define the change of measure density $\Lambda = \Lambda(\{\omega\})_{\omega \in \Omega}$, where $\Lambda(\{\omega\}) = \frac{Q(\{\omega\})}{P(\{\omega\})}$) as the *state price density*. See [241] for details of the role of these concepts.

Superhedging

We adopt a slightly more general approach (which we shall develop further in Chapter 5 and exploit more fully for continuous-time models in Chapters 7 to 10) to give an explicit justification of the 'fairness' of the option price when viewed from the different perspectives of the buyer and the seller (option writer), respectively.

Definition 2.4.5. Given a European claim $H = f(S_T)$, an (x, H)−*hedge* is an initial investment x in an admissible strategy θ such that $V_T(\theta) \geq H$ a.s.

This approach to hedging is often referred to as defining a *superhedging strategy*. This clearly makes good sense from the seller's point of view, particularly for claims of American type, where the potential liability may not always be covered exactly by replication. By investing x in the strategy θ at time 0, an investor can cover his potential liabilities whatever the stock price movements in $[0, T]$. When there is an admissible strategy θ exactly replicating H, the initial investment $x = \pi(H)$ is an example of an (x, H)−hedge. Since the strategy θ exactly covers the final liabilities, (i.e., $V_T(\theta) = H$), we call this a *minimal hedge*.

All prices acceptable to the option seller must clearly ensure that the initial receipts for the option enable him to invest in a hedge (i.e., must

ensure that there is an admissible strategy whose final value is at least H).
The *seller's price* can thus be defined as

$$\pi_s = \inf \left\{ z \geq 0 : \text{there exists } \theta \in \Theta_a \text{ with } V_T(\theta) = z + G_T(\theta) \geq H \text{ a.s.} \right\}.$$

The buyer, on the other hand, wants to pay no more than is needed to
ensure that his final wealth suffices to cover the initial outlay, or borrowings.
So his price will be the maximum he is willing to borrow, $y = -V_0$, at time
0 to invest in an admissible strategy θ, so that the sum of the option payoff
and the gains from following θ cover his borrowings. The *buyer's price* is
therefore

$$\pi_b = \sup \left\{ y \geq 0 : \text{there exists } \theta \in \Theta_a \text{ with } -y + G_T(\theta) \geq -H \text{ a.s.} \right\}.$$

In particular, θ must be self-financing, so that $\beta_T V_T(\theta) = V_0 + \beta_T G_T(\theta)$,
and since βS is a Q-martingale, we have $E_Q(\beta_T G_T(\theta)) = 0$. So the seller's
price requires that $z \geq E_Q(\beta_T H)$ for each z in (2.21)and hence $\pi_s \geq E_Q(\beta_T H)$.
Similarly, for the buyer's price, we require that $-y + E_Q(\beta_T H) \geq 0$ and
hence also $\pi_b \leq E_Q(\beta_T H)$. We have proved the following proposition.

Proposition 2.4.6. *For any integrable European claim H in a viable pricing model,*

$$\pi_b \leq E_Q(\beta_T H) \leq \pi_s. \tag{2.17}$$

If the claim H is attained by an admissible strategy θ, the minimal
initial investment z in the strategy θ that will yield final wealth $V_T(\theta) = H$
is given by $E_Q(\beta_T H)$, and conversely this represents the maximal initial
borrowing y required to ensure that $-y + G_T(\theta) + H \geq 0$. This proves the
following corollary.

Corollary 2.4.7. *If the European claim H is attainable, then the buyer's
price and seller's price are both equal to $E_Q(\beta_T H)$. Thus, in a com-
plete model, every European claim H has a unique price, given by $\pi = E_Q(\beta_T H)$, and the generating strategy θ for the claim is a minimal hedge.*

2.5 Strategies Using Contingent Claims

Our definition of arbitrage involves trading strategies that include only
primary securities (i.e., a riskless bank account which acts as numéraire
and a collection of risky assets, which we called 'stocks' for simplicity).
Our analysis assumes that these assets are traded independently of other
assets. In real markets, however, investors also have access to derivative
(or secondary) securities, whose prices depend on those of some underlying
assets. We have grouped these under the term 'contingent claim' and we
have considered how such assets should be priced. Now we need to consider
an extended concept of arbitrage since it is possible for an investor to build

a trading strategy including both primary securities and contingent claims, and we use this combination to seek to secure a riskless profit. We must therefore identify circumstances under which the market will preclude such profits.

Thus our concept of a trading strategy should be extended to include such combinations of primary and secondary securities, and we shall show that the market remains viable precisely when the contingent claims are priced according to the martingale pricing techniques for European contingent claims that we have developed. To achieve this, we need to restrict attention to trading strategies involving a bank account, stocks, and *attainable European* contingent claims.

Assume that a securities market model $(\Omega, \mathcal{F}, P, \mathbb{T}, \mathbb{F}, S)$ is given. We allow trading strategies to include attainable European claims, so that the value of the investor's portfolio at time $t \in \mathbb{T}$ will have the form

$$V_t = \theta_t \cdot S_t + \gamma_t \cdot Z_t = \sum_{i=0}^{d} \theta_t^i S_t^i + \sum_{j=1}^{m} \gamma_t^j Z_t^j, \qquad (2.18)$$

where S^0 is the bank account, $\{S_t^i : i = 1, 2, \ldots, d\}$ are the prices of d risky stocks, and $Z_t = (Z_t^j)_{j \leq m}$ are the values of m attainable European contingent claims with time T payoff functions given by $(Z^j)_{j \leq m}$. We write $S = (S^i)_{0 \leq i \leq d}$. Recall that an attainable claim Z^j can be replicated exactly by a self-financing strategy involving only the process S. The holdings of each asset are assumed to be predictable processes, so that for $t = 1, 2, \ldots, T$, θ_t^i and γ_t^j are $\mathcal{F}_{t-1}$-measurable for $i = 0, 1, \ldots, d$ and $j = 1, 2, \ldots, m$. We call our model an *extended securities market model.*

The trading strategy $\phi = (\theta, \gamma)$ is self-financing if its initial value is

$$V_0(\phi) = \theta_1 \cdot S_0 + \gamma_1 \cdot Z_0$$

and for $t = 1, 2, \ldots, T - 1$ we have

$$\theta_t \cdot S_t + \gamma_t \cdot Z_t = \theta_{t+1} \cdot S_t + \gamma_{t+1} \cdot Z_t. \qquad (2.19)$$

Note that $\cdot$ denotes the inner product in $\mathbb{R}^{d+1}$ and $\mathbb{R}^m$, respectively. A new feature of the extended concept of a trading strategy is that the final values of some of its components are known in advance since the final portfolio has value

$$V_T(\phi) = \theta_T \cdot S_T + \gamma_T \cdot Z,$$

as $Z = (Z_T^j)_{j \leq m}$ represents the m payoff functions of the European claims. Moreover, unlike stocks, we have to allow for the possibility that the values Z_t^j can be zero or negative (as can be the case with forward contracts). However, with these minor adjustments we can regard the model simply as a securities market model with one riskless bank account and $d + m$ risky assets. With this in mind, we extend the concept of arbitrage to this model.

Definition 2.5.1. An arbitrage opportunity in the extended securities market model is a self-financing trading strategy ϕ such that $V_0(\phi) = 0$, $V_T(\phi) \geq 0$, and $E_P(V_T(\phi)) > 0$. We call the model *arbitrage-free* if no such strategy exists.

As in the case of weak arbitrage in Section 2.2, we do not demand that the value process remain non-negative throughout $\mathbb{T}$. That this has no effect on the pricing of the contingent claims can be seen from the following result.

Theorem 2.5.2. *Suppose that $(\Omega, \mathcal{F}, P, \mathbb{T}, \mathbb{F}, S)$ is an extended securities market model admitting an equivalent martingale measure Q. The model is arbitrage-free if and only if every attainable European contingent claim with payoff Z has value process given by $\left\{ S_t^0 E_Q \left(\frac{Z}{S_T^0} \mid \mathcal{F}_t \right) : t \in \mathbb{T} \right\}$.*

Proof. Let $\theta = (\theta^i)_{i \leq d}$ be a generating strategy for Z. The value process of θ is then given as in equation (2.14) by $V_t(\theta) = S_t^0 E_Q \left(\frac{Z}{S_T^0} \mid \mathcal{F}_t \right)$ since the discount process is $\beta_t = \frac{1}{S_t^0}$ when S^0 is the numeraire.

We need to show that the model is arbitrage-free precisely when the value process $(Z_t)_{t \in \mathbb{T}}$ of the claim Z is equal to $(V_t(\theta))_{t \in \mathbb{T}}$. Suppose therefore that for some $u \in \mathbb{T}$ these processes differ on a set D of positive P-measure. We first assume that $D = \{Z_u > V_u(\theta)\}$, which belongs to $\mathcal{F}_u$. To construct an arbitrage, we argue as follows: do nothing for $\omega \notin D$, and for $\omega \in D$ wait until time u. At time u, sell short one unit of the claim Z for $Z_u(\omega)$, invest $V_u(\omega)$ of this in the portfolio of stocks and bank account according to the prescriptions given by strategy θ, and bank the remainder $(Z_u(\omega) - V_u(\omega))$ until time T. This produces a strategy ϕ, where

$$\phi_t = \begin{cases} 0 & \text{if } t \leq u \\ \left(\theta_t^0 + \frac{Z_u - V_u(\theta)}{S_u^0}, \theta_t^1, \ldots, \theta_t^d, -1 \right) \mathbf{1}_D & \text{if } t > u. \end{cases}$$

It is not hard to show that this strategy is self-financing; it is evidently predictable. Its value process $V(\phi)$ has $V_0(\phi) = 0$ since in fact $V_t(\phi) = 0$ for all $t \leq u$, while $V_T(\phi)(\omega) = 0$ for $\omega \notin D$. For $\omega \in D$, we have

$$(\theta_T \cdot S_T)(\omega) = V_T(\theta)(\omega) = Z(\omega)$$

since θ replicates Z. Hence

$$V_T(\phi)(\omega) = \left(\theta_T \cdot S_T + (Z_u - V_u(\theta)) \frac{S_T^0}{S_u^0} - Z \right)(\omega)$$

$$= \left((Z_u - V_u(\theta)) \frac{S_T^0}{S_u^0} \right)(\omega)$$

$$> 0.$$

This shows that ϕ is an arbitrage opportunity in the extended model since $V_T(\phi) \geq 0$ and $P(V_T(\phi) > 0) = P(D) > 0$.

To construct an arbitrage when $Z_u < V_u(\theta)$ for some $u \leq T$ on a set E with $P(E) > 0$, we simply reverse the positions described above. On E at time u, shortsell the amount $V_u(\theta)$ according to the strategy θ, buy one unit of the claim Z for Z_u, place the difference in the bank, and do nothing else. Hence, if the claim Z does not have the value process $V(\theta)$ determined by the replicating strategy θ, the extended model is not arbitrage-free.

Conversely, suppose that every attainable European claim Z has its value function given via the EMM Q as $Z_t = S_t^0 E_Q\left(\frac{Z}{S_T^0} | \mathcal{F}_t\right)$ for each $t \leq T$, and let $\psi = (\phi, \gamma)$ be a self-financing strategy, involving S and m attainable European claims $(Z^j)_{j \leq m}$, with $V_0(\psi) = 0$ and $V_T(\psi) \geq 0$. We show that $P(V_T(\psi) = 0) = 1$, so that ψ cannot be an arbitrage opportunity in the extended model. Indeed, consider the discounted value process $\overline{V}(\psi) = \frac{V(\psi)}{S^0}$ at time $t > 0$:

$$E_Q\left(\overline{V}_t(\psi) | \mathcal{F}_{t-1}\right) = E_Q\left(\sum_{i=0}^{d} \phi_t^i \overline{S}_t^i + \sum_{j=1}^{m} \gamma_t^j \frac{Z_t^j}{S_t^0} | \mathcal{F}_{t-1}\right)$$

$$= \sum_{i=0}^{d} \phi_t^i E_Q\left(\overline{S}_t^i | \mathcal{F}_{t-1}\right) + \sum_{j=1}^{m} \gamma_t^j E_Q\left(\overline{V}_t^j(\theta^j) | \mathcal{F}_{t-1}\right).$$

Here we use the fact that $\overline{S}^i = \frac{S^i}{S^0}$ is a martingale under Q and, defining $\overline{V}_t^j(\theta^j)$ as the discounted value process of the replicating strategy for the claim Z^j, we see that $\overline{V}_t^j(\theta^j) = E_Q\left(\frac{Z^j}{S_T^0} | \mathcal{F}_t\right) = \frac{Z_t^j}{S_t^0}$. Since each process $\overline{V}^j(\theta^j)$, $j \leq m$, is a Q-martingale, it follows that

$$E_Q\left(\overline{V}_t(\psi) | \mathcal{F}_{t-1}\right) = \sum_{i=0}^{d} \phi_t^i \overline{S}_{t-1}^i + \sum_{j=1}^{m} \gamma_t^j \overline{V}_{t-1}^j(\theta^j) = \overline{V}_{t-1}(\psi)$$

since the strategy $\psi = (\phi, \gamma)$ is self-financing, so that $\overline{V}(\psi)$ is also a Q-martingale. Consequently, $E_Q\left(\overline{V}_t(\psi)\right) = E_Q\left(V_0(\psi)\right) = 0$. Therefore $Q(\overline{V}_T(\psi) = 0) = 1$, and since $Q \sim P$ it follows that $P(V_T(\psi) = 0) = 1$. Therefore the extended securities market model is arbitrage-free. $\square$

This result should not come as a surprise. It remains the case that the only independent sources of randomness in the model are the stock prices $S_1, S_2, \ldots, S_d$, since the contingent claims used to construct trading strategies are priced via an equivalent measure for which their discounted versions are martingales. However, it does show that the methodology is consistent. We return to extended market models when examining possible arbitrage-free prices for claims in incomplete models in Chapter 4.

Some Consequences of Call-Put parity

In the call-put parity relation (1.3), the discount rate is given by $\beta_{t,T} = \beta^{T-t}$, where $\beta = (1 + r)$. Write (1.3) in the form

$$S_t = C_t - P_t + \beta^{T-t}K. \tag{2.20}$$

With the price of each contingent claim expressed at the expectation under the risk-neutral measure Q of its discounted final value, we show that the right-hand side of (2.20) is independent of K. Indeed,

$$S_t = \beta^{T-t}[E_Q\left((S_T - K)^+\right) - E_Q\left((K - S_T)^+\right) + K]$$

$$= \beta^{T-t}\left(\int_{\{S_T \geq K\}}(S_T - K)dQ - \int_{\{S_T < K\}}(K - S_T)dQ + K\right)$$

$$= \beta^{T-t}\left(\int_\Omega(S_T - K)dQ + K\right)$$

$$= \beta^{T-t}E_Q(S_T)$$

$$= \beta_t^{-1}E_Q\left(\beta^T S_T\right).$$

This shows that call-put parity is a consequence of the martingale property of the discounted price under Q in any market model that allows pricing of contingent claims by expectation under an equivalent martingale measure.

Remark 2.5.3. The identity also leads to the following interesting observation due to Marek Capinski, which first appeared in [35]. Recall the Modigliani-Miller theorem (see [20]), which states that the value of a firm is independent of the way in which it is financed. Since its value is represented by the sum of its equity (stock) and debt, the theorem states that the level of debt has no impact on the value of the firm. This can be interpreted in terms of options, as follows.

If the firm's borrowings at time 0 are represented by $\beta_T K$, so that it faces repayment of debt at K by time T, the stockholders have the option to buy back this debt at that time, in order to avert bankruptcy of the firm. They will only do so if the value S_T of the firm at time T is at least K. The firm's stock can therefore be represented as a European call option on S with payoff K at time T, and thus the current (time 0) value of the stock is the call option price C_0. The total current value of the firm is $S_0 = C_0 - P_0 + \beta^T K$, where P_0 is a put option on S with the same strike and horizon as the call. The calculation above shows that S_0 is independent of K, as the Modigliani-Miller theorem claims. Moreover, the current value of the debt is given via the call-put parity relation as $(\beta^T K - P_0)$. This is lower than the present value $\beta_T K$ of K, so that P_0 reflects the default risk (i.e., risk that the debt may not be recovered in full at time T).

2.6   Example: The Binomial Model

We now take another look at the Cox-Ross-Rubinstein binomial model, which provides a very simple, yet striking, example of the strength of the martingale methods developed so far.

The CRR Market Model

The Cox-Ross-Rubinstein binomial market model was described in Chapter 1. Recall that we assumed that $d = 1$. There is a single stock S^1 and a riskless bond S^0, which accrues interest at a fixed rate $r > 0$. Taking $S_0^0 = 1$, we have $S_t^0 = (1 + r)^t$ for $t \in \mathbb{T}$, and hence $\beta_t = (1 + r)^{-t}$. The ratios of successive stock values are Bernoulli random variables; that is, for all $t < T$, either $S_t^1 = S_{t-1}^1(1+a)$ or $S_t^1 = S_{t-1}^1(1+b)$, where $b > a > -1$ are fixed throughout, while S_0^1 is constant. We can thus conveniently choose the sample space

$$\Omega = \{1 + a, 1 + b\}^T$$

together with the natural filtration $\mathbb{F}$ generated by the stock price values; that is, $\mathcal{F}_0 = \{\emptyset, \Omega\}$, and $\mathcal{F}_t = \sigma(S_u^1 : u \leq t)$ for $t > 0$. Note that $\mathcal{F}_T = \mathcal{F} = 2^\Omega$ is the σ-field of all subsets of Ω.

The measure P on Ω is the measure induced by the ratios of the stock values. More explicitly, we write S for S^1 for the rest of this section to simplify the notation, and set $R_t = \frac{S_t}{S_{t-1}}$ for $t > 0$. For $\omega = (\omega_1, \omega_2, \ldots, \omega_T)$ in Ω, define

$$P(\{\omega\}) = P(R_t = \omega_t, t = 1, 2, \ldots, T). \tag{2.21}$$

For any probability measure Q on $(\Omega, \mathcal{F})$, the relation $E_Q\left(\overline{S}_t \,|\, \mathcal{F}_{t-1}\right) = \overline{S}_{t-1}$ is equivalent to

$$E_Q\left(R_t \,|\, \mathcal{F}_{t-1}\right) = 1 + r$$

since $\frac{\beta_t}{\beta_{t-1}} = 1 + r$. Hence, if Q is an equivalent martingale measure for S, it follows that $E_Q(R_t) = 1 + r$. On the other hand, R_t only takes the values $1 + a$ and $1 + b$; hence its average value can equal $1 + r$ only if $a < r < b$. We have yet again verified the following result.

Lemma 2.6.1. *For the binomial model to have an EMM, we must have*

$$a < r < b.$$

When the binomial model is viable, there is a *unique* equivalent martingale measure Q for S. We construct this measure in the following lemma.

Lemma 2.6.2. *The discounted price process $\overline{S}$ is a Q-martingale if and only if the random variables (R_t) are independent, identically distributed, and $Q(R_1 = 1 + b) = q$ and $Q(R_1 = 1 + a) = 1 - q$, where $q = \frac{r-a}{b-a}$.*

Proof. Under independence, the (R_t) satisfy

$$E_Q\left(R_t\,|\,\mathcal{F}_{t-1}\right) = E_Q\left(R_t\right) = q(1+b)+(1-q)(1+a) = q(b-a)+1+a = 1+r.$$

Hence, by our earlier discussion, $\overline{S}$ is a Q-martingale.

Conversely, if $E_Q\left(R_t\,|\,\mathcal{F}_{t-1}\right) = 1+r$, then, since R_t takes only the values $1+a$ and $1+b$, we have

$$(1 + a)Q(R_t = 1 + a\,|\,\mathcal{F}_{t-1}) + (1 + b)Q(R_t = 1 + b\,|\,\mathcal{F}_{t-1}) = 1 + r,$$

while

$$Q(R_t = 1 + a\,|\,\mathcal{F}_{t-1}) + Q(R_t = 1 + b\,|\,\mathcal{F}_{t-1}) = 1.$$

Letting $q = Q(R_t = 1 + b\,|\,\mathcal{F}_{t-1})$, we obtain

$$(1 + a)(1 - q) + (1 + b)q = 1 + r.$$

Hence $q = \frac{r-a}{b-a}$. The independence of the R_t follows by induction on $t > 0$. For $\omega = (\omega_1, \omega_2, \ldots, \omega_T) \in \Omega$, we see inductively that

$$Q\left(R_1 = \omega_1, R_2 = \omega_2, \ldots, R_t = \omega_t\right) = \prod_{i=1}^{t} q_i,$$

where $q_i = q$ when $\omega_i = 1 + b$ and equals $1 - q$ when $\omega_i = 1 + a$. Thus the (R_t) are independent and identically distributed as claimed.   □

Remark 2.6.3. Note that $q \in (0,1)$ if and only if $a < r < b$. Thus a viable binomial market model admits a *unique* EMM given by Q as in Lemma 2.6.2.

The CRR Pricing Formula

The CRR pricing formula, obtained in Chapter 1 by an explicit hedging argument, can now be deduced from our general martingale formulation by calculating the Q-expectation of a European call option on the stock. More generally, the value of the call $C_T = (S_T - K)^+$ at time $t \in \mathbb{T}$ is given by (2.14); that is,

$$V_t(C_T) = \frac{1}{\beta_t} E_Q\left(\beta_T C_T\,|\,\mathcal{F}_t\right).$$

Since $S_T = S_t \prod_{u=t+1}^{T} R_u$ (by the definition of (R_u)), we can calculate this expectation quite easily since S_t is $\mathcal{F}_t$-measurable and each R_u $(u > t)$ is independent of $\mathcal{F}_t$. Indeed,

$$V_t(C_T) = \beta_t^{-1}\beta_T E_Q\left(\left[S_t \prod_{u=t+1}^{T} R_u - K\right]^+\,|\,\mathcal{F}_t\right)$$

$$= (1+r)^{t-T} E_Q \left(\left[S_t \prod_{u=t+1}^{T} R_u - K \right]^+ | \mathcal{F}_t \right)$$

$$= v(t, S_t). \tag{2.22}$$

Here

$$v(t, x) = (1+r)^{t-T} E_Q \left(\left[x \prod_{u=t+1}^{T} R_u - K \right]^+ \right)$$

$$= (1+r)^{t-T} \sum_{u=0}^{T-t} \binom{T-t}{u} q^u (1-q)^{T-t-u} \left[x(1+b)^u (1+a)^{T-t-u} - K \right]^+$$

and, in particular, the price at time 0 of the European call option C with payoff $C_T = (S_T - K)^+$ is given by

$$v(0, S_0) = (1+r)^{-T} \sum_{u=A}^{T} \binom{T}{u} q^u (1-q)^{T-u} \left[S_0(1+b)^u (1+a)^{T-u} - K \right],$$

$$\tag{2.23}$$

where A is the first integer k for which $S_0(1+b)^k (1+a)^{T-k} > K$. The CRR option pricing formula (1.5.3) now follows exactly as in Chapter 1.

Exercise 2.6.4. Show that for the replicating strategy $\theta = (\theta^0, \theta^1)$ describing the value process of the European call C, the stock portfolio θ^1 can be expressed in terms of the differences of the value function as $\theta_t^1 = \theta(t, S_{t-1})$, where

$$\theta(t, x) = \frac{v(t, x(1+b)) - v(t, x(1+a))}{x(b-a)}.$$

Exercise 2.6.5. Derive the call-put parity relation (2.20) by describing the values of the contingent claims involved as expectations relative to Q.

2.7   From CRR to Black-Scholes

Construction of Approximating Binomial Models

The binomial model contains all the information necessary to deduce the famous Black-Scholes formula for the price of a European call option in a continuous-time market driven by Brownian motion. A detailed discussion of the mathematical tools used in that model is deferred until Chapter 6, but we now describe how the random walks performed by the steps in the binomial tree lead to Brownian motion as a limiting process when we reduce the step sizes continually while performing an ever larger number of steps within a fixed time interval $[0, T]$. From this we will see how the Black-Scholes price arises as a limit of CRR prices.

Consider a one-dimensional stock price process $S = (S_t)$ on the finite time interval $[0, T]$ on the real line, together with a European put option with payoff function $f_T = (K - S_T)^+$ on this stock. We use put options here because the payoff function f is bounded, thus allowing us to deduce that the relevant expectations (using EMMs) converge once we have shown via a central limit theorem that certain random variables converge weakly. The corresponding result for call options can then be derived using call-put parity.

We wish to construct a discrete-time binomial model beginning with the same constant stock price S_0 and with N steps in $[0, T]$. Thus we let $h_N = \frac{T}{N}$ and define the discrete timeline $\mathbb{T}_N = \{0, h_N, 2h_N, \dots, Nh_N\}$. The European put P^N with strike K and horizon T is then defined on $\mathbb{T}_N$. By (2.14), (exactly as in the derivation of (2.22)), P^N has CRR price P_0^N given by

$$P_0^N = (1 + \rho_N)^{-N} E_{Q_N}\left(\left[K - S_0 \prod_{k=1}^{N} R_k^N\right]^+\right), \qquad (2.24)$$

where, writing S_k^N for the stock price at time kh_N, the ratios $R_k^N = \frac{S_k^N}{S_{k-1}^N}$ take values $1 + b_N$ or $1 + a_N$ at each discrete time point kh_N $(k \leq N)$.

The values of a_N, b_N and the riskless interest rate ρ_N have yet to be chosen. Once they are fixed, with $a_N < \rho_N < b_N$, they will uniquely determine the risk-neutral probability measure Q_N for the Nth binomial model since by Lemma 2.6.2 the binomial random variables $(R_k^N)_{k \leq N}$ are then an independent and identically distributed sequence. We obtain, as before, that

$$Q_N(R_1^N = 1 + b_N) = q_N = \frac{\rho_N - a_N}{b_N - a_N}. \qquad (2.25)$$

We treat the parameters from the Black-Scholes model as given and adjust their counterparts in our CRR models in order to obtain convergence. To this end, we fix $r \geq 0$ and set $\rho_N = rh_N$, so that the discrete-time riskless rate satisfies $\lim_{N \to \infty}(1 + \rho_N)^N = e^{rT}$, so that r acts as the 'instantaneous' rate of return.

Fix $\sigma > 0$, which will act as the volatility per unit time of the Black-Scholes stock price, and for each fixed N we now fix a_N, b_N by demanding that the discounted logarithmic returns are given by

$$\log\left(\frac{1 + b_N}{1 + \rho_N}\right) = \sigma\sqrt{h_N} = \sigma\sqrt{\frac{T}{N}}, \quad \log\left(\frac{1 + a_N}{1 + \rho_N}\right) = -\sigma\sqrt{h_N} = -\sigma\sqrt{\frac{T}{N}},$$

so that

$$u_N = 1 + b_N = \left(1 + \frac{rT}{N}\right)e^{\sigma\sqrt{\frac{T}{N}}}, \quad d_N = 1 + a_N = \left(1 + \frac{rT}{N}\right)e^{-\sigma\sqrt{\frac{T}{N}}}.$$

Note that the discount factor at each step is $1 + \rho_N = 1 + \frac{rT}{N}$ for each $k \le N$. The random variables

$$\left\{ Y_k^N = \log\left(\frac{R_k^N}{1 + \rho_N}\right) : k \le N \right\}$$

are independent and identically distributed. We shall consider their sum

$$Z_N = \sum_{k=1}^{N} Y_k^N = \sum_{k=1}^{N} R_k^N - N\log(1 + \rho_N)$$

for each N. The discounted stock price is thus

$$\overline{S}_N^N = (1 + \rho_N)^{-N} \prod_{k=1}^{n} R_k^N = \exp\left\{ \sum_{k=1}^{N} Y_k^N \right\} = e^{Z_N},$$

so that the Nth put option price becomes

$$P_0^N = E_{Q_N}\left(\left[\left(1 + \frac{rt}{N}\right)^{-N} K - S_0 e^{Z_N} \right]^+ \right). \tag{2.26}$$

Convergence in Distribution

The values taken by Y_1^N are $\pm\sigma\sqrt{h_N}$, so its second moment is $\sigma^2 h_N = \sigma^2 \frac{T}{N}$, while its mean is given by

$$\mu_N = (2q_N - 1)\sigma\sqrt{h_N} = (2q_N - 1)\sigma\sqrt{\frac{T}{N}}.$$

Our choices will imply that q_N converges to $\frac{1}{2}$ as $N \to \infty$. We show this by checking the rate of convergence. First recall some notation: $a_N = a + o\left(\frac{1}{N}\right)$ means that $N(a_N - a) \to 0$ as $N \to \infty$.

Since $1 - q_N = \frac{u_N - \rho_N}{u_N - d_N}$, we see that $2q_N - 1$ is of order $\frac{1}{\sqrt{N}}$:

$$2q_N - 1 = 1 - 2(1 - q_N) = 1 - 2\left(\frac{e^{\sigma\sqrt{h_N}} - 1}{e^{\sigma\sqrt{h_N}} - e^{-\sigma\sqrt{h_N}}}\right)$$

$$= 1 - \frac{e^{\sigma\sqrt{h_N}} - 1}{\sinh(\sigma\sqrt{h_N})}.$$

Expanding into Taylor series the right-hand side has the form

$$1 - \frac{x + \frac{x^2}{2!} + \frac{x^3}{3!} + \cdots}{x + \frac{x^3}{3!} + \cdots} = \frac{-\frac{x^2}{2} - \frac{x^4}{4!} + \cdots}{x + \frac{x^3}{3!} + \cdots},$$

so that $2q_N - 1 = -\frac{1}{2}\sigma\sqrt{h_N} + o\left(\frac{1}{N}\right)$. Thus $\mu_N = -\frac{1}{2}\frac{\sigma^2 T}{N} + o\left(\frac{1}{N}\right)$, so that $N\mu_N \to -\frac{1}{2}\sigma^2 T$ as $N \to \infty$.

Since the second moment of Y_1^N is $\sigma^2 \frac{T}{N}$, its variance σ_N^2 therefore satisfies

$$\sigma_N^2 = \sigma^2 \frac{T}{N} + o\left(\frac{1}{N}\right). \tag{2.27}$$

We apply the central limit theorem for triangular arrays (see, e.g. , [168, VII.5.4] or [45, Corollary to Theorem 3.1.2]) in the following form to the independent and identically distributed random variables (Y_k^N) for $k \leq N$ and $N \in \mathbb{N}$.

Theorem 2.7.1 (Central Limit Theorem). *For $N \geq 1$, let $(Y_k^N)_{k \leq N}$ be an independent and identically distributed sequence of random variables, each with mean μ_N and variance σ_N^2. Suppose that there exist real μ and $\Sigma^2 > 0$ such that $N\mu_N \to \mu$ and $\sigma_N^2 = \Sigma^2 + o\left(\frac{1}{N}\right)$ as $N \to \infty$. Then the sums $Z_N = \sum_{k=1}^{N} Y_k^N$ converge in distribution to a random variable $Z \sim \mathcal{N}(\mu, \Sigma^2)$.*

We prove this by verifying the Lindeberg-Feller condition for the Y_k^N, namely that for all $\varepsilon > 0$

$$\sum_{k=1}^{N} E_{Q_N}\left((Y_k^N)^2 1_{\{|Y_k^N| > \varepsilon\}}\right) \to 0 \text{ as } N \to \infty. \tag{2.28}$$

We have seen that for fixed N and all $k \leq N$, $(Y_k^N)^2$ is constant on Ω and takes the value $\frac{\sigma^2 T}{N}$. Therefore, using Chebychev's inequality with $|Y_k^N| = \sigma\sqrt{\frac{T}{N}}$, we see that for each $k \leq N$

$$E_{Q_N}\left((Y_k^N)^2 1_{\{|Y_k^N| > \varepsilon\}}\right) = \frac{\sigma^2 T}{N} P(|Y_k^N| > \varepsilon) \leq \frac{\sigma^2 T}{N} \frac{E(|Y_k^N|)}{\varepsilon},$$

and since the right-hand side equals $\frac{\sigma^3}{\varepsilon}\left(\frac{T}{N}\right)^{\frac{3}{2}}$, the Lindeberg condition is satisfied. The Lindeberg-Feller Theorem completes the proof.

For the sequence (Y_k^N) defined above, the conditions of the theorem are satisfied with $\mu = -\frac{1}{2}\sigma^2 T$ and $\Sigma = \frac{1}{2}\sigma^2 T$ with σ as fixed above. Thus (Z_N) converges in distribution to $Z \sim \mathcal{N}(-\frac{1}{2}\sigma^2 T, \sigma^2 T)$, while $(1 + \rho_N)^{-N} \to e^{-rT}$ as $N \to \infty$. It follows that the limit of the CRR put option prices (P_0^N) is given by

$$E\left((e^{-rT}K - S_0 e^Z)^+\right), \tag{2.29}$$

where the expectation is now taken with respect to the distribution of Z.

The Black-Scholes Formula

Standardising Z, we see that the random variable $X = \frac{1}{\sigma\sqrt{T}}(Z + \frac{1}{2}\sigma^2 T)$ has distribution $N(0,1)$; that is, $Z = \sigma\sqrt{T}X - \frac{1}{2}\sigma^2 T$. The limiting value of P_0^N can be found by evaluating the integral

$$\int_{-\infty}^{\infty} \left[e^{-rT}K - S_0 e^{-\frac{1}{2}\sigma^2 T + \sigma\sqrt{T}x}\right]^+ \frac{e^{-\frac{1}{2}x^2}}{\sqrt{2\pi}} dx. \tag{2.30}$$

Observe that the integrand is non-zero only when

$$\sigma\sqrt{T}x + \left(r - \frac{1}{2}\sigma^2\right)T < \log\left(\frac{K}{S_0}\right),$$

that is, on the interval $(-\infty, \gamma)$, where

$$\gamma = \frac{\log\left(\frac{K}{S_0}\right) - (r - \frac{1}{2}\sigma^2)T}{\sigma\sqrt{T}}.$$

Thus the put option price for the limiting pricing model reduces to

$$P_0 = Ke^{-rT}(\Phi(\gamma)) - S_0 \int_{-\infty}^{\gamma} e^{-\frac{\sigma^2 T}{2}} e^{\sigma\sqrt{T}x - \frac{x^2}{2}} \frac{dx}{\sqrt{2\pi}}$$

$$= Ke^{-rT}(\Phi(\gamma)) - S_0 \int_{-\infty}^{\gamma} e^{-\frac{1}{2}(x - \sigma\sqrt{T})^2} \frac{dx}{\sqrt{2\pi}}$$

$$= Ke^{-rT}(\Phi(\gamma)) - S_0 \left(\Phi(\gamma - \sigma\sqrt{T})\right).$$

Here Φ denotes the cumulative normal distribution function.

Setting $d_- = -\gamma$ and $d_+ = d_- + \sigma\sqrt{T}$, and using the symmetry of Φ, we obtain $1 - \Phi(\gamma) = \Phi(-\gamma) = \Phi(d_-)$ and $1 - \Phi(\gamma - \sigma\sqrt{T}) = \Phi(d_+)$, where

$$d_\pm = \frac{\log\left(\frac{S_0}{K}\right) + (r \pm \frac{1}{2}\sigma^2)T}{\sigma\sqrt{T}}. \qquad (2.31)$$

By call-put parity, this gives the familiar *Black-Scholes formula* for the call option: the time 0 price of the call option $f_T = (S_T - K)^+$ is given by

$$V_0(C) = C_0 = S_0\Phi(d_+) - e^{-rT}K\Phi(d_-). \qquad (2.32)$$

Remark 2.7.2. An alternative derivation of this approximating procedure, using binomial models where for each n the probabilities of the 'up' and 'down' steps are equal to $\frac{1}{2}$, can be found in [35].

By replacing T by $T - t$ and S_0 by S_t, we can read off the value process V_t for the option similarly; in effect this treats the option as a contract written at time t with time to expiry $T - t$,

$$V_t(C) = S_t\Phi(d_{t+}) - e^{-r(T-t)}K\Phi(d_{t-}), \qquad (2.33)$$

where

$$d_{t\pm} = \frac{\log\left(\frac{S_t}{K}\right) + (r \pm \frac{1}{2}\sigma^2)(T - t)}{\sigma\sqrt{T - t}}.$$

The preceding derivation has not required us to study the dynamics of the 'limit stock price' S; it is shown in Chapter 7 that this takes the form

$$dS_t = S_t\mu dt + \sigma S_t dW_t, \qquad (2.34)$$

where W is a Brownian motion. The stochastic calculus necessary for the solution of such stochastic differential equations is developed in Chapter 6. However, we can already note one remarkable property of the Black-Scholes formula: it does not involve the mean return μ of the stock but depends on the riskless interest rate r and the volatility σ. The mathematical reason for this lies in the change to a risk-neutral measure (which underlies the martingale pricing techniques described in this chapter), which eliminates the drift term from the dynamics.

Dependence of the Option Price on the Parameters

Write $C_t = V_t(C)$ for the Black-Scholes value process of the call option; i.e.,

$$C_t = S_t \Phi(d_{t+}) - e^{-r(T-t)} K \Phi(d_{t-}),$$

where $d_{t\pm}$ is given as in (2.33). As we have calculated for the case $t = 0$, the European put option with the same parameters in the Black-Scholes pricing model is given by

$$P_t = Ke^{-r(T-t)} \Phi(-d_{t-}) - S_t \Phi(-d_{t+}).$$

We examine the behaviour of the prices C_t at extreme values of the parameters. (The reader may consider the put prices P_t similarly.)

When S_t increases, $d_{t\pm}$ grows indefinitely, so that $\Phi(d_{t\pm})$ tends to 1, and so C_t has limiting value $S_t - Ke^{-r(T-t)}$. In effect, the option becomes a forward contract with delivery price K since it is 'certain' to be exercised at time T. Similar behaviour is observed when the volatility σ shrinks to 0 since again d_{t+} become infinite, and the riskless stock behaves like a bond (or money in the bank).

When $t \to T$ (i.e., the time to expiry decreases to 0) and $S_t > K$, then $d_{t\pm}$ becomes ∞ and $e^{-r(T-t)} \to 1$, so that C_t tends to $S_t - K$. On the other hand, if $S_t < K$, $\log\left(\frac{S_t}{K}\right) < 0$ so that $d_{t\pm} = -\infty$ and $C_t \to 0$. Thus, as expected, $C_t \to (S_T - K)^+$ when $t \to T$.

Remark 2.7.3. Note finally that there is a natural 'replicating strategy' given by (2.33) since this value process is expressed as a linear combination of units of stocks S_t and bonds S_t^0 with $S_0^0 = 1$ and $S_t^0 = \beta_t^{-1} S_0^0 = e^{rt}$. Writing the value process $V_t = \theta_t \cdot S_t$ (where by abuse of notation $S = (S^0, S)$), we obtain

$$\theta_t^0 = -Ke^{-rT} \Phi(d_{t-}), \qquad \theta_t^1 = \Phi(d_{t+}). \qquad (2.35)$$

In Chapter 7, we consider various derivatives of the Black-Scholes option price, known collectively as 'the Greeks', with respect to its different parameters. This provides a sensitivity analysis with parameters that are widely used in practice.

Chapter 3

The First Fundamental Theorem

We saw in the previous chapter that the existence of a probability measure $Q \sim P$ under which the (discounted) stock price process is a martingale is sufficient to ensure that the market model is viable (i.e. , that it contains no arbitrage opportunities). We now address the converse: whether for every viable model one can construct an equivalent martingale measure for S, so that the price of a contingent claim can be found as an expectation relative to Q.

To deal with this question fully while initially avoiding difficult technical issues that can obscure the essential simplicity of the argument, we shall assume throughout Sections 3.1 to 3.4 that we are working with a *finite market model*, so each σ-field $\mathcal{F}_t$ is generated by a finite partition $\mathcal{P}_t$ of Ω. In Section 3.5, we then consider in detail the construction of equivalent martingale measures for general discrete models without any restrictions on the probability space. This requires considerably more advanced concepts and results from functional analysis.

3.1 The Separating Hyperplane Theorem in $\mathbb{R}^n$

In finite markets, the following standard separation theorem for compact convex sets in $\mathbb{R}^n$, which is a special case of the Hahn-Banach separation theorem (see [97], [264]), will suffice for our purposes.

Theorem 3.1.1 (Separating Hyperplane Theorem). *Let L be a linear subspace of $\mathbb{R}^n$ and let K be a compact convex subset in $\mathbb{R}^n$ disjoint from L. Then we can separate L and K strictly by a hyperplane containing L; (i.e., there exists a (bounded) linear functional $\phi : \mathbb{R}^n \to \mathbb{R}$ such that $\phi(x) = 0$ for all $x \in L$ but $\phi(x) > 0$ for all $x \in K$).*

The following lemma will be used in the proof but also has independent interest.

Lemma 3.1.2. *Let C be any closed convex subset of $\mathbb{R}^n$ that does not contain the zero vector. Then there is a linear functional ϕ on $\mathbb{R}^n$ that has a strictly positive lower bound on C.*

Proof. Denote by $B = B(0, r)$ the closed ball of radius r centred at the origin in $\mathbb{R}^n$, and choose $r > 0$ so that B intersects C. Then $B \cap C$ is non-empty, closed and bounded, and hence compact. Therefore the continuous map $x \mapsto |x|_n$ attains its infimum over $B \cap C$ at some $z \in B \cap C$. (Here $|x| = |x|_n$ denotes the Euclidean norm of x in $\mathbb{R}^n$.) Since $|x| > r$ when $x \notin B$, it is clear that $|x| \geq |z|$ for all $x \in C$. In particular, since C is convex, $y = \lambda x + (1 - \lambda)z$ is in C whenever $x \in C$ and $0 \leq \lambda \leq 1$, so that $|y| \geq |z|$, i.e.,

$$|\lambda x + (1 - \lambda)z|^2 \geq |z|^2. \tag{3.1}$$

Multiplying out both sides of (3.1), writing $a \cdot b$ for the scalar product in $\mathbb{R}^n$, we obtain

$$\lambda^2 x \cdot x + 2\lambda(1 - \lambda)x \cdot z + (1 - \lambda)^2 z \cdot z \geq z \cdot z,$$

which simplifies at once to

$$2(1 - \lambda)x \cdot z - 2z \cdot z + \lambda(x \cdot x + z \cdot z) \geq 0.$$

This holds for every $\lambda \in [0, 1]$. Letting $\lambda \to 0$, we obtain

$$x \cdot z \geq z \cdot z = |z|^2 > 0.$$

Defining $\phi(x) = x \cdot z$, we have found a linear functional such that $\phi(x)$ is bounded below on C by the positive number $|z|^2$. (ϕ is also bounded above, as any linear functional on $\mathbb{R}^n$ is bounded.) $\qquad\square$

Proof of Theorem 3.1.1. Let K be a compact convex set disjoint from the subspace L. Define

$$C = K - L = \{x \in \mathbb{R}^n : x = k - l \text{ for some } k \in K, l \in L\}.$$

Since K and L are convex, C is also convex.

In addition, C is closed; indeed, if $x_n = k_n - l_n$ converges to some $x \in \mathbb{R}^n$, then, as K is compact, (k_n) has a subsequence converging to some $k \in K$. Thus $x_{n_r} = k_{n_r} - l_{n_r} \to x$ as $r \to \infty$ and $k_{n_r} \to k$, so that $l_{n_r} = k_{n_r} - x_{n_r} \to k - x$ and hence $l = k - x$ belongs to L since L is closed. But then $x = k - l \in C$, so that C is closed.

As C does not contain the origin, we can therefore apply Lemma 3.1.2 to C to obtain a bounded linear functional ϕ on $\mathbb{R}^n$ such that $\phi(x) \geq |z|^2 > 0$ for z as above. In other words, writing $x = k - l$, we have $\phi(k) - \phi(l) \geq |z|^2 > 0$. This must hold for all $x \in C$. Fix k and replace l

by λl for arbitrary positive λ if $\phi(l) \geq 0$ or by λl for arbitrary negative λ if $\phi(l) < 0$. The vectors λl belong to L, as L is a linear space; since ϕ is bounded, we must have $\phi(l) = 0$ (i.e., L is a subspace of the hyperplane $\ker\phi = \{x : \phi(x) = 0\}$, while $\phi(K)$ is bounded below by $|z|^2 > 0$). The result follows. □

3.2 Construction of Martingale Measures

The above separation theorem applies to sets in $\mathbb{R}^n$. We can apply it to $\mathbb{R}^\Omega$, the space of all functions $\Omega \mapsto \mathbb{R}$, by identifying this space with $\mathbb{R}^n$ for a finite n, in view of the assumption that the σ-field $\mathcal{F}$ is finitely generated (i.e., any $\mathcal{F}$-measurable real function on Ω takes at most n distinct values, where n is the number of cells in the partition $\mathcal{P}$ that generates $\mathcal{F}$). In other words, we assume that $\Omega = D_1 \cup D_2 \cup \cdots \cup D_n$, where

$$D_i \cap D_j = \emptyset \text{ for } i \neq j, \qquad P(D_i) = p_i > 0 \text{ for } i = 1, 2, \ldots, n.$$

Without loss, we now take the (D_i) as atoms or 'points' ω_i of Ω . Thus any random variable X defined on $(\Omega, \mathcal{F})$ will be regarded as a point

$$(X(\omega_1), X(\omega_2), \ldots, X(\omega_n))$$

in $\mathbb{R}^n$. We apply this in particular to the random variables making up the value process $\{V_t(\theta)(\omega) : \omega \in \Omega\}$ and the gains process $\{G_t(\theta)(\omega) : \omega \in \Omega\}$ of a given admissible strategy $\theta \in \Theta_a$.

Recall (Definition 2.2.3) that the market model is *viable* if it contains no arbitrage opportunities (i.e., if whenever a strategy $\theta \in \Theta_a$ has initial value $V_0(\theta) = 0$, and final value $V_T(\theta) \geq 0$ a.s. (P), then $V_T(\theta) = 0$ a.s. (P)).

Denote by C the positive orthant in $\mathbb{R}^n$ with the origin removed; i.e.,

$$C = \{Y \in \mathbb{R}^n : Y_i \geq 0 \text{ for } i = 1, 2, \ldots, n, Y_i > 0 \text{ for at least one } i\}. \quad (3.2)$$

The set C is a *cone* (i.e., closed under vector addition and multiplication by non-negative scalars) and is clearly convex.

The no-arbitrage assumption means that for every admissible strategy $\theta \in \Theta_a$ we have that

$$\overline{V}_t(\theta) = \overline{G}_t(\theta) \notin C \text{ if } V_0(\theta) = 0.$$

Thus the discounted gains process $\overline{G}(\theta)$ for such a strategy θ with initial value zero cannot have a final value contained in C since otherwise it would be an arbitrage opportunity.

Recall from (2.8) that a self-financing strategy $\theta = (\theta^0, \theta^1, \theta^2, \ldots, \theta^d)$ is completely determined by the stock holdings $\hat{\theta} = (\theta^1, \theta^2, \ldots, \theta^d)$. Thus, given a predictable $\mathbb{R}^d$-valued process $\hat{\theta} = (\theta^1, \theta^2, \ldots, \theta^d)$, there is a unique predictable real-valued process θ^0 such that the augmented process $\theta =$

$(\theta^0, \theta^1, \theta^2, \ldots, \theta^d)$ has initial value $V_0(\theta) = 0$ and is self-financing. By a minor abuse of notation, we define the discounted gains process associated with $\hat{\theta}$ as

$$\overline{G}_t(\hat{\theta}) = \sum_{u=1}^{t} \theta_u \cdot \Delta \overline{S}(u) = \sum_{u=1}^{t} \left(\sum_{i=1}^{d} \theta_u^i \Delta \overline{S}_u^i \right) \text{ for } t = 1, 2, \ldots, T.$$

Suppose that $\overline{G}_t(\hat{\theta}) \in C$. Then, with β denoting the discount factor,

$$V_T(\theta) = \beta_T^{-1} \overline{V}_t(\theta) = \beta_T^{-1}(V_0(\theta) + \overline{G}_t(\theta)) = \beta_T^{-1} \overline{G}_t(\hat{\theta})$$

is non-negative and is strictly positive with positive probability. So θ is a weak arbitrage, which contradicts the viability of the model. We have proved the following result.

Lemma 3.2.1. *If the market model is viable, the discounted gains process associated with any predictable $\mathbb{R}^d$-valued process $\hat{\theta}$ cannot belong to the cone C.*

Since $\overline{G}_t(\hat{\theta})$ is a sum of scalar products $\theta_t \cdot \Delta \overline{S}_t$ in $\mathbb{R}^n$, and since any linear functional on $\mathbb{R}^n$ takes the form $x \mapsto x \cdot y$ for some $y \in \mathbb{R}^n$, the relevance of the separation theorem to these questions now becomes apparent in the proof of the next theorem, which is the main result in this section.

Theorem 3.2.2 (First Fundamental Theorem of Asset Pricing for Finite Market Models). *A finite market model is viable if and only if there exists an equivalent martingale measure (EMM) for S.*

Proof. Since we have already shown more generally (in Chapter 2) that the existence of an EMM ensures viability of the model, we need only prove the converse.

Suppose therefore that the market model is viable. We need to construct a measure $Q \sim P$ under which the price processes are martingales relative to the filtration $\mathbb{F}$. Recall that C is the convex cone of all real random variables ϕ on $(\Omega, \mathcal{F})$ such that $\phi(\omega) \geq 0$ a.s. and $\phi(\omega_i) > 0$ for at least one $\omega_i \in \Omega = \{\omega_1, \omega_2, \ldots, \omega_n\}$ (and by assumption $p_i = P(\{\omega_i\}) > 0$). We have shown that in a viable market we must have $\overline{G}_t(\hat{\theta}) \notin C$ for all predictable $\mathbb{R}^d$-valued processes $\hat{\theta}$. On the other hand, the set defined by such gains processes,

$$L = \left\{ \overline{G}_t(\hat{\theta}) : \hat{\theta} = (\theta^1, \theta^2, \ldots, \theta^d), \text{ with } \theta^i \text{ predictable for } i = 1, 2, \ldots, d \right\},$$

is a linear subspace of the vector space of all $\mathcal{F}$-measurable real-valued functions on Ω.

Since L does not meet C, we can separate L and the compact convex subset $K = \{X \in C : E_P(X) = 1\}$ of C by a linear functional f on $\mathbb{R}^n$

that is strictly positive on K and 0 on L. The linear functional has a representation of the form

$$f(x) = x \cdot q = \sum_{i=1}^{n} x_i q_i$$

for a unique $q = (q_i)$ in $\mathbb{R}^n$. Taking $\xi_i = (0, \ldots, 0, \frac{1}{p_i}, 0, \ldots, 0)$ in turn for each $i \leq n$, we see that $E_P(\xi_i) = \frac{p_i}{p_i} = 1$, so that $\xi_i \in K$, and hence $f(\xi_i) = \frac{q_i}{p_i} > 0$. Thus $q_i > 0$ for all $i \leq n$.

Now define a new linear functional $g = \frac{f}{\alpha}$, where $\alpha = \sum_{i=1}^{n} q_i > 0$. This is implemented by the vector p^* with $p_i^* = \frac{q_i}{\alpha} > 0$, so that $\sum_{i=1}^{n} p_i^* = 1$. Hence we may use the vector p^* to induce a probability measure P^* on $\Omega = \{\omega_1, \omega_2, \ldots, \omega_n\}$ by setting $P^*(\{\omega_i\}) = p_i^* > 0$, so that $P^* \sim P$.

Let $E^*(\cdot)$ denote expectation relative to P^*. Since $g(x) = \frac{1}{\alpha}f(x) = 0$ for all $x \in L$, we have $E^*\left(\overline{G}_T\left(\hat{\theta}\right)\right) = 0$ for each vector $\hat{\theta}$ of stock holdings, creating a self-financing strategy θ with $V_0(\theta) = 0$. As $\overline{V}_t(\theta) = V_0(\theta) + \overline{G}_T(\theta)$, this implies that $E^*\left(\overline{V}_T(\theta)\right) = 0$ for such θ. But by (2.8) we can generate such θ from any n-dimensional predictable process, in particular from $(0, \ldots, 0, \theta^i, 0, \ldots, 0)$, where the predictable real-valued process θ^i is given for $i \leq n$. Thus

$$E^*\left(\sum_{t=1}^{T} \theta_t^i \Delta \overline{S}_t^i\right) = 0$$

holds for every bounded predictable process $(\theta^i)_{i=1,2,\ldots,T}$. Theorem 2.3.5 now implies that each S^i is a martingale under P^*. Hence P^* is the desired EMM for the price process S. $\qquad\square$

3.3 Pathwise Description

The geometric origin of the above result is clear from the essential use that was made of the separation theorem. A geometric formulation of Theorem 3.2.2 can be based on the 'local' equivalent of the no-arbitrage condition in terms of 'one-step' changes in the value of a portfolio. In fact, although the definition of (weak) arbitrage involves only the initial and final values of a strategy, this will demonstrate that the no-arbitrage condition is an assumption about the *pathwise* behaviour of the value process. Although this discussion is somewhat detailed, it is included here for its value in providing an intuitive grasp of the ideas that underlie the more abstract proof of Theorem 3.2.2 and in giving a step-by-step construction of the equivalent martingale measure. As before, our discussion (which follows [290]) is confined to the case where $\mathcal{F}$ is finitely generated.

One-Step Arbitrage

The idea behind the construction lies in the following simple observation. Consider a market model with a single bond and stock (i.e., $d = 1$) and assume that the bond price $S^0 \equiv 1$ for all trading dates. In particular, for any self-financing strategy $\theta = (\theta^0, \theta^1)$, the value process $V_t(\theta)$ has increments $\Delta V_t = \theta_t^1 \Delta S_t^1$, as $\Delta S^0(t) = 0$. These increments will be 'concentrated' to one side of the origin precisely when the same is true for the price increments ΔS_t^1.

Now suppose we *know* at some time $(t-1) \in \mathbb{T}$ that the stock price S^1 will not decrease in the time interval $[t-1, t]$; that is, for some partition set $A \in \mathcal{P}_{t-1}$ we have $P(\{\Delta S_t^1 \geq 0\} | A) = 1$. Then we can buy stock S^1 at time $t - 1$, sell it again at time t, and invest the profit ΔS_t^1 in the riskless bond until the time horizon T. To prevent this arbitrage opportunity, we need to have $P(\{\Delta S_t^1 = 0\} | A) = 1$; i.e., that S^1 (and hence also the value process $V(\theta)$ associated with any admisssible strategy θ) is a 'one-step martingale' in the time interval $[t-1, t]$.

This idea can be extended to models with d stocks and hyperplanes in $\mathbb{R}^{d+1}$. In this case, we have

$$\Delta V_t(\theta) = \theta_t \cdot \Delta S_t = \sum_{k=1}^{d} \theta_t^k \Delta S_t^k,$$

so it is clear that condition (i) in Proposition 3.3.1 below expresses the fact that, along each sample path of the price process S, the support of the conditional distribution of the vector random variable ΔS_t, given $A \in \mathcal{P}_{t-1}$, cannot be wholly concentrated only on one 'side' of any hyperplane in $\mathbb{R}^{d+1}$.

Assume for the remainder of this section that $S^0(t) \equiv 1$ for all $t \in \mathbb{T}$.

Proposition 3.3.1. *If the finite market model $S = \left(S^0, S^1, S^2, \ldots, S^d\right)$ is viable, then, for all $\theta \in \Theta$, $t > 0$ and $A \in \mathcal{P}_{t-1}$, and with $V_t = V_t(\theta)$, the following hold:*

$$P(\Delta V_t \geq 0 | A) = 1 \text{ implies that } P(\Delta V_t = 0 | A) = 1,$$
$$P(\Delta V_t \leq 0 | A) = 1 \text{ implies that } P(\Delta V_t = 0 | A) = 1.$$

Proof. Fix $t > 0$ and $\theta \in \Theta$. Suppose that $P(\Delta V_t \geq 0 | A) = 1$ for some $A \in \mathcal{P}_{t-1}$. We define ψ with $\psi_0 = 0$ as follows for $s > 0$: let

$$\psi_s(\omega) = 0 \text{ for all } s = 1, 2, \ldots, T \text{ and } \omega \notin A,$$

while, for $\omega \in A$,

$$\psi_s(\omega) = \begin{cases} 0 & \text{if } 0 < s < t, \\ (\theta_t^0(\omega) - V_{t-1}(\theta)(\omega), \theta_t^1(\omega), \theta_t^2(\omega), \ldots, \theta_t^d(\omega))' & \text{if } s = t, \\ (V_t(\theta)(\omega), 0, \ldots, 0)' & \text{if } s > t. \end{cases}$$

Under the strategy ψ, we start with no holdings at time 0 and trade only from time t onwards, and then only if $\omega \in A$ (which we know by time $t-1$). In that case, we elect to follow the strategy θ in respect to stocks and borrow an amount equal to $(V_{t-1}(\theta) - \theta_0)$ in order to deal in stocks at $(t-1)$-prices using the strategy θ for our stock holdings. For ω in A, this is guaranteed to increase total wealth. At times $s > t$, we then maintain all wealth (i.e., our profits from these transactions) in the bond.

The strategy ψ is obviously predictable. To see that it is self-financing, we need only consider $\omega \in A$. Then we have

$$(\Delta\psi_t) \cdot S_{t-1} = (\theta_t^0 - V_{t-1}(\theta))S_{t-1}^0 + \sum_{i=1}^{d} \theta_t^i S_{t-1}^i$$

$$= \theta_t \cdot S_{t-1} - V_{t-1}(\theta)$$

$$= \theta_{t-1} \cdot S_{t-1} - V_{t-1}(\theta)$$

$$= 0$$

since $S^0 \equiv 1$ and θ is self-financing. Hence ψ is also self-financing.

With this strategy, we certainly obtain $V_T(\psi) \geq 0$. In fact, for $u \geq t$ we have

$$V_u(\psi) = \psi_t \cdot S_t = \Delta V_t(\psi) = \Delta V_t(\theta) \geq 0$$

on A and $V_u(\psi) = 0$ off A. Hence ψ defines a self-financing strategy with initial value 0 and $V_T(\psi) \geq 0$. If there is no arbitrage, we must therefore conclude that $V_T(\psi) = 0$. Since $V_T(\psi) = 0$ off A and $V_T(\psi) = \Delta V_t(\theta)$ on A, this is equivalent to

$$0 = P(V_T(\psi) > 0) = P(\{V_T(\psi) > 0\} \cap A) = P(\{\Delta V_T(\theta) > 0\}|A)P(A),$$

that is, $P(\Delta V_t = 0|A) = 1$. This proves the first assertion. The proof of the second part is similar. $\qquad\square$

The above formulation can be used to establish a further equivalent form of market model viability. Below we write $\hat{S}$ for the $\mathbb{R}^d$-valued process obtained by deleting the 0th component of S, that is; where $S = (1, \hat{S})$.

Note. For the statement and proof of the next proposition, we do *not* need the assumption that the filtration $\mathbb{F} = (\mathcal{F}_t)_{t \in \mathbb{T}}$ is finitely generated; it is valid in an arbitrary probability space $(\Omega, \mathcal{F}, P)$. It states, in essence, the 'obvious' fact that if there is an arbitrage opportunity for the model defined on the time set $\mathbb{T} = \{0, 1, \ldots, T\}$, then there is an arbitrage opportunity in at least one of the single-period markets $[t-1, t)$.

Proposition 3.3.2. *Let $(\Omega, \mathcal{F}, P, \mathbb{T}, \mathbb{F}, S)$ be an arbitrary discrete market model, where $(\Omega, \mathcal{F}, P)$ is a probability space, $\mathbb{T} = \{0, 1, \ldots, T\}$ is a discrete time set, $\mathbb{F} = (\mathcal{F}_t)_{t \in \mathbb{T}}$ is a complete filtration, and $S = (S^i)_{i=0,1,\ldots,d}$ is a price process, as defined in Section 2.1. The following are equivalent:*

(i) The model allows an arbitrage opportunity.

(ii) For some $t = 1, 2, \ldots, T$ there is an $\mathcal{F}_{t-1}$-measurable $\phi : \Omega \mapsto \mathbb{R}^{d+1}$ such that $\phi \cdot \Delta S_t \geq 0$ and $P(\phi \cdot \Delta S_t > 0) > 0$.

(iii) For some $t = 1, 2, \ldots, T$ there is an $\mathcal{F}_{t-1}$-measurable $\hat{\phi} : \Omega \mapsto \mathbb{R}^d$ such that $\hat{\phi} \cdot \Delta \hat{S}_t \geq 0$ and $P(\hat{\phi} \cdot \Delta \hat{S}_t > 0) > 0$.

Proof. The equivalence of (ii) and (iii) is clear. Now assume that (ii) holds with ϕ and $A = \{\omega : (\phi \cdot \Delta S_t)(\omega) > 0\}$. We can construct an arbitrage opportunity θ as follows: set

$$\theta_u(\omega) = 0 \text{ for all } u \in \mathbb{T} \text{ and } \omega \notin A,$$

while, for $\omega \in A$,

$$\theta_u(\omega) = \begin{cases} 0 & \text{if } u < t, \\ \left(-\sum_{i=1}^{d} \phi_i(\omega)S^i_{t-1}(\omega), \phi_1(\omega), \phi_2(\omega), \ldots, \phi_d(\omega)\right)' & \text{if } u = t, \\ (V_t(\theta)(\omega), 0, \ldots, 0)' & \text{if } u > t. \end{cases}$$

By construction, θ is predictable. (The strategy θ is in fact a special case of ψ constructed in Proposition 3.3.1.) To see that it is also self-financing, note that the value process $V(\theta)$ only changes when $\omega \in A$, and then $\Delta V_u(\theta) = 0$ unless $u = t$. Moreover,

$$\begin{aligned} \Delta V_t(\theta)(\omega) &= \theta_t \cdot S_t(\omega) - \theta_{t-1} \cdot S_{t-1}(\omega) \\ &= \theta_t \cdot S_t(\omega) \\ &= -\sum_{i=1}^{d} \phi_i(\omega)S^i_{t-1}(\omega) + \sum_{i=1}^{d} \phi_i(\omega)S^i(t)(\omega) \\ &= \theta_t \cdot \Delta S_t(\omega). \end{aligned}$$

Now $V_0(\theta) = 0$, while for $u > t$ we have $V_u(\theta) = 0$ on $\Omega \setminus A$, and, since $S^0 \equiv 1$,

$$V_u(\theta) = \Delta V_t(\theta) = \theta_t \cdot \Delta S_t = \phi_t \cdot \Delta S_t \geq 0$$

on A. Hence $V_T(\theta) \geq 0$ a.s. (P). By the definition of A,

$$\{V_T(\theta) > 0\} = \{\Delta V_t(\theta) > 0\} \cap A.$$

Hence θ is an arbitrage opportunity since $P(A) > 0$. Thus (ii) implies (i).

Conversely, assume that (i) holds. Then there is a gains process $G_T(\theta)$ that is a.s. non-negative and strictly positive with positive probability for some strategy $\theta \in \Theta$. Assume without loss of generality that $(\theta \cdot S)_0 = 0$. There must be a first index $u \geq 1$ in $\mathbb{T}$ such that $(\theta \cdot S)_u$ is a.s. non-negative and strictly positive with positive probability. Consider $(\theta \cdot S)_{u-1}$: either $(\theta \cdot S)_{u-1} = 0$ a.s. or $A = \{(\theta \cdot S)_{u-1} < 0\}$ has positive probability.

In the first case,

$$(\theta \cdot S)_u = (\theta \cdot S)_u - (\theta \cdot S)_{u-1} = \theta_u \cdot \Delta S_u \geq 0$$

since $(\theta_u - \theta_{u-1}) \cdot S_{u-1} = 0$ because θ is self-financing. For the same reason, $P[\theta_u \cdot \Delta S_u > 0] > 0$. Hence (ii) holds.

In the second case, we have

$$\theta_u \cdot \Delta S_u = (\theta \cdot S)_u - (\theta \cdot S)_{u-1} \geq -(\theta \cdot S)_{u-1} > 0$$

on A, so that the predictable random variable $\phi = 1_A \theta_u$ will satisfy (ii). This completes the proof. $\qquad\square$

This result shows that the 'global' existence of arbitrage is equivalent to the existence of 'local' arbitrage at some $t \in \mathbb{T}$. To exploit this fact geometrically, we again concentrate on the special case of finite market models. First we have the following immediate corollary.

Corollary 3.3.3. *If a finite market model is viable, then for all $t > 0$ in $\mathbb{T}$ and all (non-random) vectors $x \in \mathbb{R}^d$, we have that $x \cdot \Delta \hat{S}_t(\omega) \geq 0$ a.s. (P) implies that $x \cdot \Delta \hat{S}_t(\omega) = 0$ a.s. (P).*

Geometric Interpretation of Arbitrage

We briefly review two well-known concepts and one basic result concerning convex sets in $\mathbb{R}^d$.

First, define the *relative interior* of a subset C in $\mathbb{R}^d$ as the interior of C when viewed as a subset of its *affine hull*, where the affine hull and the *convex hull* of C are defined by

$$\text{aff}(C) = \left\{ x \in \mathbb{R}^d : x = \sum_{i=1}^{n} a_i c_i, c_i \in C, \sum_{i=1}^{n} a_i = 1 \right\},$$

$$\text{conv}(C) = \left\{ x \in \mathbb{R}^d : x = \sum_{i=1}^{n} a_i c_i, c_i \in C, a_i \geq 0, \sum_{i=1}^{n} a_i = 1 \right\}.$$

The relative interior of C is then simply the set

$$\text{ri}(C) = \left\{ x \in \text{aff}(C) : B_\epsilon(x) \cap \text{aff}(C) \subset C \text{ for some } \epsilon > 0 \right\},$$

where $B_\epsilon(x)$ is the Euclidean ϵ-ball centred at x. (See [245] for details.)

It is an easy consequence of the definitions that the existence of a hyperplane separating two non-empty convex sets is equivalent to the statement that their relative interiors are disjoint. For a proof, see [245], p.96.

In the absence of arbitrage, there is no hyperplane in $\mathbb{R}^d$ that properly separates the origin from the convex hull of $\hat{A} = \left\{ \Delta \hat{S}_t(\omega) : \omega \in A \right\}$ for any given $A \in \mathcal{P}_{t-1}$, $t > 0$. Writing $C_t(A)$ for the convex hull, we have proved the first part of the following result.

Proposition 3.3.4. *In a finite market model, the no-arbitrage condition is equivalent to the condition that, for all $t \in \mathbb{T}$ and all $A \in \mathcal{P}_{t-1}$, the* *origin belongs to the relative interior of $C_t(A)$. In other words, the finite market model allows no arbitrage opportunities if and only if, for each t and $A \in \mathcal{P}_{t-1}$, the value of S_{t-1} is a strictly convex combination of the values taken by S_t on A.*

Proof. To prove the latter equivalence, suppose that $0 \in C_t(A)$. Since $A \in \mathcal{P}_{t-1}$ and S is adapted, $\hat{S}_{t-1}(\omega) = c \in \mathbb{R}^d$ is constant for $\omega \in A$. Any vector in $C_t(A)$ thus takes the form $\sum_{i=1}^m \alpha_i(z_i - c)$, where $\alpha_i > 0$, $\sum_{i=1}^m \alpha_i = 1$, and each z_i is equal to $\hat{S}_t(\omega)$ for some $\omega \in A$. Thus $0 \in C_t(A)$ if and only if $c = \sum_{i=1}^m \alpha_i z_i$, where the vectors z_i are values of $\hat{S}_t$ on A, $\sum_{i=1}^m \alpha_i = 1$, and all $\alpha_i > 0$. □

Constructing the EMM

The last result can in turn be interpreted in terms of conditional probabilities. For each fixed $A \in \mathcal{P}_{t-1}$, we can redistribute the conditional probabilities to ensure that under this new mass distribution (probability measure) the price increment vector $\Delta \hat{S}_t$ has zero conditional expectation on A. Piecing together these conditional probabilities, we then construct an equivalent martingale measure for S.

More precisely, fix t, let $A = \bigcup_{k=1}^n A_k$ be a minimal partition of A, and let $M = (a_{ik})$ be the $d \times n$ matrix of the values taken by the price *increments* $\Delta \hat{S}_t^i$ on the cells A_k. By Proposition 3.3.4, the origin $\mathbb{R}^d$ lies in the relative interior of $C_t(A)$. Hence it can be expressed as a strictly convex combination of elements of $C_t(A)$.

This means that the equation $Mx = 0$ has a strictly positive solution $\alpha = (\alpha_k)$ in $\mathbb{R}^n$.

It is intuitively plausible that the coordinates of the vector α should give rise to an EMM for the discounted prices. To see this, we first need to derive a useful 'matrix' version of the separation theorem, for which we will also have use in Chapter 4.

Lemma 3.3.5 (Farkas (1902)). *If A is a $d \times n$ matrix and $b \in \mathbb{R}^d$, then exactly one of the following alternatives holds:*

(i) There is a non-negative solution $x \geq 0$ of $Ax = b$.

(ii) The inequalities $y'A \leq 0$ and $y \cdot b > 0$ have a solution $y \in \mathbb{R}^d$.

Proof. The columns $a_j = (a_{ij})$ $(j \leq n)$ of A define a convex polyhedral cone K in $\mathbb{R}^d$, each of whose elements is given in the form $k = \sum_{j=1}^n x_j a_j$ for scalars $x_j \geq 0$. Thus $Ax = b$ for some $x \geq 0$ if and only if the vector $b \in \mathbb{R}^d$ belongs to K. If $b \notin K$, we can separate it from K by a linear functional f on $\mathbb{R}^d$ such that $f(b) > 0$, $f(k) \leq 0$ for $k \in K$ (this is an easy adaptation of the first part of the proof of Theorem 3.1.1). Now implement f by $f(z) = y \cdot z$ for some $y \in \mathbb{R}^d$. Then $y \cdot a_j \leq 0$ for $j \leq n$. Hence $y'A \leq 0$, and $y \cdot b > 0$, as required. □

The following reformulations of Farkas' lemma follow without much difficulty and will be used in the sequel.

Lemma 3.3.6. *For a given $d \times n$ matrix M, exactly one of the following holds:*

(α) *The equation $Mx = 0$ has a solution $x \in \mathbb{R}^n$ with $x > 0$.*

(β) *There exists $y \in \mathbb{R}^d$ such that $y'M \geq 0$, and $y'M$ is not identically 0.*

For a given $d \times n$ matrix M and $b \in \mathbb{R}^d$, exactly one of the following holds:

(a) *The equation $Mx = b$ has a solution in $\mathbb{R}^n$.*

(b) *There exists $z \in \mathbb{R}^d$ with $z'M = 0$ and $z \cdot b > 0$.*

Exercise 3.3.7. Prove Lemma 3.3.6.

Applying the alternatives $(\alpha), (\beta)$ to the matrix $M = (a_{ik})$, we see that the existence of a strictly positive solution $\alpha = (\alpha_k)$ of the equation $Mx = 0$ is what precludes arbitrage: otherwise there would be a $\theta \in \mathbb{R}^d$ with $\theta'M \geq 0$ and not identically zero; such a θ would yield an arbitrage strategy.

We proceed to use the components (α_k) of this positive solution to build a one-step 'conditional EMM' for this model, restricting attention to the fixed set $A \in \mathcal{P}_{t-1}$. First denote by $\mathcal{A}_A$ the σ-field of subsets of A generated by the cells $A_1, A_2, \ldots, A_n$ of $\mathcal{P}_t$ that partition A, and let P_A be the restriction to $\mathcal{A}_A$ of the conditional probabilities $P(\cdot | A)$. Now construct a probability measure Q_A on the measurable space $(A, \mathcal{A}_A)$ by setting

$$Q_A(A_k) = \frac{\alpha_k}{|\alpha|} \text{ for } k = 1, 2, \ldots, n, \text{ where } |\alpha| = \sum_{i=1}^{n} \alpha_k.$$

Clearly $Q_A \sim P_A$. As $\mathcal{A}_A$ is generated by $(A_k)_{k \leq n}$, any $\mathcal{A}_A$-measurable vector random variable $Y : A \mapsto \mathbb{R}^d$ takes constant values $Y(\omega) = y_k \in \mathbb{R}^d$ on each of the sets A_k. Hence its expectation under Q_A takes the form

$$E_{Q_A}(Y) = \sum_{k=1}^{n} y_k Q_A(A_k) = \frac{1}{|\alpha|} \sum_{k=1}^{n} y_k \alpha_k.$$

In particular, taking $Y = \Delta \hat{S}_t$ yields $y_k = (a_{ik})_{i \leq d}$ for each $k \leq n$, where the a_{ik} are the entries of the matrix M defined above, so that $0 = M\alpha = \sum_{k=1}^{n} y_k \alpha_k$. Thus $E_{Q_A}\left(\Delta \hat{S}_t \mathbf{1}_A\right) = 0$. Since S^0 is constant by hypothesis, it follows that $E_{Q_A}(\Delta S_t \mathbf{1}_A) = 0$ (in $\mathbb{R}^{d+1}$) as well.

Conversely, suppose we are given a probability measure Q_A on $\mathcal{A}_A$ with $E_{Q_A}(\Delta S_t \mathbf{1}_A) = 0$. Setting $\alpha_k = Q_A(A_k)$ for $k \leq n$, the calculation above shows that $M\alpha = 0$, so that the zero vector in $\mathbb{R}^d$ can be expressed as a strictly convex combination of vectors in $c_t(A)$ and hence the condition of Proposition 3.3.4 is satisfied. We have proved the following proposition.

Proposition 3.3.8. *For a finitely generated filtration* $\mathbb{F}$, *the following are equivalent:*

(i) *For all* $t > 0$ *and* $A \in \mathcal{P}_{t-1}$, *the zero vector in* $\mathbb{R}^d$ *can be expressed as a strictly convex combination of vectors in the set* $c_t(A) = \left\{ \Delta \hat{S}_t(\omega) : \omega \in A \right\}$.

(ii) *For all* $t > 0$ *in* $\mathbb{T}$ *and all* $\mathcal{F}_{t-1}$-*measurable random vectors* $x \in \mathbb{R}^d$, *we have that* $x \cdot \Delta \hat{S}_t \geq 0$ *a.s.* (P) *implies that* $x \cdot \Delta \hat{S}_t = 0$ *a.s.* (P).

(iii) *There exists a probability measure* $Q_A \sim P_A$ *on* $(A, \mathcal{A}_A)$ *satisfying* $E_{Q_A}(\Delta S_t \mathbf{1}_A) = 0$.

Finally, we can put it all together to obtain three conditions, each describing the viability of the market model. Note, in particular, that condition (ii) is not affected by an equivalent change of measure. However, our proof of the steps described in Proposition 3.3.8 crucially used the fact that the filtration $\mathbb{F}$ was taken to be finitely generated.

Theorem 3.3.9. *The following statements are equivalent:*

(i) *The securities market model* $(\Omega, \mathcal{F}, P, \mathbb{T}, \mathbb{F}, S)$ *is viable.*

(ii) *For all* $t > 0$ *in* $\mathbb{T}$ *and all* $\mathcal{F}_{t-1}$-*measurable random vectors* $x \in \mathbb{R}^d$, *we have that* $x \cdot \Delta \hat{S}_t \geq 0$ *a.s.* (P) *implies that* $x \cdot \Delta \hat{S}_t = 0$ *a.s.* (P).

(iii) *There exists an equivalent martingale measure* Q *for* S.

Proof. That (i) implies (ii) was shown in Corollary 3.3.3, and that (iii) implies (i) was shown in Section 2.4. This leaves the proof that (ii) implies (iii), in which we make repeated use of Proposition 3.3.8. The family

$$\{P_A : A \in \mathcal{P}_t, \ t < T\}$$

determines P since all the σ-fields being considered are finitely generated. Thus for each $\omega \in \Omega$ we can find a unique sequence of sets $(B_t)_{t \in \mathbb{T}}$ with $B_t \in \mathcal{P}_t$ for each $t < T$ and such that

$$\Omega = B_0 \supset B_1 \supset B_2 \supset \cdots \supset B_T.$$

By the law of total probability, we can write

$$P(\{\omega\}) = P_{B_0}(B_1) P_{B_1}(B_2) \cdots P_{B_{T-1}}(\{\omega\}).$$

Now, if (ii) holds, we can use Proposition 3.3.8 successively with $t = 1$ and $A \in \mathcal{P}_0$ to construct a probability measure Q_A and then repeat for $t = 2$ and sets in $\mathcal{P}_t$, etc. In particular, this yields probability measures Q_{B_t} for each $t < T$, defined as in the discussion following Lemma 3.3.6. Setting

$$Q(\{\omega\}) = Q_{B_0}(B_1) Q_{B_1}(B_2) \cdots Q_{B_{T-1}}(\{\omega\}),$$

we obtain a probability measure $Q \sim P$ on the whole of $(\Omega, \mathcal{F})$. For any fixed $t > 0$ and $A \in \mathcal{P}_{t-1}$, the conditional probability is just

$$Q(\{\omega\} \mid A) = \mathbf{1}_A(\{\omega\}) Q_A(B_t) Q_{B_t}(B_{t+1}) \cdots Q_{B_{T-1}}(\{\omega\}).$$

Therefore, for $\omega \in A$, $E_Q\left(\Delta S_t \mid \mathcal{F}_{t-1}\right)(\omega) = 0$, and thus Q is an equivalent martingale measure for S. $\qquad\square$

3.4 Examples

Example 3.4.1. The following binomial tree example, which is adapted from [241], illustrates the stepwise construction of the EMM and also shows how viability of the market can break down even in very simple cases.

Let $\Omega = \{\omega_1, \omega_2, \omega_3, \omega_4\}$ and $T = 2$. Suppose that the evolution of a stock price S^1 is given as

$$S_0^1 = 5, \qquad S_1^1 = 8 \text{ on } \{\omega_1, \omega_2\}, \qquad S_2^1 = 9 \text{ on } \{\omega_1\},$$
$$S_1^1 = 4 \text{ on } \{\omega_3, \omega_4\}, \qquad S_2^1 = 6 \text{ on } \{\omega_2, \omega_3\},$$
$$S_2^1 = 3 \text{ on } \{\omega_4\}.$$

Note that $\mathcal{F}_0 = \{\emptyset, \Omega\}$ and that the partition $\mathcal{P}_{t-1} = \{\omega_1, \omega_2\} \cup \{\omega_3, \omega_4\}$ generates the algebra $\mathcal{F}_1 = \{\emptyset, \{\omega_1, \omega_2\}, \{\omega_3, \omega_4\}, \Omega\}$, while $\mathcal{F}_2 = \mathcal{P}(\Omega)$.

Although the stock price S_2^1 is the same in states ω_2 and ω_3, the histories (i.e., *paths*) of the price process allow us to distinguish between them. Hence the investor knows by time 2 exactly which state ω_i has been realised. For the present we shall take $S^0 \equiv 1$ (i.e., the discount rate $r = 0$).

To find an EMM $Q = \{q_1, q_2, q_3, q_4\}$ directly, we need to solve the equations $E_Q\left(S_u^1 \mid \mathcal{F}_t\right) = S_t^1$ for all t and $u > t$. This leads to the following equations:

$$
\left.
\begin{aligned}
t = 0, u = 1: \quad & 5 = 8(q_1 + q_2) + 4(q_3 + q_4), \\
t = 0, u = 2: \quad & 5 = 9q_1 + 6(q_2 + q_3) + 3q_4, \\
t = u = 1, S_1^1 = 8: \quad & 8 = \frac{1}{q_1 + q_2}(9q_1 + 6q_2), \\
t = u = 1, S_1^1 = 4: \quad & 4 = \frac{1}{q_3 + q_4}(6q_3 + 3q_4).
\end{aligned}
\right\} \tag{3.3}
$$

Solving any three of these (dependent) equations together with $\sum_{i=1}^4 q_i = 1$ yields the unique solution

$$q_1 = \frac{1}{6}, \qquad q_2 = \frac{1}{12}, \qquad q_3 = \frac{1}{4}, \qquad q_4 = \frac{1}{2}. \tag{3.4}$$

On the other hand, it is simpler to construct q_i step-by-step, as indicated in the previous section. Here this means that we must calculate the one-step conditional probabilities at each node of the tree for $t = 0$ and $t = 1$. When $S_0^1 = 5$, this requires $5 = 8p + 4(1-p)$; i.e., $p = \frac{1}{4}$.

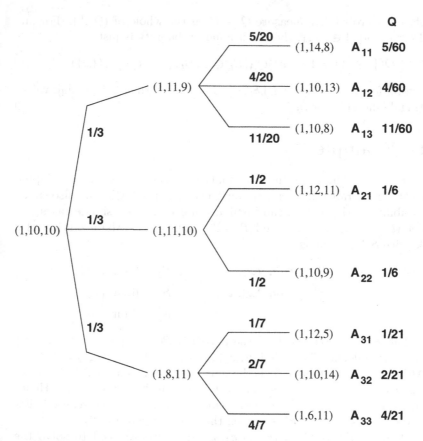

Figure 3.1: Event tree for two-stock model

For $S_1^1 = 8$ we solve $8 = 9p' + 6(1 - p')$, (i.e., $p' = \frac{2}{3}$,) while for $S_1^1 = 4$ we need $4 = 6p'' + 3(1 - p'')$, (i.e., $p'' = \frac{1}{3}$.) According to the proof of Theorem 3.3.9, this yields the q_i as

$$q_1 = \frac{1}{4} \cdot \frac{2}{3}, \qquad q_2 = \frac{1}{4} \cdot \frac{1}{3}, \qquad q_3 = \frac{3}{4} \cdot \frac{1}{3}, \qquad q_4 = \frac{3}{4} \cdot \frac{2}{3}.$$

This agrees with the values in (3.4).

It is instructive to examine the effect of discounting on this example. Suppose instead that $S^0(t) = (1+r)^t$ for each t, with $r \geq 0$. The left-hand sides of the equations (3.3) then become $5(1+r)$, $5(1+r)^2$, $8(1+r)$, and $4(1+r)$, respectively.

This yields the solution for the q_i (using the one-step method, greatly simplifying the calculation) as

$$\left. \begin{array}{ll} q_1 = \dfrac{1+5r}{4} \dfrac{2+8r}{3}, & q_3 = \dfrac{3-5r}{4} \dfrac{1+4r}{3}, \\[2ex] q_2 = \dfrac{1+5r}{4} \dfrac{1-8r}{3}, & q_4 = \dfrac{3-5r}{4} \dfrac{2-4r}{3}. \end{array} \right\} \qquad (3.5)$$

Exercise 3.4.2. Verify the solutions given in (3.5).

This time the requirement that Q be a probability measure is not automatically satisfied: when $r \geq \frac{1}{8}$, q_2 becomes non-positive. Hence Q is an EMM for $S = (S^0, S^1)$ only if $0 \leq r < \frac{1}{8}$ (i.e., if the riskless interest rate is less than 12.5%). If $r \geq \frac{1}{8}$, there is no EMM for this process, and if we observe $S_1^1 = 8$, an arbitrage opportunity can be constructed since we know in advance that the discounted stock price $\overline{S}_2^1$ will be lower than $\overline{S}_1^1 = \frac{8}{1+r}$ in each of the states ω_1 and ω_2.

Example 3.4.3. Consider a pricing model with two stocks S^1, S^2 and a riskless bond S^0 with tree structure as shown in Figure 3.1. This example is taken from [301].

The partitions giving the filtration $\mathbb{F}$ begin with $\mathcal{P}_0$ as the trivial partition and continue with

$$\mathcal{P}_1 = \{A_1, A_2, A_3\}, \quad \mathcal{P}_2 = \{A_{11}, A_{12}, A_{13}, A_{21}, A_{22}, A_{31}, A_{32}, A_{33}\}.$$

We take $T = 2$, and the various probabilities are as shown in Figure 3.1. (Note that we again keep $S^0 \equiv 1$ here.) Note that in each case the one-step transition includes both 'up' and 'down' steps, so that by Theorem 3.3.9 the model is viable and an EMM Q can be constructed for $S = (S^0, S^1, S^2)$. The calculation of Q proceeds as in the previous example (using the one-step probabilities), so that for example $Q(A_{13}) = pq$, where p is found by solving the equations

$$10 = 11p + 11p' + 8(1 - p - p'), \qquad 10 = 9p + 10p' + 11(1 - p - p'),$$

which yields $p = \frac{1}{3}$, while q must satisfy

$$11 = 10q + 10q' + 14(1 - q - q'), \qquad 9 = 8q + 13q' + 8(1 - q - q').$$

This yields $q = \frac{11}{20}$; hence $Q(A_{13}) = \frac{11}{60}$.

Exercise 3.4.4. Find the values of $Q(A)$ for all $A \in \mathcal{P}$.

To use the measure Q to calculate the price of a European call option C on stock S^2 with strike price 10, we simply find the time 0 value of C as

$$E_Q(C) = 0 \cdot \frac{5}{60} + 3 \cdot \frac{4}{60} + 0 \cdot \frac{11}{60} + 1 \cdot \frac{1}{6} + 0 \cdot \frac{1}{6} + 0 \cdot \frac{1}{21} + 4 \cdot \frac{2}{21} + 1 \cdot \frac{4}{21} = \frac{197}{210}.$$

3.5 General Discrete Models

We now turn to the construction of equivalent martingale measures for discrete market models where the underlying probability space $(\Omega, \mathcal{F}, P)$ is not necessarily finitely generated. This question has been studied intensively in recent years, both in the discrete- and continuous-time settings.

The extension from finite market models to the general case proves to be surprisingly delicate, and several different approaches have been developed since the first proof of the result by Dalang, Morton, and Willinger [68]- the interested reader should compare the expositions in [281] and [76]. We mainly follow the development in [132], which is based in turn on recent expositions in [262] and [181].

The importance of the first fundamental theorem should be clear: it provides the vital link between the economically meaningful assumption of the absence of arbitrage and the mathematical concept of the existence of equivalent measures under which the discounted stock prices are martin- gales. In generalising from the relatively simple context of finite market models, one wishes to maintain the essential aspects of the equivalence of these two conditions. In the continuous-time setting, however, the two conditions are no longer equivalent and much work has gone into reformu- lations that reflect the requirement that the market should be 'essentially' arbitrage-free while seeking to maintain a close link with the existence of equivalent martingale measures.

The general discrete-time result can be stated in the following form, which is close to that of the original paper [68].

Theorem 3.5.1 (First Fundamental Theorem of Asset Pricing). *Let* $(\Omega, \mathcal{F}, P)$ *be a probability space, and set* $\mathbb{T} = \{0, 1, \ldots, T\}$ *for some natural number* T. *Let* $\mathbb{F} = (\mathcal{F}_t)_{t \in \mathbb{T}}$ *be filtration, with* $\mathcal{F}_0$ *consisting of all* P-*null sets and their complements, and suppose the* $\mathbb{R}^{d+1}$ *-valued process* $S = (S_t^i : 0 \leq i \leq d, t \in \mathbb{T})$ *is adapted to* $\mathbb{F}$, *with* $S_t^0 > 0$ *a.s.* (P) *for each* t *in* $\mathbb{T}$. *The following are equivalent:*

(i) *There is a probability measure* $Q \sim P$ *such that the discounted price process* S/S^0 *is a* $(Q, \mathbb{F})$-*martingale.*

(ii) *The market model* $(\Omega, \mathcal{F}, P, \mathbb{T}, \mathbb{F}, S)$ *allows no arbitrage opportunities.*

If either (i) or (ii) holds, then the measure Q *can be chosen with bounded density* $\frac{dQ}{dP}$ *relative to* P.

As we have seen for finite market models, in a model with an equivalent martingale measure (i.e., when (i) holds), it is straightforward to prove the absence of arbitrage, and this has already been proved in Chapter 2 without any restrictions on $(\Omega, \mathcal{F}, P)$. Moreover, for finite market models, the task of showing that (ii) implies (i) was broken into a sequence of steps that allowed us to consider a multi-period model as a finite sequence of single-period models, where the EMM is constructed by piecing together a succession of conditional probabilities (see the steps leading to Theorem 3.3.9). The principal difficulty in extending this approach to general probability spaces, where the corresponding function spaces can no longer be identified with $\mathbb{R}^n$ for some finite n, lies in obtaining a formulation in the single-period case that allows one to avoid subtle questions of measurable selection while applying appropriate versions of the Hahn-Banach separation theorem to find the required one-step densities.

No-arbitrage in a Randomised
Single-Period Model

For the inductive procedure, we shall need to move from single-period to multi-period models, and it will not suffice to consider single-period models where the initial prices are given positive constants. We therefore need to find one-step martingale measures when the initial prices are themselves random. For this, we make the following modelling assumptions.

First, we remove, until further notice, the restriction on $\mathcal{F}_0$ stated in the theorem and instead assume as given an arbitrary σ-field $\mathcal{F}_0 \subset \mathcal{F}$. Let $S_0 = (S_0^0, S_0^1, \ldots, S_0^d) : \Omega \to \mathbb{R}^d$ be an $\mathcal{F}_0$-measurable random vector representing the bond and stock prices in a single-period market model at time 0. The prices at time 1 are given by the $\mathcal{F}$-measurable non-negative random vector $S_1 = (S_1^0, S_1^1, \ldots, S_1^d)$, so that the price process $S = (S_t^0, S_t^1, \ldots, S_t^d)_{t=0,1}$ is adapted to the filtration $(\mathcal{F}_0, \mathcal{F}_1)$, where we take $\mathcal{F}_1 = \mathcal{F}$. We take S^0 as numeraire; i.e., we assume that

$$P(S_t^0 > 0) = 1 \text{ for } t = 0, 1.$$

The discounted price increment, omitting the 0th coordinate (which is zero), is, as before, the $\mathbb{R}^d$-random vector $\Delta \widehat{S} = (\Delta \widehat{S}^i)_{1 \leq i \leq d}$, where

$$\Delta \widehat{S}^i = \frac{S_1^i}{S_1^0} - \frac{S_0^i}{S_0^0} \text{ for } 1 \leq i \leq d. \qquad (3.6)$$

The condition that this market model does not admit an arbitrage opportunity can then be stated as follows: the one-step pricing model is *viable* (also called *arbitrage-free*) if for every vector θ in $\mathbb{R}^d$ we have P-a.s. that

$$\widehat{\theta} \cdot \Delta \widehat{S} \geq 0 \text{ implies } \widehat{\theta} \cdot \Delta \widehat{S} = 0. \qquad (3.7)$$

Note that this requirement involves only the null sets of the given measure and hence is invariant under an equivalent change of measure. Moreover, since we assume that all prices are non-negative, $\Delta \widehat{S}^i$ is bounded below by $-\frac{S_0^i}{S_0^0}$, and thus $E_Q \left(\Delta \widehat{S} | \mathcal{F}_0 \right)$ is well-defined for any probability measure $Q \sim P$.

The 'martingale property' in the single-period model reduces to the requirement that

$$E_Q \left(\frac{S_1^i}{S_1^0} | \mathcal{F}_0 \right) = \frac{S_0^i}{S_0^0} \text{ a.s. } (Q) \text{ for } 1 \leq i \leq d,$$

so that we need to find an equivalent measure Q such that

$$E_Q \left(\Delta \widehat{S} | \mathcal{F}_0 \right) = 0 \text{ a.s. } (Q).$$

Notation 3.5.2. Write $\mathcal{M}_1(\Omega, \mathcal{F})$ for the space of all probability measures on $(\Omega, \mathcal{F})$, and define

$$\mathcal{P} = \left\{ Q : Q \in \mathcal{M}_1(\Omega, \mathcal{F}), \ Q \sim P \text{ and } E_Q\left(\Delta\widehat{S}\,|\mathcal{F}_0\right) = 0 \text{ a.s. } (Q) \right\}$$

and

$$\mathcal{P}_b = \left\{ Q \in \mathcal{P}, \frac{dQ}{dP} \text{ is bounded} \right\}.$$

We call elements of $\mathcal{P}$ equivalent martingale measures for the model.

We wish to analyse the geometric properties of the set of discounted gains processes arising from admissible trading strategies. By (2.8), we know that such strategies are generated by predictable $\mathbb{R}^d$-valued processes $\widehat{\theta}$, so that in the single-period model we need to consider elements of the space $L^0(\mathbb{R}^d) = L^0(\Omega, \mathcal{F}_0, P; \mathbb{R}^d)$ of all a.s. (P)-finite $\mathcal{F}_0$-measurable random vectors

$$\widehat{\theta} = \left(\theta^1, \theta^2, \ldots, \theta^d\right).$$

We then define the linear space of inner products,

$$\mathcal{K} = \left\{ \widehat{\theta} \cdot \Delta\widehat{S} : \widehat{\theta} \in L^0(\Omega, \mathcal{F}_0, P; \mathbb{R}^d) \right\}, \tag{3.8}$$

which is a subspace of $L^0 = L^0(\Omega, \mathcal{F}_0, P; \mathbb{R})$. We can now rewrite the no-arbitrage condition (3.7) as

$$\mathcal{K} \cap L^0_+ = \{0\}, \tag{3.9}$$

where L^0_+ is the convex cone of non-negative elements of L^0. (The cones L^p_+ are defined similarly for the Lebesgue spaces $L^p = L^p(\Omega, \mathcal{F}, P; \mathbb{R})$ with $1 \le p \le \infty$.) We also introduce the convex cone

$$\mathcal{C} = \mathcal{K} - L^0_+ = \left\{ Y = \widehat{\theta} \cdot \Delta\widehat{S} - U : \widehat{\theta} \in L^0(\Omega, \mathcal{F}_0, P; \mathbb{R}^d), \ U \in L^0_+(\Omega, \mathcal{F}, P) \right\}.$$

Lemma 3.5.3. $\mathcal{C} \cap L^0_+ = \{0\}$ *if and only if* $\mathcal{K} \cap L^0_+ = \{0\}$.

Proof. The first statement is clearly necessary for the second. On the other hand, if the second statement holds and Z is an element of $\mathcal{C} \cap L^0_+$, we can find $U \in L^0_+$ and $\widehat{\theta} \in L^0$ such that $Z = \widehat{\theta} \cdot \Delta\widehat{S} - U \ge 0$ a.s. (P). In particular, $\widehat{\theta} \cdot \Delta\widehat{S} \ge 0$ a.s. (P) and so is an element of $\mathcal{K} \cap L^0_+$ and hence equals 0. This forces $U = 0$; thus $Z = 0$ a.s. (P), so that the two statements are equivalent. □

Having reformulated the no-arbitrage condition, we restate the principal objective of this section as follows.

Theorem 3.5.4. *With the above definitions for the single-period model, the following are equivalent:*

(i) $\mathcal{K} \cap L^0_+ = \{0\}$,

(ii) $\mathcal{C} \cap L^0_+ = \{0\}$,

(iii) $\mathcal{P}_b \neq \emptyset$,

(iv) $\mathcal{P} \neq \emptyset$.

The equivalence of (i) and (ii) was proved in Lemma 3.5.3. Trivially, (iii) implies (iv), and the following lemma shows that (iv) implies (i).

Lemma 3.5.5. *If $\mathcal{P}$ is non-empty, then $\mathcal{K} \cap L^0_+ = \{0\}$.*

Proof. Let $Q \in \mathcal{P}$, and suppose that $\widehat{\theta} \in L^0(\mathbb{R}^d)$ has $\widehat{\theta} \cdot \Delta \widehat{S} \in \mathcal{K} \cap L^0_+$. Since $\widehat{\theta}$ is a.s. finite, we can approximate it pointwise by truncation; i.e., $\widehat{\theta}_n = \widehat{\theta} \mathbf{1}_{\{|\widehat{\theta}| < n\}}$ increases to $\widehat{\theta}$. If $\widehat{\theta} \cdot \Delta \widehat{S} > 0$ on a set of positive P-measure, the same must be true for $\widehat{\theta}_n \cdot \Delta \widehat{S}$ if n is chosen sufficiently large. Now $E_Q\left(\widehat{\theta}_n \cdot \Delta \widehat{S}\right)$ is well-defined, but since $Q \in \mathcal{P}$ we have

$$E_Q\left(\widehat{\theta}_n \cdot \Delta \widehat{S}\right) = E_Q\left(\widehat{\theta}_n \cdot E_Q\left(\Delta \widehat{S} \big| \mathcal{F}_0\right)\right) = 0.$$

This contradicts the claim that $\widehat{\theta}_n \cdot \Delta \widehat{S}$ is non-zero and in L^0_+. So $\mathcal{K} \cap L^0_+ = \{0\}$. $\qquad \square$

Remark 3.5.6. In order to complete the proof of the fundamental theorem for this single-period model, it remains to show that (ii) implies (iii) in Theorem 3.5.4 (i.e., that the reformulated no-arbitrage condition $\mathcal{C} \cap L^0_+ = \{0\}$ implies the existence of an EMM with bounded density). To do this, it will be advantageous to assume that

$$E_P\left(\frac{S^i_t}{S^0_t}\right) < \infty \text{ for } i = 0, 1, \ldots, d \text{ and } t = 0, 1.$$

In fact, we can make this assumption without loss of generality since the statement $\mathcal{C} \cap L^0_+ = \{0\}$ is invariant under equivalent changes of measure. We therefore assume that it holds for the measure P_1 whose density relative to P is given by

$$\frac{dP_1}{dP} = \frac{c}{1 + \sum_{i=0}^{d}\left(\frac{S^i_0}{S^0_0} + \frac{S^i_1}{S^0_1}\right)},$$

where c is a normalising constant chosen to make P_1 a probability measure. Clearly the P_1-expectations of the discounted prices are finite. If we find a probability measure Q with $E_Q\left(\Delta \widehat{S} \big| \mathcal{F}_0\right) = 0$ and $\frac{dQ}{dP_1}$ bounded, then $\frac{dQ}{dP} = \frac{dQ}{dP_1}\frac{dP_1}{dP}$ is bounded, so that $Q \in \mathcal{P}_b$. Henceforth we shall assume without further mention that the discounted prices are P-integrable.

We show in several steps that (ii) implies (iii), initially by adding a further assumption on the cone $\mathcal{C}$, as described below.

The following proposition presents a basic fact about the behaviour of conditional expectations under equivalent measure changes. We shall need it several times in this chapter, as well as in Chapter 9.

Proposition 3.5.7 (Bayes' Rule). *Given probability measures P, Q with $Q \ll P$ on the measurable space $(\Omega, \mathcal{F})$, a sub-σ-field $\mathcal{G}$ of $\mathcal{F}$ and a random variable $Y \geq 0$ integrable with respect to both measures, we have the identity*

$$E_Q \left(Y | \mathcal{G}\right) = \frac{E_P \left(Y \frac{dQ}{dP} | \mathcal{G}\right)}{E_P \left(\frac{dQ}{dP} | \mathcal{G}\right)} \quad a.s. (Q). \tag{3.10}$$

Proof. Let $Q \ll P$ have Radon-Nikodym derivative $\frac{dQ}{dP} = Z$. Then $Q(Z > 0) = 1$ since, for any $A \in \mathcal{F}$,

$$Q(A) = \int_A Z dP = \int_{A \cap \{Z > 0\}} Z dP = Q(A \cap \{Z > 0\}).$$

As $\mathcal{G} \subset \mathcal{F}$, the density $\frac{dQ}{dP}\big|_{\mathcal{G}}$ equals $E_P \left(Z | \mathcal{G}\right)$ since

$$Q(G) = \int_G Z dP = \int_G E_P \left(Z | \mathcal{G}\right) dP \text{ for all } G \in \mathcal{G}.$$

For the $\mathcal{F}$-measurable random variable $Y \geq 0$, let

$$W = \begin{cases} \frac{E_P(YZ|\mathcal{G})}{E_P(Z|\mathcal{G})} & \text{if } E_P \left(Z | \mathcal{G}\right) > 0, \\ 0 & \text{if } E_P \left(Z | \mathcal{G}\right) = 0. \end{cases}$$

By the above, the latter occurs only on a Q-null set.

To prove that $W = E_Q \left(Y | \mathcal{G}\right)$, we must verify that $E_Q \left(1_G W\right) = E_Q \left(1_G Y\right)$ for all $G \in \mathcal{G}$. But this follows from

$$\begin{aligned} E_Q \left(1_G W\right) &= E_P \left(1_G W Z\right) \\ &= E_P \left(E_P \left(1_G W Z | \mathcal{G}\right)\right) \\ &= E_P \left(1_G W E_P \left(Z | \mathcal{G}\right)\right) \\ &= E_P \left(1_G E_P \left(Y Z | \mathcal{G}\right)\right) \\ &= E_P \left(E_P \left(1_G Y Z\right) | \mathcal{G}\right) \\ &= E_P \left(1_G Y Z\right) \\ &= E_Q \left(1_G Y\right). \end{aligned}$$

$\square$

The role of the convex cone $\mathcal{C}$ is clarified in the following general theorem about convex cones in L^1. We use separation arguments in the Banach space L^1 to provide a normalised element of the dual space L^∞, which will act as the bounded density of the martingale measure we wish to construct.

Theorem 3.5.8 (Kreps-Yan). *Let C be a closed convex cone in L^1 containing the negative essentially bounded functions (i.e., $C \supset -L^\infty_+$) and such that $C \cap L^1_+ = \{0\}$. Then there exists $Z \in L^\infty$ such that $Z > 0$ a.s. (P) and $E_P(YZ) \leq 0$ for all $Y \in C$.*

Proof. The separation theorem we need is the analogue of the separating hyperplane theorem (Theorem 3.1.1) and follows from the Hahn-Banach theorem (see, e.g. , [264, Theorem I.9.2]): given a closed convex cone C disjoint from a compact set K in a Banach space B, we can find a continuous linear functional f in the dual space B^* and reals α, β such that $f(c) \leq \alpha < \beta < f(k)$ for all $c \in C, k \in K$.

Applying this to the convex cone C and the compact set $\{U\}$, where $0 \neq U \in L^1_+$, we find $f \in (L^1)^*$, with $f(X) = E_P(XZ)$ if $X \in L^1$, so that $Z \in L^\infty$ implements f. Thus, for all $Y \in C$,

$$E_P(YZ) \leq \alpha < \beta < E_P(UZ) \text{ for some } \alpha, \beta.$$

Since C contains 0, we must have $\alpha \geq 0$, and as C is a cone and $E_P(YZ) \leq \alpha$ for all $Y \in C$, it follows that $\alpha = 0$ ($Y \in C$ implies $\lambda Y \in C$ for all $\lambda \geq 0$, so if $E_P(YZ) > 0$ for some Y in C, $E_P(\lambda YZ) = \lambda E_P(YZ)$ cannot be bounded above as $\lambda \to \infty$). On the other hand, $E_P(-XZ) \leq 0$ holds for all $X \in L^\infty_+$ since C contains $-L^\infty_+$. Apply this with $X = 1_{\{Z<0\}}$, so that $E_P(Z^-) \leq 0$; hence $Z \geq 0$ a.s. (P).

As $E_P(UZ) > \beta > 0$, it follows that $P(Z > 0) > 0$. Note that we can replace Z by $\frac{Z}{|Z|_\infty}$ so that we can assume without loss of generality from now on that $0 \leq Z \leq 1$. Hence we have shown that for each non-zero $U \in L^1_+$ there exists a $Z_U \in L^\infty$ with $0 \leq Z_U \leq 1$, $P(Z_U > 0) > 0$, and $E_P(YZ_U) \leq 0$ for all $Y \in C$, but $E_P(UZ_U) > 0$.

However, the claim is that we can find some $Z > 0$ a.s. (P) with these properties. To construct it, we employ an exhaustion argument. First let $\sum_{k=1}^\infty \alpha_k = 1$, $\alpha_k \geq 0$, and define

$$Z = \sum_{k=1}^\infty \alpha_k Z_{U_k},$$

where each $U_k \in L^1_+$ and Z_{U_k} is as constructed above. Then, for $Y \in C$, $\sum_{k=1}^\infty |\alpha_k Z_{U_k} Y| \leq |Y|$ shows that $E_P(\sum_{k=1}^n \alpha_k Z_{U_k} Y)$ is bounded above in L^1. Therefore, by dominated convergence, we have

$$E_P(YZ) = \sum_{k=1}^\infty \alpha_k E_P(YZ_{U_k}) \leq 0.$$

Now let

$$c = \sup_{Z_U \in D} P(Z_U > 0),$$

where

$$D = \{Z_U \in L^\infty : 0 \leq Z_U \leq 1, P(Z_U > 0) > 0; E_P(YZ_U) \leq 0 \text{ if } Y \in C\}.$$

Choose a sequence (Z_{U_k}) in D such that $P(Z_{U_k} > 0) \to c$ as $k \to \infty$. The countably convex combination $Z = \sum_{k=1}^{\infty} \frac{1}{2^k} Z_{U_k}$ satisfies $E_P(YZ) \le 0$ for all Y in $\mathcal{C}$ by the above argument and hence is in D, and $\{Z > 0\} = \cup_{k=1}^{\infty} \{Z_{U_k} > 0\}$. It follows that $P(Z > 0) = c$, and it remains to show that $c = 1$.

If $c < 1$, the set $A = \{Z = 0\}$ would have $P(A) > 0$. Then $U = 1_A \in L_+^1$, $U \ne 0$, and so $E_P(UZ_U) > 0$. This would mean that $P[A \cap \{Z_U > 0\}] > 0$ and hence the function $W = \frac{1}{2}(Z + Z_U) \in D$ would have $P(W > 0) > P(Z > 0) = c$, contradicting the definition of c.

As required, we have found $Z \in L^\infty$ with $E_P(YZ) \le 0$ for all Y in $\mathcal{C}$ and $Z > 0$ a.s. (P). $\square$

Applying this to the cone $\mathcal{C} = \mathcal{K} - L_+^0$, we can now prove the following result.

Proposition 3.5.9. *If $\mathcal{C} \cap L_+^0 = \{0\}$ and $\mathcal{C} \cap L^1$ is closed in L^1, then there is a probability measure $Q \sim P$ with bounded density relative to P such that $E_Q\left(\Delta \widehat{S} \,|\, \mathcal{F}_0\right) = 0$ a.s. (Q).*

Proof. The L^1-closed cone $\mathcal{C} \cap L^1$ contains $-L_+^\infty$ since $0 \in \mathcal{K} \cap L^1$. Hence the Kreps-Yan theorem provides a $Z \in L^\infty$ with $Z > 0$ a.s. (P) and $E_P(YZ) \le 0$ for all Y in $\mathcal{C}$. Since $\mathcal{K} \cap L^1$ is a linear space, $\alpha(\widehat{\theta} \cdot \Delta \widehat{S})$ lies in $\mathcal{K} \cap L^1$ and hence in $\mathcal{C} \cap L^1$ (recall Remark 3.5.6) for any $\alpha \in \mathbb{R}$ and $\widehat{\theta}$ in $L^\infty(\mathcal{F}_0, \mathbb{R}^d)$. It follows that $E_P\left(Z(\widehat{\theta} \cdot \Delta \widehat{S})\right) = 0$ for all choices of $\widehat{\theta}$. But then

$$E_P\left(\widehat{\theta} \cdot E_P\left(Z \Delta \widehat{S} \,|\, \mathcal{F}_0\right)\right) = E_P\left(Z(\widehat{\theta} \cdot \Delta \widehat{S})\right) = 0$$

for all $\widehat{\theta} \in L^\infty(\mathcal{F}_0; \mathbb{R}^d)$, so that $E_P\left(Z \Delta \widehat{S} \,|\, \mathcal{F}_0\right) = 0$ a.s. (P). Now apply the conditional Bayes rule (3.10) so that, setting $\frac{dQ}{dP} = \frac{Z}{E_P(Z)}$, we finally obtain

$$E_Q\left(\Delta \widehat{S} \,|\, \mathcal{F}_0\right) = \frac{E_P\left(Z \Delta \widehat{S} \,|\, \mathcal{F}_0\right)}{E_P\left(Z \,|\, \mathcal{F}_0\right)} = 0 \text{ a.s. } (Q). \qquad (3.11)$$

Hence Q is an EMM for the single-period model. $\square$

We have now proved Theorem 3.5.4 under the additional assumption that the cone $\mathcal{C} \cap L^1$ is closed in the L^1-norm. The removal of this additional assumption requires a more subtle analysis, which is presented in the next section. The reader may prefer to omit this on a first reading and go directly to the proof of the fundamental theorem in a multi-period setting, which, with the above preparation, now only requires a careful backward induction procedure.

Closed Subsets of L^0

We saw that, to reformulate the no-arbitrage condition in geometric terms, we need to deal with the larger space L^0, which does not have the convenience of a norm topology. Indeed, the appropriate topology in L^0 is that of convergence in probability.

Definition 3.5.10. The random variables (X_n) in $L^0(\Omega, \mathcal{F}, P; \mathbb{R}^d)$ $(d \geq 1)$ *converge in probability* to a random variable X if

$$\lim_{n \to \infty} P(|X_n - X|_d > \varepsilon) = 0 \text{ for all } \varepsilon > 0.$$

Here $|\cdot|_d$ denotes the Euclidean norm in $\mathbb{R}^d$.

This convergence concept for $\mathbb{R}^d$-valued random vectors can of course also be defined in terms of their coordinate random variables. The topology on $L^0(\Omega, \mathcal{F}, P; \mathbb{R})$ is induced by the metric

$$d(X, Y) = E_P \left(\frac{|X - Y|}{1 + |X - Y|} \right),$$

so that with the resulting topology, $L^0(\Omega, \mathcal{F}, P; \mathbb{R}^d)$ is metrisable and the above definition suffices to describe convergence in this topology for each coordinate. It is elementary that convergence in the L^p-norm for any $p \geq 1$ implies convergence in probability. Moreover, a.s. (P)-convergence implies convergence in probability, and if $X_n \to X$ in probability, then some subsequence $(X_{n_k})_{k \geq 1}$ converges to X a.s. (P).

Our principal source of relevant information on sets in $L^0(\mathcal{F}_0; \mathbb{R}^d)$ are the $\widehat{\theta} \in L^0(\mathcal{F}_0; \mathbb{R}^d)$, which give rise to discounted gains processes whose conditional expectation relative to $\mathcal{F}_0$ vanishes a.s. (P). It is thus natural to fix vectors in $\mathbb{R}^d$ whose values are a.s. (P) orthogonal to the discounted price increments.

Write

$$N = \left\{ \phi \in L^0(\mathcal{F}_0; \mathbb{R}^d) : \phi \cdot \Delta \widehat{S} = 0 \text{ a.s. } (P) \right\},$$

$$N^\perp = \left\{ \psi \in L^0(\mathcal{F}_0; \mathbb{R}^d) : \phi \cdot \psi = 0 \text{ a.s. } (P) \text{ for all } \phi \in N \right\}.$$

It is of course by no means clear at this stage that the notation $N^\perp$ signifies any 'orthogonality' in the function space $L^0(\mathcal{F}_0; \mathbb{R}^d)$: we show below how this notation will be justified. First we note some simple properties of the linear subspaces N and $N^\perp$.

Lemma 3.5.11. N *and* $N^\perp$ *are closed subspaces of* $L^0(\mathcal{F}_0; \mathbb{R}^d)$, *and are closed under multiplication by functions in* $L^0(\mathcal{F}_0; \mathbb{R})$. *Moreover,* $N \cap N^\perp = \{0\}$.

Proof. If (ϕ_n) in N converges in probability to $\phi \in L^0(\mathcal{F}_0; \mathbb{R}^d)$ then some subsequence (ϕ_{n_k}) converges to ϕ a.s. (P). Hence

$$\left(\phi \cdot \Delta \widehat{S} \right) (\omega) = \lim_k \left(\phi_{n_k} \cdot \Delta \widehat{S} \right) (\omega) = 0 \text{ a.s. } (P),$$

so that $\phi \in N$. Hence N is closed in $L^0(\mathcal{F}_0; \mathbb{R}^d)$. An identical proof shows that $N^\perp$ is also closed.

Next, let $h : \Omega \to \mathbb{R}$ be $\mathcal{F}_0$-measurable and finite a.s. (P). Then

$$\left((h\phi) \cdot \Delta \widehat{S} \right)(\omega) = h(\omega) \left(\phi \cdot \Delta \widehat{S} \right)(\omega) = 0 \text{ a.s. } (P) \text{ for all } \phi \in N,$$

so that $h\phi \in N$. Similarly, for ψ in $N^\perp$, $((h\psi) \cdot \phi) = h(\psi \cdot \phi) = 0$ for all $\phi \in N$.

Finally, if $\phi \in N \cap N^\perp$, we have $(\phi \cdot \phi)(\omega) = |\phi(\omega)|_d^2 = 0$ a.s. (P). Hence $N \cap N^\perp = \{0\}$ as subspaces of $L^0(\mathcal{F}_0; \mathbb{R}^d)$. $\square$

The next result provides the 'orthogonal decomposition' of $L^0(\mathcal{F}_0; \mathbb{R}^d)$ indicated by the notation.

Proposition 3.5.12. *Every $\phi \in L^0(\mathcal{F}_0; \mathbb{R}^d)$ can be decomposed uniquely as $\phi = P_1\phi + P_2\phi$, where $P_1\phi \in N, P_2\phi \in N^\perp$.*

Proof. We prove this first for the *constant* functions $\omega \to e_i$, where the (e_i) form the standard ordered basis of $\mathbb{R}^d$. Any element of $L^0(\mathcal{F}_0; \mathbb{R}^d)$ can be written in the form $\phi = \sum_{i=}^d \phi_i e_i$, where the coordinate functions (ϕ_i) are $\mathcal{F}_0$-measurable real random variables.

Fix $i \leq d$, and by a minor abuse of notation write e_i for the constant function with this value. As a bounded function, e_i is in the Hilbert space $H = L^2(\mathcal{F}_0; \mathbb{R}^d)$, and $H_1 = N \cap H$ and $H_2 = N^\perp \cap H$ are linear subspaces of H. Both are closed in H since L^2-convergence implies convergence in probability. Hence the projection maps $P_i : H \to H_i$ $(i = 1, 2)$ are well-defined. Consider the element $\psi = e_i - P_1 e_i$. To show that $H_2 = H_1^\perp$, we need only prove that $\psi \in N^\perp$, which implies that $\psi = P_2 e_i$. If ψ is not in $N^\perp$, we can find $\phi \in N$ such that the inner product $(\phi \cdot \psi)(\omega) > 0$ on a set $A \in \mathcal{F}_0$ with $P(A) > 0$. Since it is possible that $E_P (\phi \cdot \psi)$ is infinite, we consider the truncations

$$\phi_n(\omega) = \begin{cases} \phi(\omega) 1_{\{|\phi| \leq n\}} & \text{if } \omega \in A, \\ 0 & \text{if } \omega \notin A. \end{cases}$$

Then each $E_P (\phi_n \cdot \psi)$ is finite, and we have $(\phi_n, \psi)_H = E_P (\phi_n \cdot \psi) > 0$ for large enough n, where $(\cdot, \cdot)_H$ is the inner product in H. As $\phi_n \in H_1 = N \cap H$, this would contradict the construction of ψ as a vector orthogonal to H_1 in H, so $\psi \in N^\perp$.

This completes the decomposition of e_i. Since $e_i = P_1 e_i + P_2 e_i$ for each $i \leq d$, with $P_1 e_i \in N \cap H$, $P_2 e_i \in N^\perp \cap H$, is a unique decomposition, we can now write $(P_1\phi)(\omega) = \sum_{i=1}^d \phi_i(\omega)(P_1 e_i)(\omega)$ and $(P_2\phi)(\omega) = \sum_{i=1}^d \phi_i(\omega)(P_2 e_i)(\omega)$ for each $\omega \in \Omega$. The function $P_1\phi$ is in N and $P_2\phi$ is in $N^\perp$ by Lemma 3.5.11, which also confirms that the decomposition is unique. $\square$

The final lemma we need provides a measurable way of selecting a convergent subsequence from a given sequence in $L^0(\mathcal{F}_0; \mathbb{R}^d)$. This is achieved by a diagonal argument on the components of the random vectors.

Lemma 3.5.13. *If $(f_n)_{n \geq 1}$ is a sequence in $L^0(\mathcal{F}_0; \mathbb{R}^d)$ with $\liminf_n |f_n|$ finite, then there is an element f in $L^0(\mathcal{F}_0; \mathbb{R}^d)$ and a strictly increasing sequence (τ_n) of $\mathcal{F}_0$-measurable random variables taking their values in $\mathbb{N}$ such that $f_{\tau_n(\omega)}(\omega) \to f(\omega)$ for P-almost all $\omega \in \Omega$.*

Proof. Write $F(\omega) = \liminf_n |f_n(\omega)|_d$, where $|\cdot|_d$ is again the Euclidean norm in $\mathbb{R}^d$. On the P-null set $B = \{F = \infty\}$, we set $\tau_m = m$ for each m. For ω in B^c we define τ_m inductively. First set

$$\sigma_m^0(\omega) = \begin{cases} 1 & \text{if } m = 1, \\ \min\left\{n > \sigma_{m-1}^0(\omega) : \left||f_n(\omega)| - F(\omega)\right|_d \leq \tfrac{1}{m}\right\} & \text{if } m \geq 2. \end{cases}$$

The first component f^1 of f is now taken as

$$f^1(\omega) = \lim_{m \to \infty} \inf f_{\sigma_m^0(\omega)}^1(\omega), \tag{3.12}$$

and at the same time we define a subsequence of random indices $(\sigma_m^1)_{m \geq 1}$ by using this 'limit value' in the construction: let $\sigma_1^1(\omega) = 1$, and for $m \geq 2$ define

$$\sigma_m^1(\omega) = \min\left\{\sigma_n^0 : \sigma_n^0(\omega) > \sigma_{m-1}^1(\omega) \text{ and } \left|f_{\sigma_{n-1}^0(\omega)}^1(\omega) - f^1(\omega)\right| \leq \frac{1}{m}\right\}.$$

Continue this inductively for $i = 2, 3, \ldots, d$, finding the second coordinate of the limit function at the next step and simultaneously constructing a subsequence (σ_m^2) of (σ_m^1) that leads to the next coordinate of f. Finally, let $\tau_m = \sigma_m^d$ for each $m \geq 1$. It is clear from the construction that $\left|f_{\tau_m(\omega)}^i(\omega) - f^i(\omega)\right| \leq \frac{1}{m}$ for each $i \leq d$ and that (τ_m) is strictly increasing, and each τ_m is $\mathcal{F}_0$-measurable by construction. $\square$

We are now ready for the final step in the proof of Theorem 3.5.4.

Proposition 3.5.14. *If $\mathcal{K} \cap L_+^0 = \{0\}$, then $\mathcal{C} = \mathcal{K} - L_+^0$ is closed in L^0.*

Proof. Let (Y_n) be a sequence in $\mathcal{C}$ converging to $Y \in L^0$ as $n \to \infty$. There is a subsequence converging to Y a.s. (P), so we can assume without loss of generality that $Y_n \to Y$ a.s. (P). Write $Y_n = \psi_n \cdot \Delta\widehat{S} - U_n$ for some $U_n \in L_+^0$ and $\psi_n \in N^\perp$ since by Proposition 3.5.12 any $\theta \in L^0(\mathcal{F}_0; \mathbb{R}^d)$ can be decomposed uniquely as $\theta = \phi + \psi$ with $\phi \in N$ and $\psi \in N^\perp$, and then $\phi \cdot \Delta\widehat{S} = 0$ so $\theta \cdot \Delta\widehat{S} = \psi \cdot \Delta\widehat{S}$.

Define $\alpha_n = (1 + |\psi_n|_d)^{-1}$ and set $f_n = \alpha_n \psi_n$. Extend this to a 'portfolio' $F_n = (\alpha_n, f_n)$ in $L^0(\mathcal{F}_0, \mathbb{R}^{d+1})$ and note that $|F_n| \leq 2$, so that we can apply Lemma 3.5.13 to provide $\mathcal{F}_0$-measurable random variables with values $\tau_1 < \tau_2 < \cdots < \tau_n < \cdots$ in $\mathbb{N}$ and a function $F \in L^\infty(\mathcal{F}_0, \mathbb{R}^{d+1})$ such that $F_{\tau_n} \to F$ P-almost surely. Since the convergence holds coordinate-wise, we can write $F = (\alpha, f)$ and then $\alpha_{\tau_n} \to \alpha$ and $f_{\tau_n} \to f$.

We show that $f_{\tau_n} \in N^\perp$ for each n. For this, let $\phi \in N$ be given. Then

$$(\phi \cdot f_{\tau_n})(\omega) = \sum_{k=1}^{\infty} \alpha_k(\omega) \mathbf{1}_{\{\tau_n(\omega)=k\}}(\omega)(\phi \cdot \psi_k)(\omega) = 0 \text{ a.s. } (P)$$

since each $\psi_k \in N^\perp$. Since $N^\perp$ is closed in $L^0(\mathcal{F}_0; \mathbb{R}^d)$, it follows that $f \in N^\perp$.

Now consider the set $A = \{\alpha = 0\}$. We claim that $P(A) = 0$. To see this, note that since $Y_n \to Y$ a.s. and $\alpha_{\tau_n} \to \alpha$ a.s. it follows that $\alpha_{\tau_n} Y_{\tau_n} = f_{\tau_n} \cdot \Delta\widehat{S} - \alpha_{\tau_n} U_{\tau_n}$ converges a.s. (P). On A the limit is obviously 0. But $f_{\tau_n} \cdot \Delta\widehat{S} \to f \cdot \Delta\widehat{S}$ a.s. (P), so we have proved that

$$\mathbf{1}_A \alpha_{\tau_n} U_{\tau_n} \to \mathbf{1}_A f \cdot \Delta\widehat{S} \text{ a.s. } (P).$$

Now each element on the left-hand side is non-negative and hence so is their limit. By the no-arbitrage condition $\mathcal{K} \cap L^0_+ = \{0\}$, it follows that $(\mathbf{1}_A f) \cdot \Delta\widehat{S} = 0$. Since $f \in N^\perp$, the same is true of $\mathbf{1}_A f$, which therefore belongs to $N \cap N^\perp = \{0\}$.

Thus $f = 0$ a.s. (P) on A. This forces $P(A) = 0$. To see this, note that by definition, $\alpha_{\tau_n(\omega)}(\omega) \to 0$ means that $(|\psi_{\tau_n(\omega)}(\omega)|_d)_n$ is unbounded above. Hence

$$\frac{|\psi_{\tau_n}|}{1 + |\psi_{\tau_n}|} \to 1,$$

so that

$$\mathbf{1}_A |f| = \mathbf{1}_A \lim_n |\alpha_{\tau_n} \psi_{\tau_n}| = \mathbf{1}_A \lim_n \frac{|\psi_{\tau_n}|}{1 + |\psi_{\tau_n}|} = \mathbf{1}_A. \tag{3.13}$$

In other words, $|f| = 1$ a.s. (P) on A, which is impossible unless $P(A) = 0$.

We therefore need only examine the convergence of $(\psi_{\tau_n})_n$ on $A^c = \{\alpha > 0\}$ since this set has full P-measure. By construction, we have

$$\alpha_{\tau_n(\omega)}(\omega) > 0 \text{ a.s. } (P).$$

Hence, as $P(A) = 0$,

$$\frac{1}{\alpha} f = \lim_n \frac{1}{\alpha_{\tau_n}} f_{\tau_n} = \lim_n \psi_{\tau_n} \text{ a.s. } (P). \tag{3.14}$$

Thus, as $U_n \geq 0$ for all n, we have

$$Y = \lim_n Y_n = \lim_n Y_{\tau_n} \leq \lim_n(\psi_{\tau_n} \cdot \Delta\widehat{S}) = \frac{1}{\alpha} f \cdot \Delta\widehat{S} \text{ a.s. } (P). \tag{3.15}$$

Thus Y has the form $\phi \cdot \Delta\widehat{S} - U$ for some $U \in L^0_+$, so that $Y \in \mathcal{C}$ as required. This completes the proof. $\qquad\square$

Remark 3.5.15. Note that we have equality in (3.15) if $Y_n = \psi_n \cdot \Delta\widehat{S}$ for all n. Therefore, if all Y_n are in $\mathcal{K}$, then so is their L^0-limit Y. Hence we have also shown that if $\mathcal{K} \cap L^0_+ = \{0\}$, then $\mathcal{K}$ is closed in L^0.

The Fundamental Theorem for a Multi-period Model

Having completed the construction of the EMM for a general single-period model with random initial prices, we can finally return to a multi-period setting to complete the proof of Theorem 3.5.1. We take as given a probability space $(\Omega, \mathcal{F}, P)$ and a time set $\mathbb{T} = \{0, 1, \ldots, T\}$ for some natural number T. We also reinstate the condition on $\mathcal{F}_0$ as in the theorem: let $\mathbb{F} = (\mathcal{F}_t)_{t \in \mathbb{T}}$ be a filtration with $\mathcal{F}_0$ consisting of all P-null sets and their complements. Suppose the $\mathbb{R}^{d+1}$-valued process $S = (S_t^i : 0 \leq i \leq d, t \in \mathbb{T})$ is adapted to $\mathbb{F}$, with $S_t^0 > 0$ a.s. (P) for each t in $\mathbb{T}$.

As usual, we take the 0th asset as numéraire and consider the discounted price processes $\overline{S}_t^i = \frac{S_t^i}{S_t^0}$ instead. This ensures that $\overline{S}^0 \equiv 1$ and that all prices are expressed in units of S^0. Given any self-financing trading strategy $\theta = \{\theta_t^i : 0 \leq i \leq d\}_{1 \leq t \leq T}$, the discounted value process $\overline{V}(\theta)$ defined as

$$\overline{V}_0(\theta) = \theta_1 \cdot \overline{S}_0, \qquad \overline{V}_t(\theta) = \theta_t \cdot \overline{S}_t(\theta) \text{ for } t = 1, 2, \ldots, T$$

satisfies

$$\overline{V}_t(\theta) = \overline{V}_0(\theta) + \overline{G}_t(\theta) \text{ for all } t = 1, 2, \ldots, T,$$

where

$$\overline{G}_t(\theta) = \sum_{u=1}^{t} \widehat{\theta}_u \cdot \Delta \widehat{S}_t,$$

with $\Delta \widehat{S}_t = \left(\frac{S_t^i}{S_t^0} - \frac{S_{t-1}^i}{S_{t-1}^0} \right)_{1 \leq i \leq d}$ and $\widehat{\theta}_t = (\theta_t^i)_{1 \leq i \leq d}$.

Proof of the Fundamental Theorem. As we have seen in Proposition 3.3.2, the no-arbitrage condition in this multi-period model can be restated as, for all t and $\theta \in L^0(\Omega, \mathcal{F}_{t-1}, P; \mathbb{R}^d)$, the requirement $\widehat{\theta}_t \cdot \Delta \widehat{S}_t \geq 0$ a.s. (P) implies that $\widehat{\theta}_t \cdot \Delta \widehat{S}_t = 0$ a.s. (P).

We therefore consider the single-period model with times $\{t-1, t\}$ instead of $\{0, 1\}$. Defining the subspace

$$\mathcal{K}_t = \left\{ \widehat{\theta}_t \cdot \Delta \widehat{S}_t : \widehat{\theta}_t \in L^0(\Omega, \mathcal{F}_{t-1}, P; \mathbb{R}^d) \right\}, \tag{3.16}$$

we have the reformulation of the no-arbitrage condition as

$$\mathcal{K}_t \cap L_+^0(\Omega, \mathcal{F}_t, P) = \{0\}. \tag{3.17}$$

This statement involves knowledge of the measure P only through its null sets and thus remains valid for any probability equivalent to P. It also allows us to apply Theorem 3.5.4 to the tth trading period for each $t \leq T$. Beginning with $t = T$, we obtain a probability measure $Q_T \sim P$ with bounded density $\frac{dQ_T}{dP}$ such that $E_{Q_T} \left(\Delta \widehat{S}_T | \mathcal{F}_{T-1} \right) = 0$. Thus we are able to start the backward induction procedure. Assume by induction that we have found a probability measure $Q_{t+1} \sim P$ that turns the process $(\widehat{S}_u)_{t+1 \leq u \leq T}$ into a martingale; i.e., that

$$E_{Q_{t+1}} \left(\Delta \widehat{S}_u \,|\, \mathcal{F}_{u-1} \right) = 0 \text{ for } u = t+1, t+2, \ldots, T.$$

Then (3.17) is valid with Q_{t+1} in place of P, and we can again apply Theorem 3.5.4 to find a probability measure $Q_t \sim Q_{t+1}$ with bounded $\mathcal{F}_t$-measurable density $\frac{dQ_t}{dQ_{t+1}}$ such that $E_{Q_t} \left(\Delta \widehat{S}_t \,|\, \mathcal{F}_{t-1} \right) = 0$. The density $\frac{dQ_t}{dP} = \frac{dQ_t}{dQ_{t+1}} \frac{dQ_{t+1}}{dP}$ remains bounded and is strictly positive a.s. (P), since $Q_t \sim Q_{t+1} \sim P$. Now apply the Bayes rule (3.10) to these measures with integrand $\Delta \widehat{S}_u$ for $t+1 \leq u \leq T$:

$$
\begin{aligned}
E_{Q_t} \left(\Delta \widehat{S}_u \,|\, \mathcal{F}_{u-1} \right) &= \frac{E_{Q_{t+1}} \left(\Delta \widehat{S}_u \frac{dQ_t}{dQ_{t+1}} \,|\, \mathcal{F}_{u-1} \right)}{E_{Q_{t+1}} \left(\frac{dQ_t}{dQ_{t+1}} \,|\, \mathcal{F}_{u-1} \right)} \\
&= E_{Q_{t+1}} \left(\Delta \widehat{S}_u \,|\, \mathcal{F}_{u-1} \right) \\
&= 0
\end{aligned}
$$

since the density $\frac{dQ_t}{dQ_{t+1}}$ is $\mathcal{F}_t$-measurable and hence $\mathcal{F}_{u-1}$-measurable for every $u \geq t+1$. Under $Q_t \sim P$, the process $(\widehat{S}_u)_{t \leq u \leq T}$ is therefore a martingale, which completes the induction step. The measure $Q_1 \sim P$ we obtain at the final step, when $t = 1$, turns $(\widehat{S}_u)_{1 \leq u \leq T}$ into a martingale. The result follows. $\square$

Equivalent Martingale Measures and Change of Numéraire

Having established the fundamental relationship between viability of the model and the existence of EMMs, it is natural to consider the impact of a change of numéraire. On the one hand, the viability of the model is not affected by a change of numéraire, since the definition of arbitrage (e.g. , as expressed in terms of the gains process at a single step, as in Proposition 3.3.2) does not involve the amount of a positive gain but only its existence. On the other hand, whether a given measure is an EMM for the model will in general depend on the choice of numéraire. At the same time, it seems plausible that there should be a simple relationship between the sets of EMMs for a given model under two different choices of numéraire: it is clear from model viability that both sets are either empty or non-empty together.

So assume that we have a viable pricing model in which the assets S^0 and S^1 are strictly positive throughout. Denote by $\mathcal{P}_i$ the non-empty set of EMMs for the model when S^i is used as numéraire ($i = 0, 1$). Recall that we write the discounted price process as $\overline{S} = \left(1, \frac{S^1}{S^0}, \frac{S^2}{S^0}, \ldots, \frac{S^d}{S^0} \right)$ when S^0

is used as numéraire. Write $\widetilde{S}$ for the discounted price process when the numéraire is S^1, so that $\widetilde{S} = \left(\frac{S^0}{S^1}, 1, \frac{S^2}{S^1}, \ldots, \frac{S^d}{S^1}\right)$. Note that $\widetilde{S}^i = \frac{S^0}{S^1}\overline{S}^i$ for $i = 0, 1, \ldots, d$. Recall that $\mathcal{M}_1(\Omega, \mathcal{F})$ denotes the space of probability measures on $(\Omega, \mathcal{F})$.

Proposition 3.5.16. *We have*

$$\mathcal{P}_1 = \left\{ \widetilde{Q} : \widetilde{Q} \in \mathcal{M}_1(\Omega, \mathcal{F}); \ \frac{d\widetilde{Q}}{dQ} = \frac{\overline{S}_t^1}{\overline{S}_0^1} \ for \ some \ Q \in \mathcal{P}_0 \right\}.$$

Proof. Denote the set of probability measures on the right by $\widetilde{\mathcal{P}}$. We first show that $\mathcal{P}_1 \subset \widetilde{\mathcal{P}}$. To do this, fix $Q \in \mathcal{P}_0$, let $t \in \mathbb{T}$ be given, and write

$$\Lambda_t = \frac{\overline{S}_t^1}{\overline{S}_0^1} = \frac{S_t^1}{S_0^0} \cdot \frac{S_0^0}{S_0^1}.$$

Then $\Lambda_0 \equiv 1$ and Λ is a Q-martingale since

$$E_Q\left(\Lambda_t \,|\, \mathcal{F}_{t-1}\right) = \frac{1}{\overline{S}_0^1} E_Q\left(\overline{S}_t^1 \,|\, \mathcal{F}_{t-1}\right) = \frac{\overline{S}_{t-1}^1}{\overline{S}_0^1} = \Lambda_{t-1} \text{ a.s. } (Q). \qquad (3.18)$$

Since $S_t^0 > 0$ and $S_t^1 > 0$ for all t by hypothesis, $\Lambda_t > 0$ a.s. (Q) for all t. In particular, $\frac{d\widetilde{Q}}{dQ} = \Lambda_t$ defines a probability measure $\widetilde{Q} \sim Q \sim P$.

It remains to show that $\widetilde{S}$ is a martingale under $\widetilde{Q}$. By Bayes' rule and the definition of Λ, we have a.s. (Q) for $u < t$ in $\mathbb{T}$ and $i = 0, 1, \ldots, d$,

$$\begin{aligned}
E_{\widetilde{Q}}\left(\widetilde{S}_t^i \,|\, \mathcal{F}_u\right) &= \frac{E_Q\left(\widetilde{S}_t^i \Lambda_t \,|\, \mathcal{F}_u\right)}{E_Q\left(\Lambda_t \,|\, \mathcal{F}_u\right)} \\
&= \frac{1}{\Lambda_u} E_Q\left(\widetilde{S}_t^i \Lambda_t \,|\, \mathcal{F}_u\right) \\
&= \frac{1}{\Lambda_u} E_Q\left(\frac{S_t^0}{S_t^1}\overline{S}_t^i \Lambda_t \,|\, \mathcal{F}_u\right) \\
&= \frac{1}{\Lambda_u} \frac{S_0^0}{S_0^1} E_Q\left(\overline{S}_t^i \,|\, \mathcal{F}_u\right) \\
&= \frac{S_u^0}{S_u^1}\overline{S}_u^i = \widetilde{S}_u^i.
\end{aligned}$$

Therefore $\widetilde{Q} \in \mathcal{P}_1$ and we have proved that $\widetilde{\mathcal{P}}$ contains $\mathcal{P}_1$. To prove the opposite inclusion, we need only reverse the roles of $\widetilde{S}$ and $\overline{S}$, so the proposition is proved. $\qquad\square$

Chapter 4

Complete Markets

Our objective in this chapter is to characterise *completeness* of the market model. First we provide a simple reformulation of completeness in terms of the representability of martingales. Although we restrict our attention (and apply the results) to *finite* market models, the more general theorems proved in the final two sections of this chapter can easily be applied to reproduce this proof for general discrete-time models.

The key result proved for finite market models states that in a viable complete model the equivalent martingale measure is unique. For finite models such as the CRR model, which is examined in detail, the fine structure of the filtrations can be identified more fully. However, we shall see later that the restriction to finite complete models is more apparent than real and that, in the discrete setting, complete models form the exception rather than the rule. To establish the desired characterisation of complete models, we also characterise the attainability of contingent claims-in the general setting, this requires the full power of the first fundamental theorem.

Let $S = (S^i : i = 0, 1, \ldots, d)$ be a non-negative $\mathbb{R}^{d+1}$-valued stochastic process representing the price vector of one riskless security with

$$S_0^0 = 1, \qquad\qquad S_t^0 = \beta_t^{-1} S_0^0,$$

and d risky securities $\{S_t^i : i = 1, 2, \ldots, d\}$ for each $t \in \mathbb{T} = \{0, 1, \ldots, T\}$.

Let X be a contingent claim (i.e., a nonnegative $\mathcal{F}$-measurable random variable $X : \Omega \mapsto \mathbb{R}$). Recall that X is said to be *attainable* if there exists an admissible trading strategy θ that *generates* X (i.e., whose value process $V(\theta) \geq 0$ satisfies $V_T(\theta) = X$ a.s. (P)).

4.1 Completeness and Martingale Representation

Let $(\Omega, \mathcal{F}, P, \mathbb{T}, \mathbb{F})$ be a complete market model with unique EMM Q. This is equivalent to the following martingale representation property: the discounted price $\overline{S}$ serves as a *basis* (under martingale transforms) for the space of $(\mathbb{F}, Q)$-martingales on $(\Omega, \mathcal{F})$. To avoid integrability issues, we restrict ourselves to finite models in the proof of the following proposition.

Proposition 4.1.1. *The viable finite market model* $(\Omega, \mathcal{F}, \mathbb{T}, \mathbb{F}, P)$ *with EMM* Q *is complete if and only if each real-valued* $(\mathbb{F}, Q)$-*martingale* $M = (M_t)_{t \in \mathbb{T}}$ *can be represented in the form*

$$M_t = M_0 + \sum_{u=1}^{t} \gamma_u \cdot \Delta \overline{S}_u = M_0 + \sum_{i=1}^{d} \left(\sum_{u=1}^{t} \gamma_u^i \Delta \overline{S}_u^i \right) \tag{4.1}$$

for some predictable process $\gamma = \{\gamma^i : i = 1, 2, \ldots, d\}$.

Proof. Suppose the model is complete, and (since every martingale is the difference of two positive martingales) assume without loss of generality that $M = (M_t)$ is a non-negative $(\mathbb{F}, Q)$-martingale. Let $C = M_T S_T^0$, and find a strategy $\theta \in \Theta_a$ that generates this contingent claim, so that $V_T(\theta) = C$, and hence $\overline{V}_T(\theta) = M_T$. Now, since the discounted value process $\overline{V}$ is a Q-martingale, we have

$$\overline{V}_t(\theta) = E_Q \left(\overline{V}_t(\theta) \,|\, \mathcal{F}_t \right) = E_Q \left(M_T \,|\, \mathcal{F}_t \right) = M_t.$$

Thus the martingale M has the form

$$M_t = \overline{V}_t(\theta) = V_0(\theta) + \sum_{u=1}^{t} \theta_u \cdot \Delta \overline{S}_u = M_0 + \sum_{u=1}^{t} \theta_u \cdot \Delta \overline{S}_u$$

for all $t \in \mathbb{T}$. Hence we have proved (4.1) with $\gamma_u = \theta_u$ for all $u \in \mathbb{T}$.

Conversely, fix a contingent claim C, and define the martingale $M = (M_t)$ by setting $M_t = E_Q \left(\beta_T C \,|\, \mathcal{F}_t \right)$. By hypothesis, the martingale M has the representation (4.1). So we define a strategy θ by setting

$$\theta_t^i = \gamma_t^i \text{ for } i \geq 1, \qquad \theta_t^0 = M_t - \gamma_t \cdot \overline{S}_t \text{ for } t \in \mathbb{T}.$$

We show that θ is self-financing by verifying that $(\Delta \theta_t) \cdot S_{t-1} = 0$. Indeed, for fixed $t \in \mathbb{T}$, we have

$$(\Delta \theta_t) \cdot S_{t-1} = S_{t-1}^0 \left(\Delta M_t - \Delta \left[\sum_{i=1}^{d} \gamma_t^i \overline{S}_t^i \right] \right) + \sum_{i=1}^{d} S_{t-1}^i \Delta \gamma_t^i$$

$$= \sum_{i=1}^{d} \left(S_{t-1}^0 \left[\gamma_t^i \Delta \overline{S}_t^i - \left(\gamma_t^i \overline{S}_t^i - \gamma_{t-1}^i \overline{S}_{t-1}^i \right) \right] + S_{t-1}^i \Delta \gamma_t^i \right)$$

$$= \sum_{i=1}^{d} S_{t-1}^i \left(\Delta \gamma_t^i - \Delta \gamma_t^i \right) = 0.$$

Moreover, $V_t(\theta) = \theta_t \cdot S_t = M_t S_t^0$ for all $t \in \mathbb{T}$. In particular, we obtain $C = V_T(\theta)$, as required. Thus the market model is complete. □

4.2 Completeness for Finite Market Models

We saw in Chapter 2 that the Cox-Ross-Rubinstein binomial market model is both viable and complete. In fact, we were able to construct the equivalent martingale measure Q for S directly and showed that in this model there is a *unique* equivalent martingale measure. We now show that this property characterises completeness in the class of viable finite market models.

Theorem 4.2.1 (Second Fundamental Theorem for Finite Market Models). *A viable finite market model is complete if and only if it admits a unique equivalent martingale measure.*

Proof. Suppose the model is viable and complete and that Q and Q' are martingale measures for S with $Q' \sim P \sim Q$. Let X be a contingent claim, and let $\theta \in \Theta_a$ generate X. Then, by (2.7), we have

$$\beta_T X = \overline{V}_T(\theta) = V_0(\theta) + \sum_{t=1}^{T} \theta_t \cdot \Delta \overline{S}_t. \tag{4.2}$$

Since each discounted price process $\overline{S}^i$ is a martingale under both Q and Q', the above sum has zero expectation under both measures. Hence

$$E_Q \left(\beta_T X \right) = V_0(\theta) = E_{Q'} \left(\beta_T X \right);$$

in particular,

$$E_Q \left(X \right) = E_{Q'} \left(X \right). \tag{4.3}$$

Equation (4.3) holds for every $\mathcal{F}$-measurable random variable X, as the model is complete. In particular, it holds for $X = \mathbf{1}_A$, where $A \in \mathcal{F}$ is arbitrary, so that $Q(A) = Q'(A)$. Hence $Q = Q'$, and so the equivalent martingale measure for this model is unique.

Conversely, suppose that the market model is viable but not complete, so that there exists a non-negative random variable X that cannot be generated by an admissible trading strategy. This implies that X cannot be generated by any self-financing strategy $\theta = \{\theta^0, \theta^1, \theta^2, \ldots, \theta^d\}$, and by (2.8) we can restrict attention to predictable processes $\{\theta^1, \theta^2, \ldots, \theta^d\}$ in $\mathbb{R}^d$, as these determine θ^0 up to constants.

Therefore, define

$$L = \left\{ c + \sum_{t=1}^{T} \theta_t \cdot \Delta \overline{S}_t : \theta \text{ predictable}, c \in \mathbb{R} \right\}.$$

Then L is a linear subspace of the vector space $L^0(\Omega, \mathcal{F}, P)$. Note that this is just $\mathbb{R}^n$, where the minimal $\mathcal{F}$-partition of Ω has n members. Since this space is finite-dimensional, L is closed.

Suppose that $\beta_T X \in L$ (i.e., $\beta_T X = c + \sum_{t=1}^T \theta_t \cdot \Delta \overline{S}_t$ for some $\mathbb{R}^d$-valued predictable process θ). By (2.8), we can always extend θ to a self-financing strategy with initial value c. However, X would be attained by this strategy. Hence we cannot have $\beta_T X \in L$, and so L is a proper subspace of L^0 and thus has a non-empty orthogonal complement $L^\perp$.

Thus, for any EMM Q, there exists a non-zero random variable $Z \in L^0$ such that

$$E_Q(YZ) = 0 \text{ for all } Y \in L. \tag{4.4}$$

As L^0 is finite-dimensional, Z is bounded. Note that $E_Q(Z) = 0$ since $Y \equiv 1$ is in L (take $\theta^i \equiv 0$ for $i \geq 1$).

Define a measure $Q' \sim Q$ by

$$\frac{Q'(\omega)}{Q(\omega)} = R(\omega),$$

where

$$R(\omega) = 1 + \frac{Z(\omega)}{2 \|Z\|_\infty}, \qquad \|Z\|_\infty = \max\{|Z(\omega)| : \omega \in \Omega\}.$$

Then Q' is a probability measure since $Q'(\{\omega\}) > 0$ for all ω and

$$Q'(\Omega) = E_Q(R) = 1,$$

as $E_Q(Z) = 0$. Moreover, for each $Y = c + \sum_{t=1}^T \theta_t \cdot \Delta \overline{S}_t \in L$, we have

$$E_{Q'}(Y) = E_Q(RY) = E_Q(Y) + \frac{1}{2 \|Z\|_\infty} E_Q(YZ) = c.$$

In particular, $E_{Q'}(Y) = 0$ when Y has $c = 0$. Thus, for any predictable process $\theta = \{\theta_t^i : t = 1, 2, \ldots, T, i = 1, 2, \ldots, d\}$, we have

$$E_{Q'}\left(\sum_{t=1}^T \theta_t \cdot \Delta \overline{S}_t\right) = 0. \tag{4.5}$$

Again using $\theta = (0, \ldots, 0, \theta^i, 0, \ldots, 0)$ successively for $i = 1, 2, \ldots, d$ in (4.5), it is clear that Theorem 2.3.5 implies that $\overline{S}$ is a Q'-martingale. We have therefore constructed an equivalent martingale measure distinct from Q. Thus, in a viable incomplete market, the EMM is not unique. This completes the proof of the theorem. $\square$

4.3 The CRR Model

Again the Cox-Ross-Rubinstein model provides a good testbed for the ideas developed above. We saw in Section 2.6 that this model is complete, by means of an explicit construction of the unique EMM as a product of one-step probabilities. We explore the content of the martingale representation result (Proposition 4.1.1) in this context and use it to provide a more precise description of the generating strategy for a more general contingent claim.

Recall that the bond price in this model is $S_t^0 = (1 + r)^t$ for $t \in \mathbb{T} = \{0, 1, \ldots, T\}$, where $r > 0$ is fixed, and that the stock price S satisfies $S_t = R_t S_{t-1}$, where

$$R_t = \begin{cases} 1 + b & \text{with probability } q = \frac{r-a}{b-a}, \\ 1 + a & \text{with probability } 1 - q = \frac{b-r}{b-a}. \end{cases}$$

Here we assume that $-1 < a < r < b$ to have a viable market model, and the sample space can be taken as $\Omega = \{1 + a, 1 + b\}^{\mathbb{T} \setminus \{0\}}$, so that the independent, identically distributed random variables $\{R_t : t = 1, 2, \ldots, T\}$ describe the randomness in the model. The unique EMM Q then takes the form

$$Q(R_t = \omega_t : s = 1, 2, \ldots, T) = \prod_{t \leq T} q_t,$$

where

$$q_t = \begin{cases} q & \text{if } \omega_t = 1 + b, \\ 1 - q & \text{if } \omega_t = 1 + a. \end{cases}$$

In such simple cases, a direct proof of the martingale representation theorem is almost obvious and does not depend on the nature of the sample space, since the (R_t) contain all the relevant information.

Proposition 4.3.1. *Suppose that $(\Omega, \mathcal{F}, Q)$ is a probability space and $(R_t,$ where $1, 2, \ldots, T$, is a finite sequence of independent and identically distributed random variables, taking the two values u, v with probabilities q and $1 - q$, respectively. Suppose further that $E(R_1) = w$, where $-1 < v < w < u$ and*

$$q = \frac{w - v}{u - v} \tag{4.6}$$

while $m_t = \sum_{s=1}^{t}(R_t - w)$, $\mathcal{F}_0 = \{\emptyset, \Omega\}$, and $\mathcal{F}_t = \sigma(R_s : s \leq t)$ for all $t = 1, 2, \ldots, T$.

Then $(m_t, \mathcal{F}_t, Q)$ is a centred martingale and every $(\mathcal{F}_t, Q)$-martingale $(M_t, \mathcal{F}_t, Q)$ with $E_Q(M_0) = 0$ can be expressed in the form

$$M_t = \sum_{s \leq t} \theta_s \Delta m_s, \tag{4.7}$$

where the process $\theta = (\theta_t)$ is $(\mathcal{F}_t)$-predictable.

Proof. We follow the proof given in [299],15.1 (see also [63], [283]). It is obvious that $m = (m_t)$ is a martingale relative to $(\mathcal{F}_t, Q)$. Since M_t is $\mathcal{F}_t$—measurable, it has the form

$$M_t(\omega) = f_t(R_1(\omega), R_2(\omega), \dots, R_t(\omega)) \text{ for all } \omega \in \Omega.$$

Now suppose that (4.7) holds. It follows that the increments of M take the form $\Delta M_t(\omega) = \theta_t(\omega)\Delta m_t(\omega)$, so that, if we set

$$f_t^u(\omega) = f_t(R_1(\omega), R_2(\omega), \dots, R_{t-1}(\omega), u),$$
$$f_t^v(\omega) = f_t(R_1(\omega), R_2(\omega), \dots, R_{t-1}(\omega), v),$$

then (4.7) results from showing that

$$f_t^u - f_{t-1} = \theta_t(u - w), \qquad\qquad f_t^v - f_{t-1} = \theta_t(v - w).$$

In other words, θ_t would need to take the form

$$\theta_t = \frac{f_t^u - f_{t-1}}{u - w} = \frac{f_t^v - f_{t-1}}{v - w}. \tag{4.8}$$

To see that this is indeed the case, we simply use the martingale property of M. Since $E_Q(\Delta M_t | \mathcal{F}_{t-1}) = 0$, we have

$$q f_t^u + (1 - q) f_t^v = f_{t-1} = q f_{t-1} + (1 - q) f_{t-1}.$$

This reduces to

$$\frac{f_t^u - f_{t-1}}{1 - q} = \frac{f_t^v - f_{t-1}}{q},$$

which is equivalent to (4.8) because of (4.6). $\square$

Valuation of General European Claims

We showed in Section 2.6 that the value process

$$V_t(C) = (1 + r)^{-(T-t)} E_Q(C | \mathcal{F}_t)$$

of a European call option C in the Cox-Ross-Rubinstein model can be expressed more concretely in the form $V_t(C) = v(t, S_t)$, where

$$v(t, x) = (1 + r)^{-(T-t)} \sum_{u=o}^{T-t} \left[\binom{T - t}{u} q^u (1 - q)^{T-t-u} \right.$$
$$\left. \times (x(1 + b)^u (1 + a)^{T-t-u} - K)^+ \right].$$

This Markovian nature of the European call (i.e., the fact that the value process depends only on the *current* price and not on the *path* taken by the process S), can be exploited more generally to provide explicit expressions for the value process and generating strategies of a European contingent claim (i.e., a claim $X = g(S_T)$). In the CRR model, we know that the evolution of S is determined by the ratios (R_t), which take only two values, $1 + b$ and $1 + a$. For any path ω, the value $S_T(\omega)$ is thus determined by the initial stock price S_0 and the *number* of 'upward' movements of the price on $\mathbb{T} = \{0, 1, \ldots, T\}$. To express this more simply, note that $R_t = (1+a) + (b-a)\delta_t$, where δ_t is a Bernoulli random variable taking the value 1 with probability q. Hence we can consider, generally, claims of the form $X = h(u_T)$, where $u_T(\omega) = \sum_{t \leq T} \delta_t(\omega)$.

Recall from Proposition 4.3.1 that the martingale $M_t = E_Q(X \mid \mathcal{F}_t)$ can be represented in the form $M_t = M_0 + \sum_{u \leq t} \theta_u \Delta m_u$. Using $v = 1 + a$, $w = 1 + r$, and $u = 1 + b$ in applying Proposition 4.3.1 in the CRR setting, we have $m_u = R_u - (1 + r)$. Therefore

$$\Delta m_u = (1+a) - (b-a)\delta_u - (1+r) = (b-a)\left(\delta_u - \frac{r-a}{b-a}\right) = (b-a)(\delta_u - q).$$

Thus the representation of M can also be written in the form

$$M_t = \sum_{u \leq t} \alpha_u (\delta_u - q),$$

where $\alpha_u = (b-a)\theta_u$.

Consider the identity $\Delta M_t = \alpha_t(\delta_t - q)$. Exactly as in the proof of Proposition 4.3.1, this leads to a description of α. Indeed,

$$\alpha_t = \frac{E_Q(M_T \mid \{\delta_u, u < t\}, \delta_t = 1) - E_Q(M_T \mid \{\delta_u, u < t\})}{1 - q}$$

$$= \frac{E_Q(h(u_T) \mid \{\delta_u, u < t\}, \delta_t = 1) - E_Q(h(u_T) \mid \{\delta_u, u < t\})}{1 - q}.$$

We now restrict our attention to the set

$$A = \{\omega : u_{t-1}(\omega) = x, \delta_t = 1\}.$$

On A, we obtain, using the independence of the (R_t),

$$E_Q(h(u_T) \mid \mathcal{F}_t) = E_Q(h(x + 1 + (u_T - u_t))),$$
$$E_Q(h(u_T) \mid \mathcal{F}_{t-1}) = E_Q(h(x + (u_T - u_{t-1})))$$
$$= qE_Q(h(x + 1 + (u_T - u_t)))$$
$$+ (1 - q)E_Q(h(x + (u_T - u_t))).$$

Thus, on the set A, we have

$$E_Q(h(u_T) \mid \mathcal{F}_t) - E_Q(h(u_T) \mid \mathcal{F}_{t-1})$$
$$= (1 - q)E_Q(h(x + 1 + (u_T - u_t)) - h(x + (u_T - u_t))),$$

and the final expectation is just

$$\sum_{s=0}^{T-t} \binom{T-t}{s} [h(x + 1 + s) - h(x + s)] q^s (1 - q)^{T-t-s}.$$

We have therefore shown that

$$\alpha_t = H_{T-t}(u_{t-1}; q),$$

where

$$H_s(x; q) = \sum_{\tau=0}^{s} \binom{s}{\tau} \left(h(x+1+\tau) - h(x+\tau)\right) q^\tau (1-q)^{s-\tau}.$$

For a European claim $X = f(S_T)$, this can be taken further using the explicit form of the martingale representation given in Proposition 4.1.1. We leave the details (which can be found in [283]) to the reader and simply note here that the function h given above now takes the form

$$h(x) = (1+r)^{-T} f(S_0(1+b)^x(1+a)^{T-x}),$$

which leads to the following ratio for the time t stock holdings:

$$\alpha_t = (1+r)^{-(T-t)} \frac{F_{T-t}(S_{t-1}(1+b); q) - F_{T-t}(S_{t-1}(1+a); q)}{S_{t-1}(b-a)}, \qquad (4.9)$$

where

$$F_t(x; p) = \sum_{s=0}^{t} \binom{t}{s} f\left(x(1+b)^s(1+a)^{t-s}\right) p^s (1-p)^{t-s}.$$

Note that for a *non-decreasing* f we obtain $\alpha_t \geq 0$ for all $t \in \mathbb{T}$. Hence the hedge portfolio can be obtained without ever having to take a short position in the stock, although clearly we may have to borrow cash to finance the position at various times.

Exercise 4.3.2. Use formula (4.8) to obtain an explicit description of the strategy that generates the European call option with strike K and expiry T in the CRR model.

4.4 The Splitting Index and Completeness

Harrison and Kreps [148] introduced the notion of the *splitting index* for viable finite market models as a means of identifying event trees that lead to complete models. This idea is closely related to the concept of extremality of a probability measure among certain convex sets of martingale measures, and in this setting, the ideas also extend to continuous-time models (see [290], [150]).

Fix a finite market model $(\Omega, \mathcal{F}, Q, \mathbb{T}, \mathbb{F}, S)$ with $S_t = (S_t^i)_{0 \leq i \leq d}$. We assume that the filtration $\mathbb{F} = (\mathcal{F}_t)$ is generated by minimal partitions $(\mathcal{P}_t)$. The *splitting index* $K(t, A)$ of a set $A \in \mathcal{P}_{t-1}$ is then the number of branches of the event tree that begin at node A; i.e.,

$$K(t, A) = \text{card}\{A' \in \mathcal{P}_t : A' \subset A\} \text{ for } t = 1, 2, \ldots, T. \qquad (4.10)$$

It is intuitively clear that this number will serve to characterise completeness of the market since we can reduce our consideration to a single-period market (as we have seen in Chapter 3) with A as the new sample space. In order to construct a hedging strategy that we use to 'span' all the possible states of the market at time t by means of a linear combination of securities (i.e., a linear combination of the prices $(S_t^i(\omega))_{0 \leq i \leq d}$) clearly the number of different possible states should not exceed $(d+1)$. Moreover, it is possible that some of the prices can be expressed as linear combinations of the remaining ones and hence are 'redundant' in the single-period market, so that, as before, what matters is the *rank* of the matrix of prices (which correspond to the price *increments* in multi-period models). Recalling finally that the bond is held constant as numéraire, the following result, for which we shall only outline the proof, becomes plausible.

Proposition 4.4.1. *A viable finite market model is complete if and only if for every $t = 1, 2, \ldots, T$ and $A \in \mathcal{P}_{t-1}$ we have*

$$\dim(\mathrm{span}\{\Delta \overline{S}_t(\omega) : \omega \in A\}) = K(t, A) - 1. \tag{4.11}$$

In particular, if the market contains no redundant securities (i.e., there is no $\alpha \neq 0$ in $\mathbb{R}^{d+1}$, $t > 0$ in $\mathbb{T}$ and $A \in \mathcal{P}_{t-1}$ such that $Q(\alpha \cdot S_t = 0 \,|\, A) = 1$), then $K(t, A) = d + 1$.

Outline of Proof. (see [290] for details) Refer to the notation introduced in the discussion following Lemma 3.3.6. We can reduce this situation to the one-step conditional probabilities as in Chapter 3 and finally 'paste together' the various steps. We also assume without loss of generality that $S^0 \equiv 1$ throughout, so that $S_t = \overline{S}_t$ for all $t \in \mathbb{T}$.

Fix $A \in \mathcal{P}_{t-1}$ and consider the set $\mathcal{M}$ of all probability measures on the space $(A, \mathcal{A}_A)$, where $\mathcal{A}_A$ is the σ-algebra generated by the sets $\{A_i, i \leq n\}$ in $\mathcal{P}_t$ that partition A. Consider an element Q_A of the convex set

$$\mathcal{M}_0 = \left\{ Q'_A \in \mathcal{M} : E_{Q'_A}(\Delta S_t \mathbf{1}_A) = 0 \right\}.$$

If Q_A is in $\mathcal{M}_0$ and assigns positive mass to $A_1, A_2, \ldots, A_m$, while giving zero mass to the other A_i, then we can write the price increment on the set A_j, $j \leq m$, as $\Delta S_t(\omega) = y_i - y$, where $S_{t-1}(\omega) = y$ is constant on A since S is adapted. The condition that Q_A cannot be expressed as a convex combination of measures in $\mathcal{M}_0$ now translates simply to the demand that the vectors $(y_i - y)$ are linearly independent. In other words, that the matrix of price increments has linearly independent columns. But we have already seen that non-singularity of the matrix of price increments is equivalent to completeness in the single-period model. The proof may now be completed by pasting together the steps to construct the unique EMM. □

Example 4.4.2. We already know that the binomial random walk model is complete by virtue of the uniqueness of the EMM. Our present interest is in the splitting index. Recall that the price process S has the form

$S_t = \prod_{u=1}^{t} r_t$, where the return process r_t takes only the values $u = 1 + b$ and $d = 1 + a$ and is independent of $\mathcal{F}_{t-1}$, so that we can describe the price dynamics by an event tree, as in Figure 1.3.

Clearly there are only two branches at each node, so that $K(t, A) = 2$, while

$$\dim(\mathrm{span}\{\Delta S_t(\omega) : \omega \in A\}) = 1$$

for each $A \in \mathcal{P}_t, t \in \mathbb{T} : \Delta S^0 \equiv 0$, and $\Delta S_t^1(\omega) = S_{t-1}^1(\omega)(R_t(\omega) - 1)$ takes the values $bS_{t-1}^1(\omega)$ and $aS_{t-1}^1(\omega)$, both of which are multiples of $S_{t-1}^1(\omega)$, which remains constant throughout A.

Example 4.4.3. For $d \geq 2$, however, the d-dimensional random walk composed of independent copies of one-dimensional walks *cannot* be complete; we have $K(t, A) = 2^d$, and this equals $d + 1$ only when $d = 1$.

We can easily construct an infinite number of EMMs for the two-dimensional (also known as *two-factor*) random walk model. In the example above, we have a price process $S = (1, S^1, S^2)$ with stock return processes R^1, R^2, which we assume to take the values $(1 \pm a_1)$ and $(1 \pm a_2)$, respectively (so that we make the 'up' and 'down' movements symmetrical in each coordinate). Suppose that $a_1 = \frac{1}{2}$ and $a_2 = \frac{1}{4}$, and define, for each $\lambda \in (0, \frac{1}{2})$, a probability measure Q_λ by fixing, at each $t = 1, 2, \ldots, T$, the return probabilities as follows:

$$Q_\lambda(R_t^1 = 1 + a_1, R_t^2 = 1 + a_2) = \lambda = Q_\lambda(R_t^1 = 1 - a_1, R_t^2 = 1 - a_2),$$

$$Q_\lambda(R_t^1 = 1 + a_1, R_t^2 = 1 - a_2) = \frac{1}{2} - \lambda = Q_\lambda(R_t^1 = 1 - a_1, R_t^2 = 1 + a_2).$$

It is straightforward to check that each Q_λ is an EMM; i.e., that

$$E_{Q_\lambda}\left(R_t^i | \mathcal{F}_{t-1}\right) = E_{Q_\lambda}\left(R_t^i\right) = 1 \text{ for all } t \geq 1.$$

It can be shown (much as we did in Chapter 2) that the multifactor Black-Scholes model is a limit of multifactor random walk models *and* is complete. Consequently, it is possible to have a complete continuous-time model that is a limit (in some sense) of incomplete discrete models. If one is interested in 'maintaining completeness' along the approximating sequence, then one is forced to use *correlated* random walks. See [63], [151] for details.

Filtrations in Complete Finite Models

The completeness requirement in finite models is very stringent. It fixes the degree of linear dependence among the values of the price increments ΔS_t on any partition set $A \in \mathcal{P}_{t-1}$ in terms of the number of cells into which $\mathcal{P}_t$ 'splits' the set A. It also ensures that the *filtration* $\mathbb{F} = (\mathcal{F}_t)$ that is determined by these partition sets is in fact the *minimal* filtration $\mathbb{F}^S$ (i.e., the σ-field $\mathcal{F}_t = \mathcal{F}_t^S = \sigma(S_u : u \leq t)$ for each t).

To see this, let Q denote the unique EMM in the complete market model and suppose that, on the contrary, the filtration $\mathbb{F} = (\mathcal{F}_t)$ strictly contains $\mathbb{F}^S$. Then there is a least $u \in \mathbb{T}$ such that $\mathcal{F}_u$ strictly contains $\mathcal{F}_u^S$. This means that some fixed $A \in \mathcal{P}_u^S$ (the minimal partition generating $\mathcal{F}_u^S$) can be split further into sets in the partition $\mathcal{P}_u$ generating $\mathcal{F}_u$ (i.e., $A = \cup_{i=1}^n A_i$ for some $A_i \in \mathcal{P}_u (n \geq 2)$).

Note that S_u is constant on $A = \cup_{i=1}^n A_i$. There is a unique set $B \in \mathcal{P}_{u-1} = \mathcal{P}_{u-1}^S$ that contains A. The partition $\mathcal{P}_u$ then contains disjoint sets $\{A_i : i = 1, 2, \ldots, m\}$ whose union is B, and since $A \subset B$, we can assume (re-ordering if needed) that $m \geq n$ and the sets $A_1, A_2, \ldots, A_n$ defined above comprise the first n of these.

Let Q^* be a probability measure on $(\Omega, \mathcal{F})$ such that $Q^*(\cdot | B)$ defines different conditional probabilities with $Q^*(A_i | B) > 0$ for all $i \leq n$ and such that

$$\sum_{i=1}^n Q^*(A_i | B) = Q(A | B), \quad Q^*(A_j | B) = Q(A | B) \text{ for } j = n+1, \ldots, m,$$

and agreeing with Q otherwise. There are clearly many choices for such Q^*.

Since ΔS_u is constant on $A = \bigcup_{i=1}^n A_i$, it follows that

$$E_{Q^*} (\Delta S_u | \mathcal{F}_{u-1}) (\omega) = E_Q (\Delta S_u | \mathcal{F}_{u-1}) (\omega) = 0$$

holds for all $\omega \in B$ and hence throughout Ω. Hence Q is not the only EMM in the model, which contradicts completeness.

Thus, in a complete *finite* market model there is no room for 'extraneous information' that does not result purely from the past behaviour of the stock prices. This severely restricts its practical applicability, as Kreps [202, p. 228] has observed: the presence of other factors (Kreps lists 'differential information, moral hazard, and individual uncertainty about future tastes' as examples) that are not fully reflected in the security prices will destroy completeness.

4.5 Incomplete Models: The Arbitrage Interval

We return to the general setup of extended securities market models that was introduced in Section 2.5. We wish to examine the set of possible prices of a European contingent claim H that preclude arbitrage. Since H is itself a tradeable asset, we need to include it in the assets that can be used to produce trading strategies. It was shown in Theorem 2.5.2 that, for any given measure Q, the only price for H consistent with the absence of arbitrage is given by the 'martingale price' $\pi(H) = E_Q (\beta_T H)$ derived in (2.15). We now consider a viable model with $\mathcal{P}$ as the set of equivalent martingale measures for the discounted price process $\overline{S}$ and augment this model by regarding H as an additional primary asset. In discounted terms, we therefore set $\overline{S}_t^{d+1} = \beta_t H$ and consider the range of possible initial prices π_H consistent with the no-arbitrage requirement in the model. We call these prices *arbitrage-free* prices for the extended model. Denote the extended (discounted) price process by $\widetilde{S} = (\overline{S}, \overline{S}^{d+1})$, where the final coordinate must satisfy the constraints

$$\overline{S}_0^{d+1} = \pi_H, \quad \overline{S}_t^{d+1} \geq 0 \text{ a.s. } (P) \text{ for } t = 1, 2, \ldots, T-1, \quad \overline{S}_T^{d+1} = H.$$

Denote by $\Pi(H)$ the set of all arbitrage-free prices for H. The first fundamental theorem immediately enables us to identify $\Pi(H)$ via the set of expectations $E_Q(\beta_T H)$ for Q in $\mathcal{P}$. However, since H cannot necessarily be generated by an admissible strategy, we do not know in advance that the integral is finite. We need the following result.

Theorem 4.5.1. *Let H be a European claim in a viable securities market model $(\Omega, \mathcal{F}, P, \mathbb{T}, \mathbb{F}, S)$ with $\mathcal{P}$ as the set of EMMs for $\overline{S}$. The set $\Pi(H)$ of arbitrage-free prices for H is given by*

$$\Pi(H) = \{E_Q(\beta_T H) : Q \in \mathcal{P}, E_Q(H) < \infty\}. \qquad (4.12)$$

The lower and upper bounds of $\Pi(H)$ are given by

$$\pi_- = \inf_{\mathcal{P}} E_Q(\beta_T H), \qquad \qquad \pi_+ = \sup_{\mathcal{P}} E_Q(\beta_T H).$$

Proof. The first fundamental theorem states that the extended model is viable if and only if it admits an EMM Q for the price process $\widetilde{S} = (\overline{S}^i : i = 0, 1, \ldots, d+1)$. This measure therefore satisfies

$$\overline{S}_t^i = E_Q\left(\overline{S}_T^i \mid \mathcal{F}_t\right) \text{ for } i = 1, 2, \ldots, d+1 \text{ and } t = 0, 1, \ldots, T.$$

Thus, in particular, $\overline{S} = (\overline{S}^i : i = 0, 1, \ldots, d)$ is a Q-martingale, so that $Q \in \mathcal{P}$, and $E_Q(\beta_T H) = E_Q\left(\overline{S}_t^{d+1}\right) < \infty$. The arbitrage-free price π_H is therefore a member of the set on the right-hand side in (4.12).

To establish the converse inclusion, let $\pi_H = E_Q(\beta_T H)$ for some $Q \in \mathcal{P}$. We need to show that π_H is an arbitrage-free price. For this, take the martingale $X = (X_t)$, where

$$X_t = E_Q(\beta_T H \mid \mathcal{F}_t) \text{ for } t \in \mathbb{T},$$

as the candidate for the 'price process' of the asset βH. This clearly satisfies the requirements in (4.12) so that, with this $\overline{S}^{d+1} = X$, the price π_H is an arbitrage-free price and Q is an EMM for the extended model, which is thus viable. Hence the two sets in (4.12) are equal.

The expectations are non-negative, so the expression for the lower bound π_- is clear. The same is true for the upper bound if the sets are bounded above.

This leaves the proof that $\pi_+ = \infty$ if $E_Q(\beta_T H) = \infty$ for some $Q \in \mathcal{P}$. This is left to the reader as an exercise in using the fact that the EMM can always be chosen to have bounded density relative to the given reference measure. □

This result allows us to characterise attainable claims as the only claims admitting a unique arbitrage-free price and further identify the possible prices of a general claim as the open interval (π_-, π_+). Our proof follows that in [132].

Theorem 4.5.2. *Let H be a European claim in a securities market model.*

 (i) *If H is attainable, then $\Pi(H)$ is a singleton and the unique arbitrage-free price for H is $\pi_- = V_0(\theta) = \pi_+$, where θ is any generating strategy for H.*

 (ii) *If H is not attainable, then either $\Pi(H) = \emptyset$ or $\pi_- < \pi_+$ and $\Pi(H)$ is the open interval (π_-, π_+).*

Proof. The first statement follows from (2.15) and Theorem 4.5.1.

For the second, note that if $\Pi(H)$ is non-empty, then it must be an interval since $\mathcal{P}$ is convex. We need to show that it is open and thus neither bound is attained. For this, we need to construct for any $\pi \in \Pi(H)$ two arbitrage-free prices π_*, π^* with $\pi_* < \pi < \pi^*$.

So fix $\pi - E_Q(\beta_T H)$, where $Q \in \mathcal{P}$. We have to construct a measure $Q^* \in \mathcal{P}$ such that
$$E_{Q^*}(\beta_T H) > E_Q(\beta_T H).$$

The given price π is the initial value of the process $V = (V_t)_{t \in \mathbb{T}}$ defined by $V_t = E_Q(\beta_T H \,|\, \mathcal{F}_t)$. Although the stochastic process V is not the value process of a generating strategy, we are nonetheless guided in our search for Q^* by what happens in that special situation. Since H is $\mathcal{F}_T$-measurable, we obtain the telescoping sum

$$V_T = E_Q(\beta_T H \,|\, \mathcal{F}_T) = \beta_T H = V_0 + \sum_{t=1}^{T}(V_t - V_{t-1}) = V_0 + \sum_{t=1}^{T} \Delta V_t. \quad (4.13)$$

By the first conclusion of this theorem, H is an attainable claim if and only if each term $\Delta V_t = E_Q(\beta_T H \,|\, \mathcal{F}_t) - E_Q(\beta_T H \,|\, \mathcal{F}_{t-1})$ has the form $\Delta G_t(\theta) = \widehat{\theta}_t \Delta \widehat{S}_t$ for some predictable process $\widehat{\theta} = (\theta^i)_{i=1,2,\dots,d}$, and by Theorem 2.3.5 this occurs for the measure $Q \in \mathcal{P}$ if and only if $E_Q(\Delta V_t) = 0$ for each $t = 1, 2, \dots, T$. Since the given claim H is not attainable, this must fail for some $t = 1, 2, \dots, T$ (i.e., for some such t, ΔV_t is *not* of the form $\widehat{\theta} \cdot \Delta \widehat{S}_t$ for any Q-integrable $\mathcal{F}_{t-1}$-measurable random vector $\widehat{\theta}$).

In other words, for this value of t, the random variable ΔV_t is disjoint from the space $\mathcal{K}_t \cap L^1(\mathcal{F}_t, Q)$, where $\mathcal{K}_t$ is as defined by (3.16). Since this is a closed subspace of $L^1(\mathcal{F}_t, Q)$, we can separate it from the compact set $\{\Delta V_t\}$ by a linear functional Z in $L^\infty(\mathcal{F}_t, Q)$. We thus obtain real numbers $\alpha < \beta$ such that for all $X \in \mathcal{K}_t \cap L^1(\mathcal{F}_t, Q)$:

$$E_Q(XZ) \le \alpha < \beta \le E_Q(\Delta V_t Z). \tag{4.14}$$

Now since $\mathcal{K}_t \cap L^1(\mathcal{F}_t, Q)$ is a subspace, $E_Q(XZ) \le \alpha$ for all $X \in \mathcal{K}_t \cap L^1(\mathcal{F}_t, Q)$ implies (as in the proof of Theorem 3.5.8) that $\alpha = 0$. But then if $E_Q(XZ) < 0$ for some X, $-X$ would violate the condition $E_Q(XZ) \le 0$. Hence $E_Q(XZ) = 0$ for all $X \in \mathcal{K}_t \cap L^1(Q)$. This means that $E_Q(\Delta V_t Z) > 0$. The same conclusion is reached if Z is replaced by $\frac{Z}{3\|Z\|_\infty}$, so that we may assume without loss of generality that $|Z| \le \frac{1}{3}$ a.s. (P).

Therefore the L^∞-function $Z^* = 1 + Z - E_Q(Z | \mathcal{F}_{t-1})$ is a.s. (P) positive and has $E_Q(Z^*) = 1$, so that $\frac{dQ^*}{dQ} = Z^*$ defines a probability measure equivalent to Q and hence to P. We calculate the Q^*-expectation of $\beta_T H$ using the fact that Z^* is $\mathcal{F}_t$-measurable:

$$
\begin{aligned}
E_{Q^*}(\beta_T H) &= E_Q(\beta_T H Z^*) \\
&= E_Q(\beta_T H) + E_Q(Z E_Q(\beta_T H | \mathcal{F}_t)) \\
&\quad - E_Q(E_Q(Z | \mathcal{F}_{t-1}) E_Q(\beta_T H | \mathcal{F}_{t-1})) \\
&= E_Q(\beta_T H) + E_Q(Z V_t) - E_Q(V_{t-1} E_Q(Z | \mathcal{F}_{t-1})) \\
&= E_Q(\beta_T H) + E_Q(Z V_t) - E_Q(E_Q(V_{t-1} Z | \mathcal{F}_{t-1})) \\
&= E_Q(\beta_T H) + E_Q(\Delta V_t Z) \\
&> E_Q(\beta_T H)
\end{aligned}
$$

by construction of Z. Therefore $\pi^* = E_{Q^*}(\beta_T H)$ will be an element of $\Pi(H)$ greater than π, provided we can show that $Q^* \in \mathcal{P}$, and thus we must show that the discounted stock prices $(\widehat{S}^i)_{i=1,2,\dots,d}$ are Q^*-martingales.

Fix $i \le d$ and $u > t$. Then, by Bayes' rule, we have

$$E_{Q^*}\left(\Delta \widehat{S}_u^i | \mathcal{F}_{u-1}\right) = \frac{E_Q\left(\Delta \widehat{S}_u^i Z^* | \mathcal{F}_{u-1}\right)}{E_Q(Z^* | \mathcal{F}_{u-1})} = E_Q\left(\Delta \widehat{S}_u^i | \mathcal{F}_{u-1}\right) = 0$$

since Z^* is $\mathcal{F}_{u-1}$-measurable for each $u > t$. On the other hand, since

$$E_Q(Z^* | \mathcal{F}_{t-1}) = E_Q(1 + Z - E_Q(Z | \mathcal{F}_{t-1}) | \mathcal{F}_{t-1}) = 1,$$

the restrictions of the measures Q and Q^* coincide on $\mathcal{F}_u$ for every $u < t$, and so

$$E_{Q^*}\left(\Delta \widehat{S}_u^i | \mathcal{F}_{u-1}\right) = E_Q\left(\Delta \widehat{S}_u^i | \mathcal{F}_{u-1}\right) = 0.$$

Thus, to show that $Q^* \in \mathcal{P}$, we need only consider $E_{Q^*}\left(\Delta \widehat{S}_t^i | \mathcal{F}_{t-1}\right)$. For this, since by construction of Z, $E_Q\left((\widehat{\theta} \cdot \Delta \widehat{S}_t) Z\right) = 0$ for all $\mathcal{F}_{t-1}$-measurable $\mathbb{R}^d$-valued random vectors θ, we have $E_Q\left(Z \Delta \widehat{S}_t^i | \mathcal{F}_{t-1}\right) =$

0 a.s. (P) for each $i \leq d$. So we can write

$$
\begin{aligned}
E_{Q*}\left(\Delta \widehat{S}_t^i \,|\mathcal{F}_{t-1}\right) &= E_Q\left(\Delta \widehat{S}_t^i Z^* \,|\mathcal{F}_{t-1}\right) \\
&= E_Q\left(\Delta \widehat{S}_t^i (1 + Z - E_Q\left(Z \,|\mathcal{F}_{t-1}\right)) \,|\mathcal{F}_{t-1}\right) \\
&= E_Q\left(\Delta \widehat{S}_t^i (1 - E_Q\left(Z \,|\mathcal{F}_{t-1}\right)) \,|\mathcal{F}_{t-1}\right) + E_Q\left(Z \Delta \widehat{S}_t^i \,|\mathcal{F}_{t-1}\right).
\end{aligned}
$$

The final term is a.s. (P) zero, as was shown above, while the first is, a.s. (P),

$$
E_Q\left(\Delta \widehat{S}_t^i \,|\mathcal{F}_{t-1}\right)[1 - E_Q\left(Z \,|\mathcal{F}_{t-1}\right)] = 0
$$

since $E_Q\left(Z \,|\mathcal{F}_{t-1}\right)$ is $\mathcal{F}_{t-1}$-measurable and $Q \in \mathcal{P}$. So we have verified that $Q^* \in \mathcal{P}$ and hence $\pi^* \in \Pi(H)$.

The construction of a suitable $\pi_* < \pi$ in $\Pi(H)$ is now straightforward. For example, we can use the probability measure Q_* with density

$$
\frac{dQ_*}{dQ} = 2 - Z^*,
$$

a choice that ensures that $E_{Q_*}\left(2 - Z^*\right) = 1$ and that $0 < 2 - Z^* \leq \frac{5}{3}$ since $|Z| \leq \frac{1}{3}$. Since $2 - Z^* = 1 - Z + E_Q\left(Z \,|\mathcal{F}_{t-1}\right)$, it follows as in (4.16) that

$$
E_{Q_*}\left(\beta_T H\right) = E_Q\left(\beta_T H\right) - E_Q\left(\Delta V_t Z\right) < E_Q\left(\beta_T H\right).
$$

Also $Q_* \in \mathcal{P}$:

$$
E_Q\left(\Delta \widehat{S}_t^i (2 - Z^*) \,|\mathcal{F}_{t-1}\right) = 2 E_Q\left(\Delta \widehat{S}_t^i \,|\mathcal{F}_{t-1}\right) - E_Q\left(\Delta \widehat{S}_t^i Z^* \,|\mathcal{F}_{t-1}\right) = 0,
$$

as $Q, Q^* \in \mathcal{P}$. $\qquad\square$

Remark 4.5.3. Note that Theorems 4.5.1 and 4.5.2 together imply that if $\Pi(H)$ is empty, then there is no EMM in the model for which the claim H has finite expectation.

4.6 Characterisation of Complete Models

We saw in Theorem 4.5.2 that a viable finite market model is complete if and only if the set $\mathcal{P}$ of its EMMs is a singleton. We could not establish this result in greater generality until we had dealt with the first fundamental theorem in the general setting (i.e., shown that the model is viable if and only if $\mathcal{P} \neq \emptyset$). Having done this, and also having characterised the attainability of claims, we can now go much further in identifying the class of complete market models more fully. We shall demonstrate, after the fact, that the argument provided to prove Theorem 4.5.2 will suffice in general since every complete model in the discrete-time setting must actually be a finite market model.

Theorem 4.6.1. *A viable securities market model* $(\Omega, \mathcal{F}, P, \mathbb{T}, \mathbb{F}, S)$ *is complete if and only if it allows a unique equivalent martingale measure. When $\mathcal{P}$ is a singleton, the underlying probability space Ω is finitely generated and its generating partition has at most $(d+1)^T$ atoms.*

Proof. With the more advanced tools now at our disposal, the proof of this far-reaching result is elementary for general market models. Throughout, we only need to work with bounded claims: in a finite-dimensional space of random variables all elements are automatically bounded.

First consider the single-period case (i.e., let $T = 1$). Completeness of the market means that every European contingent claim, that is, every non-negative function in $L^0(\mathcal{F})$ is attainable by some generating strategy based on the price processes $(S)_{0 \le i \le d}$. In particular, as already observed in Chapter 2, the indicator $\mathbf{1}_A$ of any set in $\mathcal{F}$ is an attainable claim. Theorem 4.5.2 shows that the unique arbitrage-free price $E_Q\left(\beta_T \mathbf{1}_A\right)$ is independent of the choice of EMM Q. Hence $Q(A)$ is also uniquely determined for each A, so that $\mathcal{P}$ is a singleton.

Conversely, if Q is the unique EMM in the model, any bounded claim H is Q-integrable, and its price is given uniquely by $E_Q\left(\beta_T H\right)$. Again by Theorem 4.5.2, it follows that H is attainable. Thus every element $H \in L^\infty(\mathcal{F})$ is of the form $\theta_t \cdot S_t$ for some $\mathbb{R}^{d+1}$-valued random vector θ. In other words, the collection of possible portfolio values $\left\{\theta \cdot S : \theta \in \theta \in \mathbb{R}^{d+1}\right\}$ contains $L^\infty(\mathcal{F})$. This is only possible if $L^\infty(\mathcal{F})$ has dimension at most $d+1$ and thus the σ-field $\mathcal{F}$ is generated by a finite partition with at most $d+1$ atoms (see Lemma 4.6.2 below, whose proof is an easy exercise). Thus every contingent claim is automatically bounded, hence attainable.

Turning now to the multi-period case, we argue by induction on T. Note first that if every $\mathcal{F}$-measurable bounded claim is attainable then $\mathcal{F} = \mathcal{F}_T$ since for any $A \in \mathcal{F}$ the generating value process is by construction $\mathcal{F}_T$-measurable. We know that, when $T = 1$, the probability space Ω of a complete model has at most $d+1$ atoms. Assume that, for every complete model with $T-1$ trading periods, the underlying probability space has at most $(d+1)^{T-1}$ atoms, and consider a complete model with T trading periods. Thus every $\mathcal{F}_T$-measurable non-negative bounded random variable can be written in the form $V_{T-1} + \theta_T \cdot \Delta S_T$ for some $\mathcal{F}_{T-1}$-measurable functions V_{T-1} and θ_T. These functions are constant on each of the (at most $(d+1)^{T-1}$) atoms of $(\Omega, \mathcal{F}_{T-1}, P)$. For each such atom A, we can consider the conditional probability $P(\cdot | A)$ since $P(A) > 0$. The vector space $L^\infty(\Omega, \mathcal{F}_T, P(\cdot | A))$ has dimension at most $(d+1)$, so by Lemma 4.6.2 it follows that $(\Omega, \mathcal{F}_T, P(\cdot | A))$ has at most $(d+1)^T$ atoms.

Since the vector spaces L^p are therefore all finite-dimensional, all contingent claims in the given model are bounded. Thus the value of each claim H is given by the unique element of $\Pi(H)$, and by Theorem 4.5.2 it follows that H is attainable. $\qquad\square$

Lemma 4.6.2. *For $0 \le p \le \infty$, the dimension of the space $L^p(\Omega, \mathcal{F}, P)$*

equals

$$\sup \{ n \geq 1 : \exists \ partition \ \{ F_i \in \mathcal{F} : i \leq n \} \ of \ \Omega \ with \ P(A_i) > 0 \ for \ i \leq n \}.$$

The dimension n *of* $L^p(\mathcal{F})$ *is finite if and only if there is a partition of* Ω *into* n *atoms.*

Remark 4.6.3. There are various other characterisations of completeness, notably in terms of the set of extreme points of $\mathcal{P}$, which are better adapted to their continuous-time analogues. We refer to [132] and [280] for details. Note, however, that the characterisation given above illustrates that, in mathematical terms, completeness will hold only for a very restricted subset of the class of viable market models, since all complete models must in fact be finite market models. Finance theorists, on the other hand, might argue that realistic market models are necessarily finite.

Chapter 5

Discrete-time American Options

American options differ fundamentally from their European counterparts since the exercise date is now at the holder's disposal and not fixed in advance. The only constraint is that the option ceases to be valid at time T and thus cannot be exercised *after* the expiry date T. The pricing problem for American options is more complex than those considered up to now, and we need to develop appropriate mathematical concepts to deal with it. As in the preceding chapters, we shall model discrete-time options on a given securities market model $(\Omega, \mathcal{F}, P, \mathbb{T}, \mathbb{F}, S)$.

5.1 Hedging American Claims

Random Exercise Dates

First, we require a concept of 'random exercise dates' to reflect that the option holder can choose different dates at which to exercise the option depending on her perception of the random movement of the underlying stock price. The exercise date τ is therefore no longer the constant T but becomes a function on Ω with values in $\mathbb{T}$, that is, a *random variable* $\tau : \Omega \mapsto \mathbb{T}$. It remains natural to assume that investors are not prescient, so that the decision whether to exercise at time t when in state ω depends only on information contained in the σ-field $\mathcal{F}_t$. Hence our exercise dates should satisfy the requirement that $\{\tau = t\} \in \mathcal{F}_t$.

Exercise 5.1.1. Show that the following requirements on a random variable $\tau : \Omega \mapsto \mathbb{T}$ are equivalent:

a) For all $t \in \mathbb{T}$, $\{\tau = t\} \in \mathcal{F}_t$.

b) For all $t \in \mathbb{T}$, $\{\tau \leq t\} \in \mathcal{F}_t$.

Hint: Recall that the $(\mathcal{F}_t)$ increase with t.

We briefly review relevant aspects of martingale theory and optimal stopping. These often require care about measurability problems. The greater technical complexity is offset by wider applicability of our results, and they provide good practice for the unavoidable technicalities that we encounter in the continuous-time setting. Throughout, however, it is instructive to focus on the underlying ideas, and it may be advantageous in this and the following chapters to skip lightly over some technical matters at a first reading.

Hedging Constraints

Hedge portfolios also require a little more care than in the European case since the writer may face the liability inherent in the option at any time in $\mathbb{T}$. More generally, an American contingent claim is a function of the whole *path* $t \mapsto S_t(\omega)$ of the price process under consideration, for each $\omega \in \Omega$, not just a function of $S_T(\omega)$. We again assume that $S = \{S_t^i : i = 0, 1, \ldots, d; t \in \mathbb{T}\}$, where $S_t^0 = \beta_t^{-1}$ is a (non-random) riskless bond, and the stock price S^i is a random process indexed by $\mathbb{T}$ for each $i = 1, 2, \ldots, d$.

Accordingly, let $f = (f_t(S))_{t \in \mathbb{T}}$ denote an American contingent claim, so that f is a *sequence* of non-negative random variables, each depending, in general, on $\{S^i(\omega) : 0 \le i \le d\}$ for every $\omega \in \Omega$. As considered in Section 2.4, the *hedge portfolio* with initial investment $x > 0$ for this claim will now be a self-financing strategy $\theta = \{\theta_t^i : i = 0, 1, \ldots, d; t \in \mathbb{T}\}$, producing a value process $V(\theta)$ that satisfies the hedging constraints

$$V_0(\theta) = \theta_1 \cdot S_0 = x,$$
$$V_t(\theta)(\omega) \ge f_t(S_0(\omega), S_1(\omega), \ldots, S_T(\omega)) \text{ for all } \omega \in \Omega \text{ and } t > 0.$$

The hedge portfolio θ is now described as *minimal* if, for some random variable τ with $\{\omega : \tau(\omega) = t\} \in \mathcal{F}_t$ for all $t \in \mathbb{T}$, we have

$$V_{\tau(\omega)}(\theta)(\omega) = f_{\tau(\omega)}(S_0(\omega), \ldots, S_T(\omega)). \tag{5.1}$$

Since the times at which the claim f takes its greatest value may vary with ω, the hedge portfolio θ must enable the seller (writer) of the claim to cover his losses in all eventualities since the buyer has the freedom to exercise his claim at any time. The hedge portfolio will thus no longer 'replicate' the value of the claim in general, but it may never be less than this value; that is, it must 'superhedge' or super-replicate the claim. This raises several questions for the given claim f:

(i) Do such self-financing strategies exist for a given value of the initial investment $x > 0$?

(ii) Do minimal self-financing strategies always exist for such x?

(iii) What is the optimal choice of the random exercise time τ?

(iv) How should the 'rational' time-0 *price* of the option be defined?

These questions are examined in this chapter. To deal with them, however, we first need to develop the necessary mathematical tools.

5.2 Stopping Times and Stopped Processes

The preceding considerations lead us to study 'random times', which we call *stopping times*, more generally for (discrete) stochastic processes. While our applications often have a finite time horizon, it is convenient to take the study further, to include stopping times that take values in the set $\bar{\mathbb{N}} = \{0, 1, \dots, \infty\}$. This extension requires us to establish results about martingale convergence, continuous-time versions of which will also be needed in later chapters. The well-known martingale convergence theorems are discussed briefly; we refer to other texts (e.g. ,[109], [199], [299]) for detailed development and proofs of these results.

The idea of stopping times for stochastic processes, while intuitively obvious, provides perhaps the most distinguishing feature of the techniques of probability theory that we use in this book. At its simplest level, a stopping time τ should provide a gambling strategy for a gambler seeking to maximise his winnings; since martingales represent 'fair' games, such a strategy should not involve prescience, and therefore the decision to 'stop' the adapted process $X = (X_t)$ representing the gambler's winnings at time t should only involve knowledge of the progress of the winnings up to that point; that is, if state ω occurs, the choice $\tau(\omega) = t$ should depend only on $\mathcal{F}_t$. Generally, suppose we are given a filtration $\mathbb{F} = (\mathcal{F}_t)_{t \in \mathbb{N}}$ on $(\Omega, \mathcal{F}, P)$ with $\mathcal{F} = \mathcal{F}_\infty = \sigma \left(\bigcup_{t=0}^{\infty} \mathcal{F}_t \right)$ and such that $\mathcal{F}_0$ contains all P-null sets. We have the following definition.

Definition 5.2.1. A *stopping time* is a random variable $\tau : (\Omega, \mathcal{F}) \to \bar{\mathbb{N}}$ such that for all $t \in \mathbb{N}$, $\{\tau \leq t\} \in \mathcal{F}_t$.

Remark 5.2.2. Exercise 5.1.1 shows that we could equally well have used the condition: for all $t \in \mathbb{N}$, $\{\tau = t\} \in \mathcal{F}_t$. Note, however, that this depends on the countability of $\mathbb{N}$. For continuous-time models, the time set $\mathbb{T}$ is a finite or infinite interval on the positive halfline, and we have to use the condition $\{\tau \leq t\} \in \mathcal{F}_t$ for all $t \in \mathbb{T}$ in the definition of stopping times. In discrete-time models, the condition $\{\tau = t\}$ is often much simpler to verify.

Nevertheless, many of the basic results about stopping times, and their proofs, are identical in both setups, and the exceptions become clear from the following examples and exercises.

Example 5.2.3.   (i) Observe that if $\tau = t_0$ a.s., then $\{\tau = t_0\} \in \mathcal{F}_0 \subset \mathcal{F}_{t_0}$, so that each 'constant time' is a stopping time. Similarly, it is easy to see that $\tau + t_0$ is a stopping time for each stopping time τ and constant t_0.

(ii) Suppose that σ and τ are stopping times. Then $\sigma \vee \tau = \max\{\sigma, \tau\}$ and $\sigma \wedge \tau = \min\{\sigma, \tau\}$ are both stopping times. Indeed, consider

$$\{\sigma \vee \tau \leq t\} = \{\sigma \leq t\} \cap \{\tau \leq t\}, \quad \{\sigma \wedge \tau \leq t\} = \{\sigma \leq t\} \cup \{\tau \leq t\}.$$

In both cases, the sets on the right-hand side are in $\mathcal{F}_t$ since σ and τ are stopping times.

(iii) Let $(X_t)_{t \in \mathbb{N}}$ be an $\mathbb{F}$-adapted process and let B be a Borel set. We now show that $\tau_B : \Omega \to \mathbb{N}$ defined by $\tau_B(\omega) = \inf\{s \geq 1 : X_s \in B\}$ (where $\inf \emptyset = \infty$) is an $\mathbb{F}$-stopping time. (We call τ_B the *hitting time* of B.)

To see this, note that each $X_s^{-1}(B)$ is in $\mathcal{F}_s$ since X_s is $\mathcal{F}_s$-measurable. Moreover, since $\mathbb{F}$ is increasing, $\mathcal{F}_s \subset \mathcal{F}_t$ when $s \leq t$. Hence, for any $t \geq 0$, $\{\tau_B = t\} \in \mathcal{F}_t$ since

$$\{\tau_B = t\} = \bigcap_{s=0}^{t-1} \{\tau_B > s\} \cap X_t^{-1}(B) = \bigcap_{s=0}^{t-1} (\Omega \setminus X_s^{-1}(B)) \cap X_t^{-1}(B).$$

The continuous-time counterpart of this result is rather more difficult in general and involves delicate measurability questions; in special cases, such as when B is an open set and $t \mapsto X_t(\omega)$ is continuous, it becomes much simpler (see, e.g. ,[199]).

Exercise 5.2.4. Suppose that (τ_n) is a sequence of stopping times. Extend the argument in the second example above to show that $\bigvee_{n \geq 1} \tau_n = \sup(\tau_n : n \geq 1)$ and $\bigwedge_{n \geq 1} \tau_n = \inf(\tau_n : n \geq 1)$ are stopping times. (Note that this uses the requirement that the σ-fields $\mathcal{F}_t$ are closed under *countable* unions and intersections.)

Fix a stochastic basis $(\Omega, \mathcal{F}, \bar{\mathbb{N}}, \mathbb{F}, P)$ with $\mathcal{F} = \mathcal{F}_\infty = \sigma(\cup_{t=0}^\infty \mathcal{F}_t)$. Recall that we assume throughout that the σ-fields $\mathcal{F}_t$ are *complete*. First we consider random processes 'stopped' at a *finite* stopping time τ, as most of our applications assume a finite trading horizon T.

Definition 5.2.5. If $X = (X_t)$ is an adapted process and τ is any a.s. *finite* stopping time, then we define the map $\omega \mapsto X_{\tau(\omega)}(\omega)$, giving the values of X at the stopping time τ, by the random variable

$$X_\tau = \sum_{t \geq 0} X_t \mathbf{1}_{\{\tau = t\}}.$$

To see that X_τ is $\mathcal{F}$-measurable, note that, for any Borel set B in $\mathbb{R}$,

$$\{X_\tau \in B\} = \bigcup_{t \geq 0} (\{X_t \in B\} \cap \{\tau = t\}) \in \mathcal{F}. \qquad (5.2)$$

Moreover, if we define the σ-field of *events prior to* τ by

$$\mathcal{F}_\tau = \{A \in \mathcal{F} : A \cap \{\tau = t\} \in \mathcal{F}_t \text{ for all } t \geq 1\}, \qquad (5.3)$$

then (5.2) shows that X_τ is $\mathcal{F}_\tau$-measurable since $\{X_t \in B\}$ is in $\mathcal{F}_t$ for each t, so that $\{X_\tau \in B\} \in \mathcal{F}_\tau$. Trivially, τ itself is $\mathcal{F}_\tau$-measurable.

Exercise 5.2.6. Let σ and τ be stopping times.

(i) Suppose that $A \in \mathcal{F}_\sigma$. Show that $A \cap \{\sigma \leq \tau\}$ and $A \cap \{\sigma = \tau\}$ belong to $\mathcal{F}_\tau$. Deduce that if $\sigma \leq \tau$ then $\mathcal{F}_\sigma \subset \mathcal{F}_\tau$. (*Hint:* The continuous-time analogue of this result is proved in Theorem 6.1.8. Convince yourself that a virtually identical statement and proof applies here.)

Deduce that, for any σ, τ, $\mathcal{F}_{\sigma \wedge \tau} \subset \mathcal{F}_\sigma \subset \mathcal{F}_{\sigma \vee \tau}$.

(ii) Show that the sets $\{\sigma < \tau\}$, $\{\sigma = \tau\}$, and $\{\sigma > \tau\}$ belong to both $\mathcal{F}_\sigma$ and $\mathcal{F}_\tau$.

The next two results, which we will extend considerably later, use the fact that stopping a martingale is essentially a special case of taking a martingale transform. They are used extensively in the rest of this chapter.

Theorem 5.2.7 (Optional Sampling for Bounded Stopping Times).
Let X be a supermartingale and suppose that σ and τ are bounded stopping times with $\sigma \leq \tau$ a.s. Then

$$E\left(X_\tau \mid \mathcal{F}_\sigma\right) \leq X_\sigma \text{ a.s.} \tag{5.4}$$

If X is a martingale, then $E\left(X_\tau \mid \mathcal{F}_\sigma\right) = X_\sigma$ a.s.

Proof. Consider the process $\phi = (\phi_t)$, where $\phi_t = 1_{\{\sigma < t \leq \tau\}}$. The random variable ϕ_t is $\mathcal{F}_{t-1}$-measurable for $t > 0$ since

$$\{\sigma < t \leq \tau\} = \{\sigma < t\} \cap (\Omega \setminus \{\tau < t\}).$$

Thus ϕ is predictable and non-negative. We consider the transform $\phi \cdot X$. Since τ is assumed to be bounded (by some $k \in \mathbb{N}$, say), we have

$$|(\phi \cdot X)_t| \leq |X_0| + \cdots + |X_k| \text{ for all } t,$$

so that each $Z_t = (\phi \cdot X)_t$ is integrable. Thus Z is a supermartingale with $Z_0 = 0$ and $Z_k = X_\tau - X_\sigma$. Hence

$$0 = E\left(Z_0\right) \geq E\left(Z_k\right) = E\left(X_\tau - X_\sigma\right).$$

Now consider $A \in \mathcal{F}_\sigma$ and apply the preceding equation to the bounded stopping times σ' and τ', where σ' equals σ on A, and k otherwise, with a similar definition for τ'.

Exercise 5.2.8. Check carefully, using (5.3) and Exercise 5.2.6, that σ' and τ' are indeed stopping times.

This yields

$$\int_A X_\tau dP \leq \int_A X_\sigma dP.$$

Hence the result follows, again using Exercise 5.2.6. □

Definition 5.2.9. Let X be a stochastic process on $(\Omega, \mathcal{F}, P, \mathbb{T}, \mathbb{F})$, and let σ be any stopping time. Define the *process X^σ stopped at time σ* by $X_t^\sigma = X_{\sigma \wedge t}$ for all $t \in \mathbb{T}$.

Remark 5.2.10. Note carefully that $(t, \omega) \mapsto X_t^{\sigma(\omega)}(\omega) = X_{t \wedge \sigma(\omega)}(\omega)$ is a *random process*, while $\omega \mapsto X_{\sigma(\omega)}(\omega)$ is a *random variable*.

Then X^σ is again a transform $\phi \cdot X$, with $\phi_t = \mathbf{1}_{\{\sigma \geq t\}}$. To complement Theorem 5.2.7, we have the following result.

Theorem 5.2.11 (Optional Stopping Theorem). *Suppose that X is a (super-)martingale and let σ be a bounded stopping time. Then X^σ is again a (super-)martingale for the filtration $\mathbb{F}$.*

Proof. We deal with the supermartingale case. For $t \geq 1$,

$$X_{t \wedge \sigma} = X_0 + \sum_{s \leq t} \phi_s \Delta X_s,$$

where we have set $\phi_s = \mathbf{1}_{\{s \leq \sigma\}}$, which is predictable. Hence X^σ is adapted to $\mathbb{F}$ and $\phi_s \geq 0$. Hence X^σ is a supermartingale. The martingale case is then obvious. $\qquad\square$

5.3   Uniformly Integrable Martingales

In order to deal with unbounded stopping times, we need to develop a little of the convergence theory for a particularly important class of martingales indexed by $\mathbb{N}$, namely uniformly integrable (UI) martingales. The counterparts of these results in the continuous-time setting are outlined in Chapter 6.

Definition 5.3.1. A family $\mathcal{C}$ of random variables is *uniformly integrable (UI)* if, given $\epsilon > 0$, there exists $K > 0$ such that

$$\int_{\{|X| > K\}} |X| \, dP < \epsilon \text{ for all } X \in \mathcal{C}. \tag{5.5}$$

In other words, $\sup_{X \in \mathcal{C}} \int_{\{|X| > K\}} |X| \, dP \to 0$ as $K \to \infty$, which explains the terminology. Such families are easy to find.

Examples of UI Families

First of all, if $\mathcal{C}$ is bounded in $L^p(\Omega, \mathcal{F}, P)$ for some $p > 1$, then $\mathcal{C}$ is UI. To see this, choose A such that $E(|X|^p) < A$ for all $X \in \mathcal{C}$ and fix $X \in \mathcal{C}$, $K > 0$. Write $Y = |X| \mathbf{1}_{\{|X| > K\}}$. Then $Y(\omega) \geq K > 0$ for all $\omega \in \Omega$, and since $p > 1$ it is clear that $Y \leq K^{1-p} Y^p$. Thus

$$E(Y) \leq K^{1-p} E(Y^p) \leq K^{1-p} E(|X|^p) \leq K^{1-p} A.$$

But K^{1-p} decreases to 0 when $K \to \infty$, so (5.5) holds.

Exercise 5.3.2. Prove that if $\mathcal{C}$ is UI, then it is bounded in L^1, but the converse is false.

A useful additional hypothesis is domination in L^1: if there exists $Y \geq 0$ in L^1 such that $|X| \leq Y$ for all $X \in \mathcal{C}$, then $\mathcal{C}$ is UI. (See, e.g. ,[299] for a simple proof.)

To illustrate why uniform integrability is so important for martingales, consider the following.

Proposition 5.3.3. *Let $X \in L^p$, $p \geq 1$. The family*

$$\mathcal{U} = \{E(X \,|\, \mathcal{G}) : \mathcal{G} \text{ is a sub-}\sigma\text{-field of } \mathcal{F}\}$$

is UI.

We prove this for the case $p > 1$ (which is all we need in the sequel) and refer to [299, Theorem 13.4] for the case $p = 1$. First we need an important inequality, which we will use frequently.

Proposition 5.3.4 (Jensen's Inequality). *Suppose that $X \in L^1$. If $\phi : \mathbb{R} \to \mathbb{R}$ is convex and $\phi(X) \in L^1$, then*

$$E(\phi(X) \,|\, \mathcal{G}) \geq \phi(E(X \,|\, \mathcal{G})). \tag{5.6}$$

Proof. Any convex function $\phi : \mathbb{R} \mapsto \mathbb{R}$ is the supremum of a family of affine functions, so there exists a sequence (ϕ_n) of real functions with $\phi_n(x) = a_n x + b_n$ for each n, such that $\phi = \sup_n \phi_n$. Therefore $\phi(X) \geq a_n X + b_n$ holds a.s. for each (and hence all) n. So by the positivity of $E(\cdot \,|\, \mathcal{G})$, we have

$$E(\phi(X) \,|\, \mathcal{G}) \geq \sup_n (a_n E(X \,|\, \mathcal{G}) + b_n) = \phi(E(X \,|\, \mathcal{G})) \text{ a.s.}$$

$\square$

Proof of Proposition 5.3.3. With $\phi(x) = |x|^p$, Jensen's inequality implies that $|E(X \,|\, \mathcal{G})|^p \leq E(|X|^p \,|\, \mathcal{G})$, and taking expectations and p^{th} roots on both sides, we obtain

$$\|E(X \,|\, \mathcal{G})\|_p \leq \|X\|_p \quad \text{for all } \mathcal{G} \subset \mathcal{F}.$$

Thus the family $\mathcal{U}$ is L^p-bounded and hence UI. $\square$

Remark 5.3.5. Jensen's inequality shows that the conditional expectation operator is a *contraction* on L^p. The same is true for L^1. Taking $\phi(x) = |x|$, we obtain $|E(X \,|\, \mathcal{G})| \leq E(|X| \,|\, \mathcal{G})$, and hence $\|E(X \,|\, \mathcal{G})\|_1 \leq \|X\|_1$.

Jensen's inequality also shows that, given $p > 1$ and an L^p-bounded martingale $(M_t, \mathcal{F}_t)_{t \in \mathbb{T}}$, the sequence $(|M_t|^p, \mathcal{F}_t)$ is a *sub*martingale. This follows upon taking $\phi(x) = |x|^p$, so that by (5.6), with $t \geq s$, we have

$$E(|M_t|^p \,|\, \mathcal{F}_s) \geq |E(M_t \,|\, \mathcal{F}_s)|^p = |M_s|^p.$$

Here the integrability of N_t, which is required for the application of (5.6), follows from the L^p-boundedness of M_t. Similar results follow upon applying (5.6) with $\phi(x) = x^+$ or $\phi(x) = (x - K)^+$ with suitable integrability assumptions.

Martingale Convergence

We now review briefly the principal limit theorems for martingales. The role of uniform integrability is evident from the following proposition.

Proposition 5.3.6. *Suppose* (X_n) *is a sequence of integrable random variables and* X *is integrable. The following are equivalent:*

a) *The sequence* (X_n) *converges to* X *in the* L^1-*norm; i.e. ,* $\|X_n - X\|_1 = E(|X_n - X|) \to 0$.

b) *The sequence* (X_n) *is UI and converges to* X *in probability.*

See [109] or [299] for the proof of this standard result. Since a.s. convergence implies convergence in probability, we also have the following.

Corollary 5.3.7. *If* (X_n) *is UI and* $X_n \to X$ *a.s., then* $X \in L^1$ *and* $X_n \to X$ *in* L^1-*norm.*

Thus, to prove that a UI martingale converges in L^1-norm, the principal task is showing a.s. convergence. Doob's original proof of this result remains instructive and has been greatly simplified by the use of martingale transforms. We outline here the beautifully simple treatment given in [299], to which we refer for details.

Let $t \mapsto M_t(\omega)$ denote the sample paths of a random process M defined on $\mathbb{N} \times \Omega$ and interpret $\Delta M_t = M_t - M_{t-1}$ as 'winnings' per unit stake on game t. The total winnings ('gains process') can be represented by the martingale transform $Y = C \cdot M$ given by a playing strategy C in which we stake one unit as soon as M has taken a value below a, continue placing unit stakes until M reaches values above b, after which we do not play until M is again below a, and repeat the process indefinitely. It is 'obvious' (and can be shown inductively) that C is predictable.

Let $U_T[a, b](\omega)$ denote the number of 'upcrossings' of $[a, b]$ by the path $t \mapsto M_t$, that is, the maximal $k \in \mathbb{N}$ such that there are $0 \le s_1 < t_1 < s_2 < \cdots < t_k < T$ for which $M_{s_i}(\omega) < a$ and $M_{t_i}(\omega) > b$ $(i = 1, 2, \ldots, k)$. Then

$$Y_T(\omega) \ge (b - a)U_T[a, b](\omega) - (M_T(\omega) - a)^- \qquad (5.7)$$

since Y increases by at least $(b - a)$ during each upcrossing, while the final term overestimates the potential loss in the final play.

Now suppose that M is a supermartingale. Since C is bounded and non-negative, the transform Y is again a supermartingale (the results of Chapter 2 apply here as everything is restricted to the finite time set $\{0, 1, \ldots, T\}$. Thus $E(Y_T) \le E(Y_0) = 0$. Then (5.7) yields

$$(b - a)E(U_T[a, b]) \le E(M_T - a)^-. \qquad (5.8)$$

If, moreover, $M = (M_t)_{t \in \mathbb{N}}$ is L^1-bounded, $K = \sup_t \|M_t\|_1$ is finite, so that

$$(b - a)E(U_T[a, b]) \le |a| + K.$$

The bound is independent of T, so monotone convergence implies that

$$(b-a) E\left(U_\infty[a,b]\right) < \infty,$$

where $U_\infty[a,b] = \lim_{T \to \infty} U_T[a,b]$.

Hence $\{U_\infty[a,b] = \infty\}$ is a P-null set; that is, every interval is 'up-crossed' only finitely often by almost all paths of M. Now the set $D \subset \Omega$ on which $M_t(\omega)$ does *not* converge to a finite *or* infinite limit can be written as

$$D = \bigcup_{\{a,b \in \mathbb{Q} : a < b\}} D_{a,b},$$

where

$$D_{a,b} = \left\{\omega : \liminf_t M_t(\omega) < a < b < \limsup_t M_t(\omega)\right\}.$$

Since $D_{a,b} \subset \{\omega : U_\infty[a,b] = \infty\}$, it follows that D is also P-null.

Thus the a.s. limit M_∞ exists a.s. (P) in $[-\infty, \infty]$. By Fatou's lemma,

$$\|M_\infty\|_1 = E\left(\liminf |M_t|\right) \le \liminf \|M_t\|_1 \le K,$$

so that M_∞ is in L^1 and consequently a.s. finite.

Finally, if the family $(M_t)_{t \in \mathbb{N}}$ is a martingale and is also UI (we simply say that M is a UI martingale), then it follows at once from Corollary 5.3.7 that $M_t \to M_\infty$ in L^1-norm. Moreover, the martingale property 'extends to the limit'; that is, for all t,

$$M_t = E\left(M_\infty \,|\, \mathcal{F}_t\right). \tag{5.9}$$

To see this, note that for $A \in \mathcal{F}_t$ and $u \ge t$, the martingale property yields $\int_A M_u \, dP = \int_A M_s \, dP$, while

$$\left|\int_A M_t \, dP - \int_A M_\infty \, dP\right| \le \int_A |M_t - M_\infty| \, dP \le \|M_t - M_\infty\|_1 \to 0$$

as $t \to \infty$. This proves (5.9). Whenever (5.9) holds, we say that the limit random variable M_∞ *closes* the martingale M.

To summarise, we have the following.

Theorem 5.3.8 (Martingale Convergence). *(i) If the supermartingale M is bounded in L^1, then $M_\infty(\omega) = \lim_{t \to \infty} M_t(\omega)$ exists a.s. (P) and the random variable M_∞ is integrable.*

(ii) If M is a UI martingale, $M_t \to M_\infty$ a.s. and in L^1, and M_∞ closes the martingale M.

(iii) If $X \in L^1$ and $M_t = E\left(X \,|\, \mathcal{F}_t\right)$ for all $t \in \mathbb{N}$, then M is a UI martingale and $M_t \to E\left(X \,|\, \mathcal{F}_\infty\right)$ a.s. and in L^1.

Proof. Only the final statement still requires proof; this can be found in [299, 14.2]. Note that if $\mathcal{F} = \mathcal{F}_\infty$ (as we assume), then $M_t \to X$. □

One immediate consequence of the convergence theorems is that for UI martingales we can extend Definition 5.2.5 to general stopping times. Given a UI martingale M and any stopping time τ, the random variable $M_\tau(\omega) = M_{\tau(\omega)}(\omega)$ is now also well-defined on the set $\{\tau = \infty\}$, on which we set $M_\tau = M_\infty$.

We extend Theorems 5.2.7 (optional sampling) and 5.2.11 (optional stopping) to general stopping times when M is a UI martingale. In the first place, we have the following result.

Theorem 5.3.9. *Let M be a UI martingale and τ a stopping time. Then*

$$E\left(M_\infty \,|\, \mathcal{F}_\tau\right) = M_\tau \quad a.s.\,(P).$$ (5.10)

Proof. As M_∞ closes M, we have $M_t = E\left(M_\infty \,|\, \mathcal{F}_t\right)$ for all t. In addition, as $\tau \wedge t$ is a bounded stopping time, Theorem 5.2.7 yields $M_{\tau \wedge t} = E\left(M_t \,|\, \mathcal{F}_{\tau \wedge t}\right)$. Hence $E\left(M_\infty \,|\, \mathcal{F}_{\tau \wedge t}\right) = M_{\tau \wedge t}$.

Let $A \in \mathcal{F}_\tau$. The set $B_t = A \cap \{\tau \le t\}$ is in $\mathcal{F}_t$ by definition and in $\mathcal{F}_\tau$ since τ is $\mathcal{F}_\tau$-measurable. Hence $B_t \in \mathcal{F}_{\tau \wedge t}$ and so

$$\int_{B_t} M_\infty \, dP = \int_{B_t} M_{\tau \wedge t} \, dP = \int_{B_t} M_\tau \, dP.$$ (5.11)

Assume without loss of generality that M_∞ (and hence each M_t) is non-negative, and let $t \uparrow \infty$. Then (5.11) shows that

$$\int_{A \cap \{\tau < \infty\}} M_\infty \, dP = \int_{A \cap \{\tau < \infty\}} M_\tau \, dP.$$

Since $M_\tau = M_\infty$ trivially on $\{\tau = \infty\}$, the result follows. □

Corollary 5.3.10. *Let M be a UI martingale.*

(i) *(Optional Sampling) If $\sigma \le \tau$ are stopping times, then*

$$E\left(M_\tau \,|\, \mathcal{F}_\sigma\right) = M_\sigma \quad a.s.\,(P).$$ (5.12)

(ii) *(Optional Stopping) If τ is a stopping time, then $M_\tau \in L^1$ and M^τ is a UI martingale. In particular, $E\left(M_\tau\right) = E\left(M_0\right)$.*

Doob Decomposition and Quadratic Variation

Again let $(\Omega, \mathcal{F}, P, \mathbb{N}, \mathbb{F})$ be a stochastic basis, and let $X = (X_t)_{t \in \mathbb{T}}$ be an $\mathbb{F}$-adapted process. Since martingales describe what we might call 'purely random' behaviour, it is natural to ask to what extent the 'martingale part' of X can be isolated from the 'long-term trends' that X exhibits. In discrete time, this is easily accomplished; remarkably there is also such a decomposition in continuous time (the *Doob-Meyer decomposition*, see [109], [199]). This fact underlies the success of general stochastic integration and the success of martingale methods in continuous-time finance.

Definition 5.3.11. Given an adapted sequence $X = (X_t)$ of random variables on $(\Omega, \mathcal{F}, P)$, construct the stochastic processes M and A by setting

$$M_0 = 0, \quad \Delta M_t = M_t - M_{t-1} = X_t - E\left(X_t \,|\, \mathcal{F}_{t-1}\right) \text{ for } t > 0,$$
$$A_0 = 0, \quad \Delta A_t = A_t - A_{t-1} = E\left(X_t \,|\, \mathcal{F}_{t-1}\right) - X_{t-1} \text{ for } t > 0.$$

Adding terms for $s \leq t$, it is clear that

$$X_t = X_0 + M_t + A_t \text{ for all } t \geq 0. \tag{5.13}$$

We call this the *Doob decomposition* of the adapted process X.

It is clear that A_t is $\mathcal{F}_{t-1}$-measurable, so that A is predictable. The process M is a martingale null at 0 since $E\left(\Delta M_t \,|\, \mathcal{F}_{t-1}\right) = 0$. Thus we have

$$E\left(\Delta X_t \,|\, \mathcal{F}_{t-1}\right) = \Delta A_t \text{ for all } t > 0. \tag{5.14}$$

By construction, $\Delta M_t + \Delta A_t = \Delta X_t$ for all $t > 0$.

The Doob decomposition is unique in the following sense. If we also have $X - X_0 = M' + A'$ for some martingale M' and predictable process A', then $M + A = X - X_0 = M' + A'$, so that $M - M' = A' - A$ is a predictable martingale. Such a process must be constant, as we saw in Chapter 2. Hence (up to some fixed P-null set N, for all $t \in \mathbb{N}$) equation (5.13) is the *unique* decomposition of an adapted process X into the sum of its initial value, a martingale, and a predictable process A, both null at 0.

When X is a submartingale, equation (5.14) shows that $\Delta A_t \geq 0$, so that $t \mapsto A_t(\omega)$ is *increasing* in t, for almost all $\omega \in \Omega$. This increasing predictable process A therefore has an a.s. limit A_∞ (which can take the value $+\infty$ in general).

Now consider the special case where $X = M^2$ and M is an L^2-bounded martingale with $M_0 = 0$; then M^2 is a submartingale by Jensen's inequality (5.6) (see Remark 5.3.5). The Doob decomposition $M^2 = N + A$ consists of a UI martingale N and a predictable increasing process A, both null at 0. Define $A_\infty = \lim_{t \uparrow \infty} A_t$ a.s. We have

$$E\left(M_t^2\right) = E\left(N_t\right) + E\left(A_t\right) = E\left(A_t\right) \text{ for all } t \in \mathbb{N},$$

and these quantities are bounded precisely when $A_\infty \in L^1$.

Observe, using (5.14), that, since M is a martingale,

$$\Delta A_t = E\left(\left(M_t^2 - M_{t-1}^2\right) |\mathcal{F}_{t-1}\right) = E\left((\Delta M_t)^2 \,|\, \mathcal{F}_{t-1}\right). \tag{5.15}$$

For this reason, we call A the *quadratic variation of M* and write $A = \langle M \rangle$. We have shown that *an L^2-bounded martingale has integrable quadratic variation.*

Remark 5.3.12. In Chapters 6 to 8, we make fuller use of the preceding results in the continuous-time setting. The translation of the convergence theorems so that they apply to continuous-time UI martingales is straight-forward (though somewhat tedious) once one has established that such a martingale M, with time set $[0, T]$ or $[0, \infty)$, always possesses a 'version' almost all of whose paths $t \mapsto M_t(\omega)$ are right-continuous and have left limits. This enables one to use countable dense subsets to approximate the path behaviour and use the results just presented; see [109], [199] for details. With the interpretation of $\mathbb{T}$ as an interval in $\mathbb{R}^+$, the convergence theorems and the optional sampling and optional stopping results proved in the foregoing go over verbatim to the continuous-time setting. We assume this in Chapter 6 and beyond.

Of particular importance in continuous time is the analogue of the Doob decomposition, the *Doob-Meyer decomposition* of a sub- (or super-) mar-tingale; we briefly outline its principal features without proof (see [109] for a full treatment). We discuss the Doob-Meyer decomposition further when introducing Itô processes in Chapter 6; we will make essential use of the decomposition when analysing American put options in Chapter 8.

If $\mathbb{T} = [0, \infty)$ and $X = (X_t)$ is a supermartingale with right-continuous paths $t \mapsto X_t(\omega)$ for P-almost all $\omega \in \Omega$, then we say that X *is of class* D if the family $\{X_\tau : \tau \text{ is a stopping time}\}$ is UI. If X is a UI martingale, this is automatic from Theorem 5.3.9, but this is not generally the case for supermartingales. Every such supermartingale has decomposition

$$X_t = M_t - A_t,$$

where M is a UI martingale and the increasing process A has $A_0 = 0$ and is *predictable*. In continuous time, this definition requires that A be measurable with respect to the σ-field $\mathcal{P}$ on $[0, \infty) \times \Omega$ that is generated by the continuous processes. The Doob-Meyer decomposition is unique up to indistinguishability (see Definition 6.1.12), and the process A is integrable.

Given an L^2-bounded (hence UI) martingale M, the decomposition again defines a quadratic variation for the submartingale $M^2 = N + A$, and we write $A = \langle M \rangle$. Note that since M is a martingale, (5.15) also holds in this setting, which justifies the terminology. Of particular interest to us are martingales whose quadratic variation is non-random; we shall find (Chapter 6) that Brownian motion W is a martingale such that $\langle W \rangle_t = t$.

5.4   Optimal Stopping: The Snell Envelope

American Options

We return to our consideration of American options on a finite discrete time set. Consider a price process $S = (S^0, S^1)$ consisting of a riskless bond $S_t^0 = (1+r)^t$ and a single risky stock $(S_t^1)_{t \in \mathbb{T}}$, where $\mathbb{T} = \{0, 1, \dots, T\}$

for finite $T > 0$ and $r > 0$, defined on a probability space $(\Omega, \mathcal{F}, P)$. We have seen that the holder's freedom to choose the exercise date (without prescience) requires the option writer (seller) of an American call option with strike K to hedge against a liability of $(S_\tau^1 - K)^+$ at a (random) stopping time $\tau : \Omega \mapsto \mathbb{T}$. Thus, if the system is in state $\omega \in \Omega$, and if $\tau(\omega) = t$, the liability is $\left(S_t^1(\omega) - K\right)^+$. In general, both the stopping time and the liability *vary* with ω. We write $\mathcal{T} = \mathcal{T}_\mathbb{T}$ for the class of all $\mathbb{T}$-valued stopping times. Since $\mathbb{T}$ is assumed finite, we can restrict attention to *bounded* stopping times for the present, and hence Theorems 5.2.7 and 5.2.11 apply to this situation.

Suppose that the writer tries to construct a hedging strategy $\theta = (\theta^0, \theta^1)$ to guard against the potential liability. This will generate a value process $V(\theta)$ with

$$V_t(\theta) = V_0(\theta) + \sum_{u \le t} \theta_u \cdot \Delta S_u = V_0(\theta) + \sum_{u \le t} (\theta_u^0 \Delta S_u^0 + \theta_u^1 \Delta S_0^1).$$

The strategy should be self-financing, so we also demand that $(\Delta \theta_t) \cdot S_{t-1} = 0$ for $t \ge 1$.

We assume that the model is viable and that Q is an EMM for S. Then the discounted value process $M = \overline{V}(\theta)$ is a martingale under $(\mathbb{F}, Q)$ and by Theorem 5.2.7 we conclude that

$$V_0(\theta) = M_0 = E_Q\left(\overline{V}_\tau(\theta)\right) = E_Q\left((1+r)^{-\tau} V_\tau(\theta)\right). \tag{5.16}$$

Note that, since τ is a random variable, we cannot now take the term $(1 + r)^{-\tau}$ outside the expectation as in the case of European options.

Hence, if the writer is to hedge successfully against the preceding liability, the initial capital required for this portfolio is $E_Q\left((1+r)^{-\tau} V_\tau(\theta)\right)$. This holds for every $\tau \in \mathcal{T}$. But since we need $V_\tau(\theta) \ge (S_\tau - K)^+$, the initial outlay x with which to form the strategy θ must satisfy

$$x \ge \sup_{\tau \in \mathcal{T}} E_Q\left((1+r)^{-\tau}(S_\tau^1 - K)^+\right). \tag{5.17}$$

More generally, given an American option, we saw in Section 5.1 that its payoff function is a random *sequence* $f_t = f_t(S^1)$ of functions that (in general) depend on the *path* taken by S^1. The initial capital x needed for a hedging strategy satisfies

$$x \ge \sup_{\tau \in \mathcal{T}} E_Q\left((1+r)^{-\tau} f_\tau\right).$$

If we can find a self-financing strategy θ *and* a stopping time $\tau^* \in \mathcal{T}$ such that $V_{\tau^*}(\theta) = f_{\tau^*}$ almost surely, then the initial capital required is exactly

$$x = \sup_{\tau \in \mathcal{T}} E_Q\left((1+r)^{-\tau} f_\tau\right) = E_Q\left((1+r)^{-\tau^*} f_{\tau^*}\right). \tag{5.18}$$

Recall from Section 5.1 that a hedging strategy (or simply a *hedge*) is a self-financing strategy θ that generates a value process $V_t(\theta) \geq f_t$ a.s. (Q) for all $t \in \mathbb{T}$, and we say that the hedge θ is *minimal* if there exists a stopping time τ^* with $V_{\tau^*}(\theta) = f_{\tau^*}$ a.s. (Q). Thus (5.18) is *necessary* for the existence of a minimal hedge θ, and we show that it is also *sufficient*. This justifies calling x the *rational price* of the American option with payoff function f.

To see how the value process $V(\theta)$ changes in each underlying single-period model, we again consider the problem faced by the option writer but work backwards in time from the expiry date T. Since f_T is the value of the option at time T, the hedge must yield at least $V_T = f_T$ in order to cover exercise at that time. At time $T - 1$, the option holder has the choice either to exercise immediately or to hold the option until time T. The time $T - 1$ value of the latter choice is

$$(1 + r)^{-1} f_T = S^0_{T-1} E_Q \left(\overline{f}_T \,|\, \mathcal{F}_{T-1} \right);$$

recall that we write $\overline{Y}_t = \beta_t Y_t = (S^0_t)^{-1} Y_t$ for the discounted value of any quantity Y_t. Thus the option writer needs income from the hedge to cover the potential liability $\max\left\{ f_{T-1}, S^0_{T-1} E_Q \left(\overline{f}_T \,|\, \mathcal{F}_{T-1} \right) \right\}$, so this quantity is a rational choice for $V_{T-1}(\theta)$. Inductively, we obtain

$$V_{t-1}(\theta) = \max\left\{ f_{t-1}, S^0_{t-1} E_Q \left(V_t \,|\, \mathcal{F}_{t-1} \right) \right\} \text{ for } t = 1, 2, \ldots, T. \tag{5.19}$$

In particular, if $\beta_t = (1+r)^t$ for some constant interest rate $r > 0$, equation (5.19) simplifies to

$$V_{t-1}(\theta) = \max\left\{ f_{t-1}, (1+r)^{-1} E_Q \left(V_t \,|\, \mathcal{F}_{t-1} \right) \right\} \text{ for } t = 1, 2, \ldots, T. \tag{5.20}$$

The option writer's problem is to construct such a hedge.

The Snell Envelope

Adapting the treatment given in [236], we now solve this problem in a more abstract setting in order to focus on its essential features; given a *finite* adapted sequence $(X_t)_{t \in \mathbb{T}}$ of *non-negative* random variables on $(\Omega, \mathcal{F}, Q)$, we show that the optimisation problem of determining $\sup_{\tau \in \mathcal{T}} E_Q(X_\tau)$ can be solved by the inductive procedure suggested previously and that the optimal stopping time $\tau^* \in \mathcal{T}$ can be described in a very natural way.

Definition 5.4.1. Given $(X_t)_{t \in \mathbb{T}}$ with $X_t \geq 0$ a.s. for all t, define a new adapted sequence $(Z_t)_{t \in \mathbb{T}}$ by backward induction by setting

$$Z_T = X_T, \quad Z_{t-1} = \max\left\{ X_{t-1}, E_Q \left(Z_t \,|\, \mathcal{F}_{t-1} \right) \right\} \text{ for } t = 1, 2, \ldots, T. \tag{5.21}$$

We call Z the *Snell envelope* of the finite sequence (X_t).

The sequence $(Z_t)_{t \in \mathbb{T}}$ is clearly adapted to the filtration $\mathbb{F} = (\mathcal{F}_t)_{t \in \mathbb{T}}$. In the following, we give a more general definition, applicable also to infinite sequences.

Note that (Z_t) is defined 'backwards in time'. It is instructive to read the definition with a 'forward' time variable using the *time to maturity* $s = T - t$. Then the definitions (5.21) become

$$Z_T = X_T, \quad Z_{T-s} = \max \left\{ X_{T-s}, E_Q \left(Z_{T-s+1} | \mathcal{F}_{T-s} \right) \right\} \text{ for } s = 1, 2, \ldots, T.$$

We now examine the properties of the process Z.

Proposition 5.4.2. *Let $(Z_t)_{t \in \mathbb{T}}$ be the Snell envelope of a process $(X_t)_{t \in \mathbb{T}}$ with $X_t \geq 0$ a.s. for all t.*

(i) *The process Z is the smallest $(\mathbb{F}, Q)$-supermartingale dominating X.*

(ii) *The random variable $\tau^* = \min \{t \geq 0 : Z_t = X_t\}$ is a stopping time, and the stopped process Z^{τ^*} defined by $Z_t^{\tau^*} = Z_{t \wedge \tau^*}$ is an $(\mathbb{F}, Q)$-martingale.*

Proof. From (5.21) we deduce that $Z_t \geq X_t$ for $t \in \mathbb{T}$; hence Z dominates X. Since

$$Z_{t-1} \geq E_Q \left(Z_t | \mathcal{F}_{t-1} \right) \text{ for all } t = 1, 2, \ldots, T,$$

the process Z is also a supermartingale.

To see that it is the *smallest* such supermartingale, we argue by backward induction. Suppose that $Y = (Y_t)$ is any supermartingale with $Y_t \geq X_t$ for all $t \subset \mathbb{T}$. Clearly, $Y_T \geq X_T = Z_T$. Now if $Y_t \geq Z_t$ for a fixed $t \in \mathbb{T}$, then we have $Y_{t-1} \geq E_Q \left(Y_t | \mathcal{F}_{t-1} \right)$ since Y is a supermartingale. It follows from the positivity of the conditional expectation operator that $Y_{t-1} \geq E_Q \left(Z_t | \mathcal{F}_{t-1} \right)$. On the other hand, Y dominates X; hence $Y_{t-1} \geq X_{t-1}$. Therefore

$$Y_{t-1} \geq \max \left\{ X_{t-1}, E_Q \left(Z_t | \mathcal{F}_{t-1} \right) \right\} = Z_{t-1},$$

which completes the induction step. The first assertion of the proof follows.

For the second claim, note that $Z_0 = \max \{X_0, E_Q \left(Z_1 | \mathcal{F}_0 \right) \}$, and $\{\tau^* = 0\} = \{Z_0 = X_0\} \in \mathcal{F}_0$ since X_0 and Z_0 are $\mathcal{F}_0$-measurable. By the definition of τ^*, we have

$$\{\tau^* = t\} = \bigcap_{s=0}^{t-1} \{Z_s > X_s\} \cap \{Z_t = X_t\} \text{ for } t = 1, 2, \ldots, T.$$

This set belongs to $\mathcal{F}_t$ since X and Z are adapted. Thus τ^* is a stopping time. Note that $\tau^*(\omega) \leq T$ a.s.

To see that the stopped process $Z_t^{\tau^*} = Z_{t \wedge \tau^*}$ defines a martingale, we again use a martingale transform, as in the proof of Theorem 5.2.11. Define

$$\phi_t = \mathbf{1}_{\{\tau^* \geq t\}} \text{ for } t = 1, 2, \ldots, T;$$

the process ϕ is predictable since $\{\tau^* \geq t\} = \Omega \setminus \{\tau^* < t\}$. Moreover,

$$Z_t^{\tau^*} = Z_0 + \sum_{u=1}^{t} \phi_u \Delta Z_u \text{ for } t = 1, 2, \ldots, T.$$

Now, for $t = 1, 2, \ldots, T$, we have

$$Z_t^{\tau^*} - Z_{t-1}^{\tau^*} = \phi_t(Z_t - Z_{t-1}) = \mathbf{1}_{\{\tau^* \geq t\}}(Z_t - Z_{t-1});$$

if $\tau^*(\omega) \geq t$, then $Z_{t-1}(\omega) > X_{t-1}(\omega)$, so that $Z_{t-1}(\omega) = E_Q(Z_t | \mathcal{F}_{t-1})(\omega)$ on this set. For all $t = 1, 2, \ldots, T$, we therefore have

$$E_Q\left(\left(Z_t^{\tau^*} - Z_{t-1}^{\tau^*}\right) | \mathcal{F}_{t-1}\right) = \mathbf{1}_{\{\tau^* \geq t\}} E_Q\left((Z_t - E_Q(Z_t | \mathcal{F}_{t-1})) | \mathcal{F}_{t-1}\right) = 0.$$

Thus the stopped process Z^{τ^*} is a martingale on $(\Omega, \mathbb{F}, Q)$. Recall that we assume that the σ-field $\mathcal{F}_0$ is trivial, so that it contains only Q-null sets and their complements (in the case of a finite market model, this reduces to $\mathcal{F}_0 = \{\emptyset, \Omega\}$). Therefore X_0 and Z_0 are a.s. constant since both are $\mathcal{F}_0$-measurable. $\qquad \square$

Definition 5.4.3. We call a stopping time $\sigma \in \mathcal{T} = \mathcal{T}_{\mathbb{T}}$ optimal for $(X_t)_{t \in \mathbb{T}}$ if

$$E_Q(X_\sigma) = \sup_{t \in \mathcal{T}} E_Q(X_\tau). \tag{5.22}$$

Proposition 5.4.4. Let $(Z_t)_{t \in \mathbb{T}}$ be the Snell envelope of a process $(X_t)_{t \in \mathbb{T}}$ with $X_t \geq 0$ a.s. for all t. The stopping time $\tau^* = \min\{t \geq 0 : Z_t = X_t\}$ is optimal for X, and

$$Z_0 = E_Q(X_{\tau^*}) = \sup_{\tau \in \mathcal{T}} E_Q(X_\tau). \tag{5.23}$$

Proof. Since Z^{τ^*} is a martingale, we have

$$Z_0 = Z_0^{\tau^*} = E_Q\left(Z_T^{\tau^*}\right) = E_Q(Z_{\tau^*}) = E_Q(X_{\tau^*}),$$

where the final equality follows from the definition of τ^*. On the other hand, given any $\tau \in \mathcal{T}$, we know from Proposition 5.4.2 that Z^τ is a supermartingale. Hence

$$Z_0 = E_Q(Z_0^\tau) \geq E_Q(Z_\tau) \geq E_Q(X_\tau)$$

since Z dominates X. $\qquad \square$

Characterisation of Optimal Stopping Times

We are now able to describe how the martingale property characterises optimality more generally. Let $(Z_t)_{t \in \mathbb{T}}$ be the Snell envelope of a process $(X_t)_{t \in \mathbb{T}}$ with $X_t \geq 0$ a.s. for all t.

Proposition 5.4.5. *The stopping time $\sigma \in \mathcal{T}$ is optimal for X if and only if the following two conditions hold.*

(i) $Z_\sigma = X_\sigma$ *a.s.* (Q).

(ii) Z^σ *is an* $(\mathbb{F},Q)$*-martingale.*

Proof. If Z^σ is a martingale, then

$$Z_0 = E_Q\left(Z_0^\sigma\right) = E_Q\left(Z_T^\sigma\right) = E_Q\left(Z_\sigma\right) = E_Q\left(X_\sigma\right),$$

where the final step uses condition 1. On the other hand, Z^τ is a super-martingale for $\tau \in \mathcal{T}$. Hence

$$Z_0 = E_Q\left(Z_0^\tau\right) \geq E_Q\left(Z_T^\tau\right) = E_Q\left(Z_\tau\right) \geq E_Q\left(X_\tau\right),$$

as Z dominates X. Since $\sigma \in \mathcal{T}$, it follows that σ is optimal.

Conversely, suppose that σ is optimal for X. By Proposition 5.4.4, we have $Z_0 = \sup_{\tau \in \mathcal{T}} E_Q\left(X_\tau\right)$; it follows that $Z_0 = E_Q\left(X_\sigma\right) \leq E_Q\left(Z_\sigma\right)$ since Z dominates X. However, Z^σ is a supermartingale, so $E_Q\left(Z_\sigma\right) \leq Z_0$. In other words, for any optimal σ, $E_Q\left(X_\sigma\right) = Z_0 = E_Q\left(Z_\sigma\right)$. But Z dominates X, and thus $X_\sigma = Z_\sigma$ a.s. (Q). This proves condition 1 above.

Now observe that we have $Z_0 = E_Q\left(Z_\sigma\right)$ as well as

$$Z_0 \geq E_Q\left(Z_{\sigma \wedge t}\right) \geq E_Q\left(Z_\sigma\right)$$

since Z^σ is a supermartingale. Hence

$$E_Q\left(Z_{\sigma \wedge t}\right) = E_Q\left(Z_\sigma\right) = E_Q(E_Q\left(Z_\sigma \mid \mathcal{F}_t\right)).$$

Again because Z is a supermartingale, we also have, by Theorem 5.2.7, that

$$Z_{\sigma \wedge t} \geq E_Q\left(Z_\sigma \mid \mathcal{F}_t\right),$$

so that again $Z_{\sigma \wedge t} = E_Q\left(Z_\sigma \mid \mathcal{F}_t\right)$. This means that Z^σ is in fact a martingale. $\square$

From Proposition 5.4.5, it is clear that τ^* is the *smallest* optimal stopping time for X since by definition it is the smallest stopping time such that $Z_{\tau^*} = X_{\tau^*}$ a.s. (Q). To find the *largest* optimal stopping time for X, we look for the first time that the increasing process A in the Doob decomposition of Z 'leaves zero'; that is, the time ν at which the stopped process Z^ν ceases to be a martingale.

Since Z is a supermartingale, its Doob decomposition $Z = Z_0 + N + B$ has N as a martingale and B as a predictable *decreasing* process, both null at 0. Let $M = Z_0 + N$, which is a martingale, since Z_0 is a.s. constant, and set $A = -B$, so that $A = (A_t)_{t \in \mathbb{T}}$ is increasing, with $A_0 = 0$, and $Z = M - A$.

Define a random variable $\nu : \Omega \mapsto \mathbb{T}$ by

$$\nu(\omega) = \begin{cases} T & \text{if } A_T(\omega) = 0, \\ \min\{t \geq 0 : A_{t+1} > 0\} & \text{if } A_T(\omega) > 0. \end{cases} \tag{5.24}$$

To see that $\nu \in \mathcal{T}$, simply observe that

$$\{\nu = t\} = \bigcap_{s \leq t} \{A_s = 0\} \cap \{A_{t+1} > 0\}$$

is in $\mathcal{F}_t$ because A_{t+1} is $\mathcal{F}_t$-measurable. Thus ν is a stopping time. It is clearly $\mathbb{T}$-valued and therefore bounded.

Proposition 5.4.6. *The stopping time ν in (5.24) is the largest optimal stopping time for X.*

Proof. For $s \leq \nu(\omega)$, $Z_s(\omega) = M_s(\omega) - A_s(\omega)$. Hence Z^ν is a martingale, so that the second condition in Proposition 5.4.5 holds for ν. To verify the first condition (i.e. , $Z_\nu = X_\nu$), let us write Z_ν in the form

$$Z_\nu = \sum_{s=0}^{T} \mathbf{1}_{\{\nu=s\}} Z_s = \sum_{s=0}^{T-1} \mathbf{1}_{\{\nu=s\}} \max\{X_s, E(Z_{s+1}|\mathcal{F}_s)\} + \mathbf{1}_{\{\nu=T\}} X_t.$$

Now

$$E(Z_{s+1}|\mathcal{F}_s) = E(M_{s+1} - A_{s+1}|\mathcal{F}_s) = M_s - A_{s+1}.$$

On the set $\{\nu = s\}$, we have $A_s = 0$ and $A_{s+1} > 0$; hence $Z_s = M_s$. This means that, on this set, $E(Z_{s+1}|\mathcal{F}_s) < Z_s$ a.s., and therefore that $Z_s = \max\{X_s, E(Z_{s+1}|\mathcal{F}_s)\} = X_s$. Thus $Z_\nu = X_\nu$ a.s.; hence ν is optimal.

It is now clear that ν is the *largest* optimal time for (X_t). Indeed, if $\tau \in \mathcal{T}$ has $\tau \geq \nu$ and $Q(\tau > \nu) > 0$, then

$$E(Z_\tau) = E(M_\tau) - E(A_\tau) = E(Z_0) - E(A_\tau) < E(Z_0) = Z_0.$$

By (5.23), the stopping time τ cannot be optimal. $\qquad\square$

Extension to Unbounded Stopping Times

We need to consider value processes at *arbitrary* times $t \in \mathbb{T}$ since the holder's possible future actions from time t onwards will help to determine those processes. So let $\mathcal{T}_t$ denote the set of stopping times $\tau : \Omega \mapsto \mathbb{T}_t = \{t, t+1, \ldots, T\}$, and consider instead the optimal stopping problem $\sup_{\tau \in \mathcal{T}_t} E(X_\tau)$. Although the stopping times remain bounded, an immediate difficulty in attempting to transfer the results we have for $t = 0$ to more general $t \in \mathbb{T}$ is that we made use in our proofs of the fact that Z_0 was a.s. constant. This followed from our assumption that $\mathcal{F}_0$ contained only null sets and their complements, and it led us to establish (5.23), which we used throughout.

In the general case, we are obliged to replace expectations $E_Q(Z_\tau)$ by *conditional* expectations $E_Q(Z_\tau | \mathcal{F}_t)$, thus facing the problem of defining the supremum of a family of random variables rather than real numbers. We need to ensure that we obtain this supremum as an $\mathcal{F}$-measurable function, even for an uncountable family. We use this opportunity to extend the definition of the Snell envelope in preparation for a similar extension to continuous-time situations needed in Chapter 8.

Proposition 5.4.7. *Let $(\Omega, \mathcal{F}, P)$ be a probability space. Let $\mathcal{L}$ be a family of $\mathcal{F}$-measurable functions $\Omega \mapsto [-\infty, \infty]$. There exists a unique $\mathcal{F}$-measurable function $g : \Omega \mapsto [-\infty, \infty]$ with the following properties:*

(i) $g \geq f$ a.s. for all $f \in \mathcal{L}$.

(ii) If an $\mathcal{F}$-measurable function h satisfies $h \geq f$ a.s. for all $f \in \mathcal{L}$, then $h \geq g$ a.s.

We call g the *essential supremum* of $\mathcal{L}$ and write $g = \operatorname{ess\,sup}_{f \in \mathcal{L}} f$. There exists a sequence (f_n) such that $g = \sup_n f_n$. If $\mathcal{L}$ is upward filtering (i.e., if for given f', f'' in $\mathcal{L}$ there exists $f \in \mathcal{L}$ with $f \geq \max\{f', f''\}$), then the sequence (f_n) can be chosen to be increasing, so that $f = \lim_n f_n$.

Proofs of this result can be found in [199], [236]. The idea is simple: identify the closed intervals $[0, 1]$ and $[-\infty, \infty]$, for example, via the increasing bijection $x \mapsto e^x$. Any countable family $\mathcal{C}$ in $\mathcal{L}$ has a well-defined $\mathcal{F}$-measurable ($[0, 1]$-valued) $f_\mathcal{C}$, which thus has finite expectation under P. Define

$$\alpha = \sup\{E(f_\mathcal{C}) : \mathcal{C} \subset \mathcal{L}, \mathcal{C} \text{ countable}\}$$

and choose a sequence $(f'_n, \mathcal{C}_n)$ with $E(f'_n) \to \alpha$. Since $\mathcal{K} = \bigcup_n \mathcal{C}_n$ is countable and $E(f'_\mathcal{K}) = \alpha$, we can set $g = f'_\mathcal{K}$. The sequence (f'_n) serves as an approximating sequence, and $f_0 = f'_0$, $f_{n+1} \geq f_n \vee f'_{n+1}$ will make it increasing with n.

Definition 5.4.8. Let $(\Omega, \mathcal{F}, \mathbb{T}, \mathbb{F}, P)$ be a stochastic base with $\mathbb{T} = \mathbb{N}$. Given an adapted process $(X_t)_{t \in \mathbb{T}}$ such that $X^* = \sup_t X_t \in L^1$, define $\mathcal{T}_t$ as the family of $\mathbb{F}$-stopping times τ such that $t \leq \tau < \infty$. We call $\tau \in \mathcal{T}_t$ a *t-stopping rule*. The *Snell envelope* of (X_t) is the process Z defined by

$$Z_t = \operatorname*{ess\,sup}_{\tau \in \mathcal{T}_t} E(X_\tau | \mathcal{F}_t) \text{ for } t \in \mathbb{T}. \tag{5.25}$$

This definition allows *unbounded* (but a.s. finite) stopping times. When X is UI, we can still use the optional stopping results proved earlier in this context. The martingale characterisation of optimal stopping times can be extended as well; see [199] or [236] for details.

5.5   Pricing and Hedging American Options

Existence of a Minimal Hedge

Return to the setup at the beginning of Section 5.4 and assume henceforth that the market model $(\Omega, \mathcal{F}, P, \mathbb{T}, \mathbb{F}, S)$ is viable and complete, with Q as the unique EMM.

Given an American option (f_t) in this model (e.g. , an American call with strike K, where $f_t = (S_t^1 - K)^+$), we saw that a hedging strategy θ would need to generate a value process $V(\theta)$ that satisfies (5.19); that is,

$$V_T(\theta) = f_T,$$
$$V_{t-1}(\theta) = \max\left\{ f_{t-1}, (1+r)^{-1} E_Q\left(V_t \,|\mathcal{F}_{t-1}\right)\right\} \quad \text{for } t = 1, 2, \ldots, T,$$

since $S_t^0 = (1+r)^{-t}$ for all $t \in \mathbb{T}$. Moving to discounted values, $Z = \left(\overline{V}_t(\theta)\right)$ is then the Snell envelope of the discounted option price $\overline{f}_t = (1+r)^{-t} f_t$, so that

$$Z_T = \overline{f}_T, \qquad Z_{t-1} = \max\left\{ \overline{f}_{t-1}, E_Q\left(\overline{V}_t \,|\mathcal{F}_{t-1}\right)\right\} \quad \text{for } t = 1, 2, \ldots, T.$$

In particular, the results of the previous section yield

$$Z_t = \sup_{\tau \in \mathcal{T}_t} E_Q\left(\overline{f}_\tau \,|\mathcal{F}_t\right) \text{ for } t \in \mathbb{T}, \tag{5.26}$$

and the stopping time $\tau_t^* = \min\left\{ s \geq t : Z_s = \overline{f}_s\right\}$ is optimal, so that the supremum in (5.26) is attained by τ_t^*. (We developed these results for $t = 0$, but with the extended definition of the Snell envelope, they hold for general t.)

For $\tau^* = \tau_0^*$ and $\mathcal{T} = \mathcal{T}_0$, we have, therefore,

$$Z_0 = \sup_{\tau \in \mathcal{T}} E_Q\left(\overline{f}_\tau\right) = E_Q\left(\overline{f}_{\tau^*}\right). \tag{5.27}$$

This *defines* the rational price of the option at time 0 and thus also the initial investment needed for the existence of a hedging strategy.

Now write the Doob decomposition of the supermartingale Z as $Z = \overline{M} - \overline{A}$, where $\overline{M}$ is a martingale and $\overline{A}$ a predictable increasing process. Also write $M_t = S_t^0 \overline{M}_t$ and $A_t = S_t^0 \overline{A}_t$.

Since the market is complete, we can attain the contingent claim M_T by a self-financing strategy θ (e.g. , we could use the strategy constructed by means of the martingale representation in the proof of Proposition 4.1.1) and we may assume that θ is admissible. Thus $\overline{V}_t(\theta) = \overline{M}_t$, and as $\overline{V}(\theta)$ is a martingale under the EMM Q,

$$\overline{V}_t(\theta) = \overline{M}_t = Z_t + \overline{A}_t \text{ for all } t \in \mathbb{T}.$$

Hence also

$$Z_t S_t^0 = V_t(\theta) - A_t \text{ for } t \in \mathbb{T}. \tag{5.28}$$

From the results of the previous section, we know that on the set

$$C = \{(t, \omega) : 0 \leq t < \tau^*(\omega)\},$$

the Snell envelope Z is a martingale and $\overline{A}_t(\omega) = 0$ on this set. Hence

$$V_t(\theta)(\omega) = \sup_{t \leq \tau \leq T} E_Q\left((1+r)^{-(\tau-t)} f_\tau \,|\, \mathcal{F}_t\right) \text{ for all } (t, \omega) \in C. \quad (5.29)$$

We saw that τ^* is the *smallest* optimal exercise time and that $\overline{A}_{\tau^*(\omega)}(\omega) = 0$. Hence (5.28) and (5.29) imply that

$$V_{\tau^*(\omega)}(\theta)(\omega) = Z_{\tau^*(\omega)}(\omega) S^0_{\tau^*(\omega)}(\omega) = f_{\tau^*(\omega)}(\omega). \quad (5.30)$$

Thus the hedge θ with initial capital investment

$$V_0(\theta) = x = \sup_{\tau \in \mathcal{T}} E_Q\left((1+r)^{-\tau} f_\tau\right) \quad (5.31)$$

is minimal. Thus we have verified that this condition is *sufficient* for the existence of a minimal hedge for the option.

The Rational Price and Optimal Exercise

Hedging requires an initial investment x of at least $\sup_{\tau \in \mathcal{T}} E_Q\left((1+r)^{-\tau} f_\tau\right)$, and the supremum is attained at the optimal time τ^*. It follows that x is the minimum initial investment for which a hedging strategy can be constructed. Thus (5.31) provides a natural choice for the 'fair' or *rational* price of the American option.

The optimal exercise time need not be uniquely defined, however; any optimal stopping time (under Q) for the payoff function f_t will be an optimal exercise time. In fact, the holder of the option (the buyer) has no incentive to exercise the option while $Z_t S^0_t > f_t$ since using the option price as initial investment he could create a portfolio yielding greater payoff than the option at time τ by using the hedging strategy θ. Thus the buyer would wait for a stopping time σ for which $\overline{Z}_\sigma = \overline{f}_\sigma$; that is, until the optimality criterion in Proposition 5.4.5 is satisfied. However, he would also choose $\sigma \leq \nu$, where ν is the largest optimal stopping time defined in (5.24), since otherwise the strategy θ would, at times greater than $t > \nu$, yield value $V_t(\theta) > Z_t S^0_t$ by (5.28). Thus, for any optimal exercise time σ, we need to have $Z_{t \wedge \sigma} = V_{t \wedge \sigma}$, so that Z^σ is a martingale. This means that the second condition in Proposition 5.4.5 holds, so that σ is optimal for the stopping problem solved by the Snell envelope. (Note that the same considerations apply to the option writer: if the buyer exercises at a non-optimal time τ, the strategy θ provides an arbitrage opportunity for the option writer since either $A_\tau > 0$ or $Z_\tau > \overline{f}_\tau$, so that $V_\tau(\theta) - f_\tau = Z_\tau S^0_\tau + A_\tau - f_\tau > 0$.)

We have proved the following theorem.

Theorem 5.5.1. *A stopping time* $\hat{\tau} \in \mathcal{T}$ *is an optimal exercise time for the American option* $(f_t)_{t \in \mathbb{T}}$ *if and only if*

$$E_Q\left((1+r)^{-\hat{\tau}} f_{\hat{\tau}}\right) = \sup_{\tau \in \mathcal{T}} E_Q\left((1+r)^{-\tau} f_\tau\right). \tag{5.32}$$

Remark 5.5.2. We showed by an arbitrage argument in Chapter 1 that American options are more valuable than their European counterparts in general but that for a simple call option there is no advantage in early exercise, so that the American and European call options have the same value. Using the theory of optimal stopping, we can recover these results from the martingale properties of the Snell envelope. Indeed, if $f_t = (S_t^1 - K)^+$ is an American call option with strike K on $\mathbb{T}$, then its discounted value process is given by the Q-supermartingale (Z_t) as in (5.26). Now if $\overline{C}_t$ is the discounted time t value of the European option $C_T = (S_T^1 - K)^+$, then $C_T = f_T$, so that

$$Z_t \geq E_Q\left(Z_T \,|\, \mathcal{F}_t\right) = E_Q\left(\overline{f}_T \,|\, \mathcal{F}_t\right) = E_Q\left(\overline{C}_T \,|\, \mathcal{F}_t\right) = \overline{C}_t. \tag{5.33}$$

This shows that the value process of the American call option dominates that of the European call option.

On the other hand, for these call options, we have $C_t \geq f_t = (S_t^1 - K)^+$, as we saw in (1.23), and hence the Q-martingale $(\overline{C}_t)$ dominates $(\overline{f}_t)$. It is therefore a supermartingale dominating $(\overline{f}_t)$ and, by the definition of the Snell envelope, (Z_t) is the *smallest* supermartingale with this property. We conclude that $\overline{C}_t \geq Z_t$ for all $t \in \mathbb{T}$. Hence $\overline{C}_t = Z_t$, and so the value processes of the two options coincide.

5.6   Consumption-Investment Strategies

Extended 'Self-Financing' Strategies

In the study of American options in Chapter 8, and especially in studying continuous-time consumption-investment problems in Chapter 10, we shall extend the concept of 'self-financing' strategies by allowing for potential consumption. In the present discrete-time setting, the basic concepts appear more transparent, and we outline them briefly here in preparation for the technically more demanding discussion in Chapter 10.

Assume that we are given a price process $\left\{S_t^i : i = 0, 1, \ldots, d\right\}_{t=0,1,\ldots,T}$ on a stochastic basis $(\Omega, \mathcal{F}, P, \mathbb{T}, \mathbb{F})$. For any process X, the discounted version is denoted by $\overline{X}$, where $\overline{X}_t = \beta_t X_t$ as usual.

If $c = (c_t)_{t \in \mathbb{T}}$ denotes a 'consumption process' (which, if c_t is negative, equates to additional investment at time t), then the self-financing constraint for strategies (i.e., $(\Delta \theta_t) \cdot S_{t-1} = 0$) should be amended to read

$$(\Delta \theta_t) \cdot S_{t-1} + c_t = 0. \tag{5.34}$$

An investment-consumption strategy is a pair (θ, c) of *predictable* processes that satisfies (5.34), and their associated value or *wealth process* V is given by $V_t = \theta_t \cdot S_t$ as before. Also define the cumulative consumption process C by $C_t = \sum_{u=1}^{t} c_u$. The constraint (5.34) is trivially equivalent to each of the following (for all $t > 0$):

$$\Delta V_t = \theta_t \cdot \Delta S_t - c_t, \tag{5.35}$$

$$V_t = V_0 + \sum_{u=1}^{t} \theta_u \cdot \Delta S_u - C_t, \tag{5.36}$$

$$\overline{V}_t = V_0 + \sum_{u=1}^{t} \theta_u \cdot \Delta \overline{S}_u - \overline{C}_t. \tag{5.37}$$

Assume from now on that the market model $(\Omega, \mathcal{F}, P, \mathbb{T}, \mathbb{F}, S)$ is viable and complete and that Q is the unique EMM for $\overline{S}$. Assume further that C is a pure consumption process; that is, $\Delta C_t = c_t \geq 0$ for all $t \in \mathbb{T}$. Then for a strategy (θ, c) as previously, the discounted value process $\overline{V}$ satisfies, for $t \in \mathbb{T}$,

$$E_Q\left(\Delta \overline{V}_t \,|\, \mathcal{F}_{t-1}\right) = E_Q\left((\theta_t \cdot \Delta \overline{S}_t - \overline{c}_t)\,|\,\mathcal{F}_{t-1}\right) = -\overline{c}_t \leq 0$$

since $\overline{S}$ is a Q-martingale and $\overline{c}_t \geq 0$. In summary, we have the following.

Proposition 5.6.1. *For every consumption strategy (θ, c) satisfying (5.34), the discounted value process $\overline{V}$ is a Q-supermartingale.*

Construction of Hedging Strategies

Suppose that $U = (U_t)$ is an adapted process whose discounted version $\overline{U}$ is a Q-supermartingale. Then we can use the increasing process in its Doob decomposition to define a consumption process c and a self-financing strategy θ such that the pair (θ, c) satisfies (5.34) and has value process U.

To do this, write $\overline{U} = \overline{M} - \overline{A}$ for the Doob decomposition of $\overline{U}$, so that $\overline{A}_0 = 0$ and $\overline{M}$ is a Q-martingale. As the market is complete, the contingent claim $M_T = S_T^0 \overline{M}_T$ can be generated by a unique self-financing strategy θ, so that $\theta_T \cdot S_T = M_T$; that is, $\theta_T \cdot \overline{S}_t T = \overline{M}_T$. As $\overline{M}$ is a martingale, we have $\overline{M}_t = E_Q\left(\theta_T \cdot \overline{S}_T \,|\, \mathcal{F}_t\right)$ for all $t \in \mathbb{T}$. Thus

$$\overline{U}_t = E_Q\left(\theta_T \cdot \overline{S}_T \,|\, \mathcal{F}_t\right) - \overline{A}_t \text{ for } t \in \mathbb{T},$$

so that

$$U_t = S_t^0 E_Q\left(\theta_T \cdot \overline{S}_T \,|\, \mathcal{F}_t\right) - A_t \text{ for } t \in \mathbb{T},$$

where the process $A_t = S_t^0 \overline{A}_t$ is increasing and has $A_0 = 0$. Since θ is self-financing, the final portfolio value has the form

$$\theta_T \cdot \overline{S}_T = \theta_0 \cdot S_0 + \sum_{u=1}^{T} \theta_u \cdot \Delta \overline{S}_u \text{ for } t \in \mathbb{T},$$

so that

$$E_Q \left(\theta_T \cdot \overline{S}_T \,|\mathcal{F}_t \right) = \theta_0 \cdot S_0 + \sum_{u=1}^{t} \theta_u \cdot \Delta \overline{S}_u \text{ for } t \in \mathbb{T}. \qquad (5.38)$$

Choosing C so that $\overline{A}_t = \sum_{u=1}^{t} \overline{c}_u$ and $C_0 = 0 = \overline{A}_0$, we see that $c_u = S_{u-1}^0(\Delta \overline{A}_u)$ meets the requirement and that C is predictable and non-negative (as A is increasing). Inductively, $\overline{A}_t = \sum_{u=1}^{t} \overline{C}_u$ yields

$$\overline{A}_{t+1} = A_t + \Delta \overline{A}_{t+1} = \sum_{u=1}^{t+1} \overline{c}_u,$$

and by (5.37) we obtain $\overline{V}_t = \overline{U}_t$; that is, $V_t = U_t$ for the value process associated with (θ, c).

Guided by our discussion of American options, we now call a consumption strategy (θ, c) a *hedge* for a given claim (i.e., an adapted process) $X = (X_t)$ if $V_t(\theta) \geq X_t$ for all $t \in \mathbb{T}$. Writing Z for the Snell envelope of $\overline{X}$, the supermartingale Z dominates $\overline{X}$ and can be used as the process $\overline{U}$ in the previous discussion. Thus we can find a hedging strategy (θ, c) for X and obtain

$$V_t(\theta) = U_t = S_t^0 Z_t \geq X_t \text{ for } t \in \mathbb{T}, \qquad V_T(\theta) = S_T^0 Z_T = X_T.$$

As Z is the smallest supermartingale dominating $\overline{X}$, it follows that *any* hedge (θ', c') for X must have a value process dominating $S^0 Z$.

Financing Consumption

Suppose an investor is given an initial endowment $x > 0$ and follows a consumption strategy $c = (c_t)_{t \in \mathbb{T}}$ (a non-negative predictable process). How can this consumption be *financed* by a self-financing investment strategy utilising the endowment x?

It seems natural to say that c *can be financed* (or is budget-feasible) from the endowment x provided that there is a predictable process $\theta = (\theta^0, \theta^1, \theta^2, \ldots, \theta^d)$ for which (θ, c) is a consumption strategy with $V_0(\theta) = x$ and $V_t(\theta) \geq 0$ for all $t \in \mathbb{T}$. By (5.37), we require that

$$\overline{V}_t(\theta) = x + \sum_{u=1}^{t} \theta_u \cdot \Delta \overline{S}_u - \sum_{u=1}^{t} \overline{c}_u \geq 0 \qquad (5.39)$$

if such a strategy θ exists. But $\overline{S}$ is a Q-martingale, so, taking expectations, (5.39) becomes, with $C = \sum_{u=1}^{t} c_u$ as cumulative consumption,

$$E_Q \left(\overline{C}_t \right) = E_Q \left(\sum_{u=1}^{t} \overline{c}_u \right) \leq x. \qquad (5.40)$$

The budget constraint (5.40) is therefore necessary if the consumption C is to be financed by the endowment x. It is also sufficient as shown in the following.

Given a consumption process C with $c_t = \Delta C_t$, define the process $\overline{U}_t = x - \overline{C}_t$. Since C is predictable and $c_{t+1} \geq 0$,

$$\overline{U}_{t+1} = E_Q \left(\overline{U}_{t+1} \,|\, \mathcal{F}_t \right) \leq \overline{U}_t,$$

so that $\overline{U}$ is a supermartingale. By (5.40), $E_Q \left(\overline{U}_t \right) \geq 0$ for all $t \in \mathbb{T}$. But then we can find a hedging strategy θ for the claim $X = 0$ with $V_0(\theta) = x$ and $V_t(\theta) \geq 0$ for all t. We have proved the following.

Theorem 5.6.2. *The consumption process C can be financed by an initial endowment x if and only if the constraint (5.40) is satisfied.*

Chapter 6

Continuous-Time Stochastic Calculus

6.1 Continuous-Time Processes

In this and the succeeding chapters, the time parameter takes values in either a finite interval $[0, T]$ or the infinite intervals $[0, \infty)$, $[0, \infty]$. We denote the time parameter set by $\mathbb{T}$ in each case.

Filtrations and Stopping Times

Suppose $(\Omega, \mathcal{F}, P)$ is a probability space. As before, we use the concept of a filtration on $(\Omega, \mathcal{F}, P)$ to model the acquisition of information as time evolves. The definition of a filtration is as in Chapter 2 and now takes account of the change in the time set $\mathbb{T}$.

Definition 6.1.1. A *filtration* $\mathbb{F} = (\mathcal{F}_t)_{t \in \mathbb{T}}$ is an increasing family of sub-σ-fields of $\mathcal{F}$(i.e., $\mathcal{F}_t \subset \mathcal{F}$ and if $s \leq t$, then $\mathcal{F}_s \subset \mathcal{F}_t$).

We assume that $\mathbb{F}$ satisfies the 'usual conditions'. This means the filtration $\mathbb{F}$ is:

(a) *complete*; that is, every null set in $\mathcal{F}$ belongs to $\mathcal{F}_0$ and thus to each $\mathcal{F}_t$, and

(b) *right continuous*; that is, $\mathcal{F}_t = \bigcap_{s > t} \mathcal{F}_s$ for $t \in \mathbb{T}$.

Remark 6.1.2. Just as in the discrete case, $\mathcal{F}_t$ represents the history of some process or processes up to time t. However, all possible histories must be allowed. If an event $A \in \mathcal{F}$ is $\mathcal{F}_t$-measurable, then it only depends on what has happened to time t. Unlike the situation we discussed in Chapter 2, new information can arrive at any time $t \in [0, T]$ (or even $t \in [0, \infty)$), and the filtration consists of an uncountable collection of σ-fields. The right continuity assumption is specific to this situation.

Definition 6.1.3. Suppose the time parameter $\mathbb{T}$ is $[0, \infty]$ (or $[0, \infty)$, or $[0, T]$). A random variable τ taking values in $\mathbb{T}$ is a *stopping time* if, for every $t \geq 0$,

$$\{\tau \leq t\} \in \mathcal{F}_t.$$

Remark 6.1.4. Consequently, the event $\{\tau \leq t\}$ depends only on the history up to time t. The first time a stock price reaches a certain level is a stopping time, as is, say, the first time the price reaches a certain higher level after dropping by a specified amount. However, the last time, before some given date, at which the stock price reaches a certain level is not a stopping time because to say it is the 'last time' requires information about the future. Note that in the continuous-time setting it does not make sense to replace the condition $\{\tau \leq t\} \in \mathcal{F}_t$ by $\{\tau = t\} \in \mathcal{F}_t$. Many of the properties of stopping times carry over to this setting, however.

Just as in Chapter 5, a constant random variable, $T(\omega) = t$ for all $\omega \in \Omega$, is a stopping time. If T is any stopping time, then $T + s$ is also a stopping time for $s \geq 0$.

We continue with some basic properties of stopping times.

Proposition 6.1.5. *If S and T are stopping times, then $S \wedge T$ and $S \vee T$ are also stopping times. Consequently, if $(T_n)_{n \in \mathbb{N}}$ is a sequence of stopping times, then $\wedge_n T_n = \inf_n T_n$ and $\vee_n T_n = \sup_n T_n$ are stopping times.*

Proof. The proof is identical to that given in Example 5.2.3 for the discrete case. □

Definition 6.1.6. Suppose T is a stopping time with respect to the filtration $(\mathcal{F}_t)$. Then the σ-field $\mathcal{F}_T$ of events occurring up to time T is the collection of events $A \in \mathcal{F}$ satisfying

$$A \cap \{T \leq t\} \in \mathcal{F}_t \text{ for all } t \in \mathbb{T}.$$

Exercise 6.1.7. Prove that $\mathcal{F}_T$ is a σ-field.

One then can establish the following (compare with Exercise 5.2.6 for the discrete case).

Theorem 6.1.8. *Suppose S, T are stopping times.*

a) If $S \leq T$, then $\mathcal{F}_S \subset \mathcal{F}_T$.

b) If $A \in \mathcal{F}_S$, then $A \cap \{S \leq T\} \in \mathcal{F}_T$.

Proof. (a) Suppose that $B \in \mathcal{F}_S$. Then, for $t \in \mathbb{T}$,

$$B \cap \{T \leq t\} = B \cap \{S \leq t\} \cap \{T \leq t\} \in \mathcal{F}_t.$$

(b) Suppose that $A \in \mathcal{F}_S$. For $t \in \mathbb{T}$, we have

$$A \cap \{S \leq T\} \cap \{T \leq t\} = (A \cap \{S \leq t\}) \cap \{T \leq t\} \cap \{S \wedge t \leq T \wedge t\}.$$

Each of the three sets on the right-hand side is in $\mathcal{F}_t$: the first because $A \in \mathcal{F}_S$, the second because T is a stopping time, and the third because $S \wedge t$ and $T \wedge t$ are $\mathcal{F}_t$-measurable random variables. □

Definition 6.1.9. A continuous-time *stochastic process* X taking values in a measurable space $(E, \mathcal{E})$ is a family of random variables $\{X_t\}$ defined on $(\Omega, \mathcal{F}, P)$, indexed by t, that take values in $(E, \mathcal{E})$.

That is, for each t, we have a random variable $X_t(\cdot)$ with values in E. Alternatively, for each ω (i.e., fixing ω and letting t vary), we have a sample path $X.(\omega)$ of the process.

Remark 6.1.10. X could represent the evolution of the price of oil or the price of a stock over time.

For some (future) time t, $X_t(\omega)$ is a random quantity, a random variable. Each ω represents a 'state of the world' corresponding to which there is a price $X_t(\omega)$. Conversely, fixing ω means one realization of the world, as time evolves, is considered. This gives a realization, or *path*, of the price $X.(\omega)$ as a function of time t.

Equivalence of Processes

A natural question is to ask when two stochastic processes model the same phenomenon. We discuss several possible definitions for stochastic processes defined on a probability space $(\Omega, \mathcal{F}, P)$ and taking values in the measurable space $(E, \mathcal{E})$.

The weakest notion of equivalence of processes reflects the fact that in practice one can only observe a stochastic process at *finitely* many instants. Assume for simplicity that $E = \mathbb{R}$ and $\mathcal{E}$ is the Borel σ-field $\mathcal{B}$ on $\mathbb{R}$. Then we can form the family of *finite-dimensional distributions* of the process $X = (X_t)_{t \geq 0}$ by considering the probability that for $n \in \mathbb{N}$, times $t_1, t_2, \ldots, t_n \in \mathbb{T}$ and a Borel set $A \subset \mathbb{R}^n$, the random vector $(X_{t_1}, X_{t_2}, \ldots, X_{t_n})$ takes values in A. Indeed, set

$$\phi^X_{t_1, t_2, \ldots, t_n}(A) = P\left(\{\omega \in \Omega : (X_{t_1}(\omega), X_{t_2}(\omega), \ldots, X_{t_n}(\omega)) \in A\}\right).$$

For each family $\{t_1, t_2, \ldots, t_n\}$, this defines $\phi^X_{t_1, t_2, \ldots, t_n}$ as a measure on $\mathbb{R}^n$. We say that two processes X and Y are *equivalent* (or have the same *law*) if their families of finite-dimensional distributions coincide, and then we write $X \sim Y$.

Note that the preceding does *not* require Y to be defined on the *same* probability space as X. This means that we can avoid complicated questions about the 'proper' probability space for a particular problem since only the finite-dimensional distributions and not the full realisations of the process (i.e., the various random 'paths' it traces out) are relevant for our description of the probabilities concerned. It turns out that if we consider the process as a map $X : \Omega \mapsto \mathbb{R}^\mathbb{T}$ (i.e., $\omega \mapsto X(\cdot, \omega)$) and we stick to *Borel sets* A in $\mathbb{R}^\mathbb{T}$, then the finite-dimensional distributions give us sufficient information to identify a canonical version of the process, up to equivalence. (This is the famous *Kolmogorov extension theorem*; see [194, Theorem 2.2]).

However, at least when $\mathbb{T}$ is uncountable, most of the interesting sets in $\mathbb{R}^{\mathbb{T}}$ are not Borel sets, so that we need a somewhat stronger concept of 'equivalence' that 'fixes' the paths of our process X tightly enough. Two such definitions are now given; each of them requires the two processes concerned to be defined on the same probability space.

Definition 6.1.11. Suppose $(X_t)_{t\geq 0}$ and $(Y_t)_{t\geq 0}$ are two processes defined on the same probability space $(\Omega, \mathcal{F}, P)$ and taking values in $(E, \mathcal{E})$. The process $\{Y_t\}$ is said to be a *modification* of (X_t) if

$$X_t = Y_t \text{ a.s. for all } t \in \mathbb{T};$$

i.e.,

$$P(X_t = Y_t) = 1 \text{ for all } t \in \mathbb{T}.$$

Definition 6.1.12. The processes (X_t) and (Y_t) defined as in Definition 6.1.11 are said to be *indistinguishable* if, for almost every $\omega \in \Omega$,

$$X_t(\omega) = Y_t(\omega) \text{ for all } t \in \mathbb{T}. \tag{6.1}$$

The difference between Definition 6.1.11 and Definition 6.1.12 is that in Definition 6.1.11 the set of zero measure on which X_t and Y_t may differ may depend on t, whereas in Definition 6.1.12 there is a single set of zero measure outside of which (6.1) holds. When the time index set is countable, the two definitions coincide.

Exercise 6.1.13. A process X is *right-continuous* if for almost every ω the map $t \mapsto X_t(\omega)$ is right-continuous. Show that if the processes X and Y are right-continuous and one is a modification of the other, then they are indistinguishable.

Definition 6.1.14. Suppose $A \subset [0, \infty] \times \Omega$ and that $\mathbf{1}_A(t, \omega) = \mathbf{1}_A$ is the indicator function of A; that is,

$$\mathbf{1}_A(t, \omega) = \begin{cases} 1 & \text{if } (t, \omega) \in A, \\ 0 & \text{if } (t, \omega) \notin A. \end{cases}$$

Then A is called *evanescent* if $\mathbf{1}_A$ is indistinguishable from the zero process.

Exercise 6.1.15. Show that A is evanescent if the projection

$$\{\omega \in \Omega : \text{there exists } t \text{ with } (t, \omega) \in A\}$$

of A onto Ω is a set of measure zero.

Finally, we recall the following.

Definition 6.1.16. A process $(t, \omega) \mapsto X_t(\omega)$ from $([0, T] \times \Omega, \mathcal{B}([0, T] \times \mathcal{F}))$ to a measurable space $(E, \mathcal{E})$ is said to be *progressively measurable*, or *progressive*, if for every $t \in [0, T]$ the map $(s, \omega) \to X_s(\omega)$ of $[0, T] \times \Omega$ to E is measurable with respect to the σ-field $\mathcal{B}([0, \mathbb{T}]) \times \mathcal{F}_t$.

6.2 Martingales

Definition 6.2.1. Suppose $(\mathcal{F}_t)_{t\geq 0}$ is a filtration of the measurable space $(\Omega, \mathcal{F})$ and (X_t) is a stochastic process defined on $(\Omega, \mathcal{F})$ with values in $(E, \mathcal{E})$. Then X is said to be *adapted* to $(\mathcal{F}_t)$ if X_t is $\mathcal{F}_t$-measurable for each t.

The random process that models the concept of randomness in the most fundamental way is a martingale; we now give the continuous-time definition for $t \in [0, \infty]$; the discrete-time analogue was discussed in Chapters 2 through 5.

Definition 6.2.2. Suppose $(\Omega, \mathcal{F}, P)$ is a probability space with a filtration $(\mathcal{F}_t)_{t\geq 0}$. A real-valued adapted stochastic process (M_t) is said to be a *martingale* with respect to the filtration $(\mathcal{F}_t)$ if $E\,|M_t| \leq \infty$ for all t and for all $s \leq t$

$$E\,(M_t\,|\mathcal{F}_s\,) = M_s.$$

If the equality is replaced by $\leq$ (resp. $\geq$), then (M_t) is said to be a supermartingale (resp. submartingale).

Remark 6.2.3. A martingale is a purely random process in the sense that, given the history of the process so far, the expected value of the process at some later time is just its present value. Note that in particular

$$E\,(M_t) = E\,(M_0) \text{ for all } t \geq 0.$$

Brownian Motion

The most important example of a continuous-time martingale is a Brownian motion. This process is named for Robert Brown, a Scottish botanist who studied pollen grains in suspension in the early nineteenth century. He observed that the pollen was performing a very random movement and thought this was because the pollen grains were alive. We now know this rapid movement is due to collisions at the molecular level.

Definition 6.2.4. A *standard Brownian motion* $(B_t)_{t\geq 0}$ is a real-valued stochastic process that has continuous sample paths and stationary independent Gaussian increments. In other words,

a) $B_0 = 0$ a.s.

b) $t \to B_t(\omega)$ is continuous a.s.

c) For $s \leq t$, the increment $B_t - B_s$ is a Gaussian random variable that has mean 0, variance $t-s$, and is independent of $\mathcal{F}_s = \sigma\,\{B_u : u \leq s\}$.

We can immediately establish the following.

Theorem 6.2.5. *Suppose (B_t) is a standard Brownian motion with respect to the filtration $(\mathcal{F}_t)_{t\geq 0}$. Then*

a) $(B_t)_{t\geq0}$ is an $\mathcal{F}_t$-martingale.

b) $\left(B_t^2 - t\right)_{t\geq0}$ is an $\mathcal{F}_t$-martingale.

c) $\left(e^{\sigma B_t - \frac{\sigma^2}{2}t}\right)_{t\geq0}$ is an $\mathcal{F}_t$-martingale.

Proof.   a) Since $B_t - B_s$ is independent of $\mathcal{F}_s$ for all $s \leq t$, we have

$$E\left(B_t - B_s \,|\mathcal{F}_s\right) = E\left(B_t - B_s\right) = 0.$$

Consequently, $E\left(B_t \,|\mathcal{F}_s\right) = B_s$ a.s.

b) For $\left(B_t^2 - t\right)$, we have

$$E\left(B_t^2 - B_s^2 \,|\mathcal{F}_s\right) = E\left((B_t - B_s)^2 + 2B_s(B_t - B_s)\,|\mathcal{F}_s\right)$$
$$= E\left((B_t - B_s)^2 \,|\mathcal{F}_s\right) + 2B_s E\left((B_t - B_s)\,|\mathcal{F}_s\right). \qquad (6.2)$$

The second term in (6.2) is zero by the first part. Independence implies that

$$E\left((B_t - B_s)^2 \,|\mathcal{F}_s\right) = E\left((B_t - B_s)^2\right) = t - s.$$

Therefore $E\left(B_t^2 - t\,|\mathcal{F}_s\right) = B_s^2 - s$.

c) If Z is a standard normal random variable, with density $\frac{1}{\sqrt{2\pi}}e^{-\frac{x^2}{2}}$, then

$$E\left(e^{\lambda Z}\right) = \frac{1}{\sqrt{2\pi}}\int_{-\infty}^{\infty} e^{\lambda x}e^{-\frac{x^2}{2}}\,dx = e^{\frac{\lambda^2}{2}} \text{ for } \lambda \in \mathbb{R}.$$

For $s < t$, by independence and stationarity, we have

$$E\left(e^{\sigma B_t - \frac{\sigma^2}{2}t}\,|\mathcal{F}_s\right) = e^{\sigma B_s - \frac{\sigma^2}{2}t}E\left(e^{\sigma(B_t - B_s)}\,|\mathcal{F}_s\right)$$
$$= e^{\sigma B_s - \frac{\sigma^2}{2}t}E\left(e^{\sigma(B_t - B_s)}\right)$$
$$= e^{\sigma B_s - \frac{\sigma^2}{2}t}E\left(e^{\sigma B_{t-s}}\right).$$

Now σB_{t-s} is $N\left(0, \sigma^2(t - s)\right)$; that is, if Z is $N(0,1)$ as previously, the random variable σB_{t-s} has the same law as $\sigma\sqrt{t - s}Z$ and

$$E\left(e^{\sigma B_{t-s}}\right) = E\left(e^{\sigma\sqrt{t-s}Z}\right) = e^{\frac{\sigma^2}{2}(t-s)}.$$

Therefore

$$E\left(e^{\sigma B_t - \frac{\sigma^2}{2}t}\,|\mathcal{F}_s\right) = e^{\sigma B_s - \frac{\sigma^2}{2}s} \text{ a.s. for } s < t.$$

$\square$

Conversely, we prove in Theorem 6.4.16 that a continuous process that satisfies the first two statements in Theorem 6.2.5 is, in fact, a Brownian motion. (Indeed, the third statement characterises a Brownian motion.)

Uniform Integrability and Limit Theorems

We discussed the role of uniform integrability in some detail in the discrete-time setting of Chapter 5. Here we review briefly how these ideas carry over to continuous-time martingales. Definition 5.3.1 immediately prompts the following.

Definition 6.2.6. A martingale $\{M_t\}$ with $t \in [0, \infty)$ (or $t \in [0, T]$) is said to be uniformly integrable if the set of random variables $\{M_t\}$ is uniformly integrable.

Remark 6.2.7. If $\{M_t\}$ is a uniformly integrable martingale on $[0, \infty)$, then $\lim M_t = M_\infty$ exists a.s. as we proved for the discrete case in Chapter 5.

Again, a consequence of $\{M_t\}$ being a uniformly integrable martingale on $[0, \infty)$ is that $M_\infty = \lim M_t$ in the L^1 norm; i.e., $\lim_t \|M_t - M_\infty\|_1 = 0$. In this case, $\{M_t\}$ is a martingale on $[0, \infty]$ and

$$M_t = E\left(M_\infty \,|\, \mathcal{F}_t\right) \quad \text{a.s. for all } t.$$

We again say that M is *closed* by the random variable M_∞.

Recall from the examples following Definition 5.3.1 that if a class $\mathcal{K}$ of random variables is in $L^1(\Omega, \mathcal{F}, P)$ and is L^p-bounded for some $p > 1$, then $\mathcal{K}$ is uniformly integrable.

Notation 6.2.8. Write $\mathcal{M}$ for the set of uniformly integrable martingales.

An important concept is that of 'localization'. If $\mathcal{C}$ is a class of processes, then $\mathcal{C}_{\ell oc}$ is the set of processes defined as follows. We say that $X \in \mathcal{C}_{\ell oc}$ if there is an increasing sequence $\{T_n\}$ of stopping times such that $\lim T_n = \infty$ a.s. and $X_{t \wedge T_n} \in \mathcal{C}$. For example, $\mathcal{C}$ might be the bounded processes, or the processes of bounded variation.

Notation 6.2.9. $\mathcal{M}_{\ell oc}$ denotes the set of local martingales.

The defining relation for martingales, $E\left(M_t \,|\, \mathcal{F}_s\right) = M_s$, can again be extended to stopping times. This result, which is the analogue of Theorem 5.3.9, is again known as Doob's optional stopping theorem since it says the martingale equality is preserved even if (non-anticipative) random stopping rules are allowed. A complete proof of this result in continuous time can be found in [109, Theorem 4.12, Corollary 4.13]. Note that our discussion of the discrete case in Chapter 5 showed how the extension from bounded to more general stopping times required the martingale convergence theorem and conditions under which a supermartingale is closed by an L^1-function. This condition is also required in the following.

Theorem 6.2.10. *Suppose $(M_t)_{t \geq 0}$ is a right-continuous supermartingale (resp. submartingale) with respect to the filtration $(\mathcal{F}_t)$. If S and T are two $(\mathcal{F}_t)$-stopping times such that $S \leq T$ a.s., then*

$$E\left(M_T \,|\, \mathcal{F}_S\right) \leq M_S \quad a.s. \qquad (resp., \ E\left(M_T \,|\, \mathcal{F}_S\right) \geq M_S \ a.s.).$$

Corollary 6.2.11. *In particular, if $(M_t)_{t \geq 0}$ is a right-continuous martingale and S, T are $(\mathcal{F}_t)$-stopping times with $S \leq T$, then*

$$E\left(M_T \mid \mathcal{F}_S\right) = M_S \quad \text{a.s.}$$

Remark 6.2.12. Note that, if T is any $(\mathcal{F}_t)$-stopping time, then $E\left(M_T\right) = E\left(M_0\right)$.

The following is a consequence of the optional stopping theorem. Note that we write $x^+ = \max\{x, 0\}$ and $x^- = \max\{-x, 0\}$.

Lemma 6.2.13. *Suppose $X_t, t \in [0, \infty]$ is a supermartingale. Then*

$$\alpha P\left(\left(\inf_t X_t\right) \leq -\alpha\right) \leq \sup_t E\left(X_t^-\right) \quad \text{for all } \alpha \geq 0.$$

Proof. Write

$$S(\omega) = \inf\{t : X_t(\omega) \leq -\alpha\}$$

and $S_t = S \wedge t$. Using the optional stopping theorem (Theorem 6.2.10), we have

$$E\left(X_{S_t}\right) \geq E\left(X_t\right).$$

Therefore

$$E\left(X_t\right) \leq -\alpha P\left\{\inf_{s \leq t} X_s \leq -\alpha\right\} + \int_{\{\inf_{s \leq t} X_s > -\alpha\}} X_t dP;$$

that is,

$$\alpha P\left\{\inf_{s \leq t} X_s \leq -\alpha\right\} \leq E\left(-X_t\right) + \int_{\{\inf_{s \leq t} X_s > -\alpha\}} X_t dP$$

$$= \int_{\{\inf_{s \leq t} X_s \leq -\alpha\}} -X_t dP$$

$$\leq E\left(X_t^-\right). \tag{6.3}$$

Letting $t \to \infty$ in (6.3), the result follows. $\qquad\square$

As a consequence, we can deduce Doob's maximal theorem.

Theorem 6.2.14. *Suppose $(X_t)_{t \in [0, \infty]}$ is a martingale. Then*

$$\alpha P\left\{\sup_t |X_t| \geq \alpha\right\} \leq \sup_t \|X_t\|_1 \quad \text{for all } \alpha \geq 0.$$

Proof. From Jensen's inequality (see Proposition 5.3.4), if X is a martingale, then $Y_t = -|X_t|$ is a (negative) supermartingale with $\|Y_t\|_1 = \|X_t\|_1 = E\left(Y_t^-\right)$. In addition,

$$\left\{\inf_t Y_t \leq -\alpha\right\} = \left\{\sup_t |X_t| \geq \alpha\right\},$$

so the result follows from Lemma 6.2.13. $\qquad\square$

Before proving Doob's L^p inequality, we first establish the following result.

Theorem 6.2.15. *Suppose X and Y are two positive random variables defined on the probability space $(\Omega, \mathcal{F}, P)$ such that $X \in L^p$ for some $p, 1 < p < \infty$, and*

$$\alpha P(\{Y \geq \alpha\}) \leq \int_{\{Y \geq \alpha\}} X dP \text{ for all } \alpha > 0.$$

Then

$$\|Y\|_p \leq q \|X\|_p \,,$$

where $\frac{1}{p} + \frac{1}{q} = 1$.

Proof. Let $F(\lambda) = P(Y > \lambda)$ be the complementary distribution function of Y. Using integration by parts,

$$
\begin{aligned}
E\left(Y^p\right) &= -\int_0^\infty \lambda^p dF(\lambda) \\
&= \int_0^\infty F(\lambda) d(\lambda^p) - \lim_{h \to \infty} \left(\lambda^p F(\lambda)\right)_0^h \\
&\leq \int_0^\infty F(\lambda) d(\lambda^p) \\
&\leq \int_0^\infty \lambda^{-1} \left(\int_{\{Y \geq \lambda\}} X dP\right) d(\lambda^p) \qquad \text{by hypothesis} \\
&= E\left(X \int_0^Y \lambda^{-1} d(\lambda^p)\right) \qquad \text{by Fubini's theorem} \\
&= \left(\frac{p}{p-1}\right) E\left(XY^{p-1}\right) \\
&\leq q \|X\|_p \left\|Y^{p-1}\right\|_q \qquad \text{by Hölder's inequality.}
\end{aligned}
$$

That is,

$$E\left(Y^p\right) \leq q \|X\|_p \left(E\left(Y^{pq-q}\right)\right)^{\frac{1}{q}} .$$

If $\|Y\|_p$ is finite, the result follows immediately because $pq - q = p$. Otherwise, consider the random variable

$$Y_k = Y \wedge k, k \in N.$$

Then $Y_k \in L^p$ and Y_k also satisfies the hypotheses. Therefore

$$\|Y_k\|_p \leq q \|X\|_p .$$

Letting $k \to \infty$, the result follows. $\qquad \square$

Theorem 6.2.16. *Suppose $(X_t)_{t \geq 0}$ is a right-continuous positive submartingale. Write $X^*(\omega) = \sup_t X_t(\omega)$. Then, for $1 < p \leq \infty$, $X^* \in L^p$ if and only if*

$$\sup_t \|X_t\|_p < \infty.$$

Also, for $1 < p < \infty$ and $q^{-1} = 1 - p^{-1}$,

$$\|X^*\|_p \leq q \sup_t \|X_t\|_p.$$

Proof. When $p = \infty$, the first part of the theorem is immediate because

$$\sup_t \|X_t\|_\infty = B < \infty$$

implies that $X_t \leq B$ a.s. for all $t \in [0, \infty]$. The right-continuity is required to ensure there is a single set of measure zero outside which this inequality is satisfied for all t. Also, for $1 < p < \infty$, if $X^* \in L^p$, then

$$\sup_t \|X_t\|_p \leq \|X^*\|_p < \infty.$$

As in Section 5.3, the random variables (X_t) are uniformly integrable, so from [109, Corollaries 3.18 and 3.19] we have that

$$X_\infty(\omega) = \lim_{t \to \infty} X_t(\omega)$$

exists a.s. Using Fatou's lemma, we obtain

$$E\left(\lim_t X_t^p\right) \leq \liminf_t E\left(X_t^p\right) \leq \sup_t E\left(X_t^p\right) < \infty.$$

Therefore $X_\infty \in L^p$ and $\|X_\infty\|_p \leq \sup_t \|X_t\|_p$.

Write $X_t^*(\omega) = \sup_{s \leq t} X_s(\omega)$. Then $\{-X_t\}$ is a supermartingale, so from inequality (6.3) in Lemma 6.2.13, for any $\alpha > 0$,

$$\alpha P\left(\inf_{s \leq t}(-X_s) \leq -\alpha\right) = \alpha P\left(X_t^* \geq \alpha\right) \leq \int_{\{X_t^* \geq \alpha\}} X_t dP \leq \int_{\{X^* \geq \alpha\}} X_t dP.$$

Letting $t \to \infty$, we have for any $\alpha > 0$,

$$\alpha P\left(X^* \geq \alpha\right) \leq \int_{\{X^* \geq \alpha\}} X_\infty dP.$$

Therefore, Theorem 6.2.15 can be applied with $Y = X^*$ and $X = X_\infty$ to obtain

$$\|X^*\|_p \leq q \|X_\infty\|_p$$

and the result follows.   □

The following important special case arises when $p = q = 2$ and the time interval is taken as $[0, T]$.

Corollary 6.2.17 (Doob's Inequality). *Suppose $(M_t)_{t \geq 0}$ is a continuous martingale. Then*

$$E\left(\sup_{0 \leq t \leq T} |M_t|^2\right) \leq 4E\left(|M_T|^2\right).$$

6.3   Stochastic Integrals

In discrete time the discounted value of a portfolio process having initial value V_0 and generated by a self-financing strategy $(H_t)_{t \geq 0}$ is given by

$$V_0 + \sum_{j=1}^{n} H_j(\overline{S}_j - \overline{S}_{j-1}).$$

Recall that under an equivalent measure the discounted price process $\overline{S}$ is a martingale. Consequently, the preceding value process is a martingale transform. The natural extension to continuous time of such a martingale transform is the stochastic integral $\int_0^t H_s d\overline{S}_s$. However, $d\overline{S} = \overline{S}\sigma dW_t$, where (W_t) is a Brownian motion. Almost all sample paths $W.(\omega)$ of Brownian motion are known to be of unbounded variation. They are therefore certainly not differentiable. The integral $\int H d\overline{S}$ cannot be defined as $\int H \frac{dS}{dt} \cdot dt$ or even as a Stieltjes integral. It can, however, be defined as the limit of suitable approximating sums in $L^2(\Omega)$.

We work initially on the time interval $[0, T]$. Suppose (W_t) is an $(\mathcal{F}_t)$-Brownian motion defined on $(\Omega, \mathcal{F}, P)$ for $t \in [0, T]$; that is, W is adapted to the filtration $(\mathcal{F}_t)$.

Simple Processes

Definition 6.3.1. A real-valued *simple process* on $[0, T]$ is a function H for which

a) there is a partition $0 = t_0 < t_1 < \ldots t_n = T$; and

b) $H_{t_0} = H_0(\omega)$ and $H_t = H_i(\omega)$ for $t \in (t_i, t_{i+1}]$, where $H_i(\cdot)$ is $\mathcal{F}_{t_i}$-measurable and square integrable. That is,

$$H_t = H_0(\omega) + \sum_{i=0}^{n-1} H_i(\omega)\mathbf{1}_{(t_i, t_{i+1}]} \text{ for } t \in [0, T].$$

Definition 6.3.2. If H is a simple process, the *stochastic integral* of H with respect to the Brownian motion (W_t) is the process defined for $t \in (t_k, t_{k+1}]$, by

$$\int_0^t H_s dW_s = \sum_{i=0}^{k-1} H_i(W_{t_{i+1}} - W_{t_i}) + H_k(W_t - W_{t_k}).$$

This can be written as a martingale transform:

$$\int_0^t H_s dW_s = \sum_{i=0}^{n} H_i(W_{t_{i+1} \wedge t} - W_{t_i \wedge t}).$$

We write $\int_0^t H \, dW = \int_0^t H_s \, dW_s$.

Note that, because $W_0 = 0$, there is no contribution to the integral at $t = 0$.

Theorem 6.3.3. *Suppose H is a simple process. Then:*

a) $\left(\int_0^t H_s \, dW_s \right)$ *is a continuous $\mathcal{F}_t$-martingale.*

b) $E \left(\left[\int_0^t H_s \, dW_s \right]^2 \right) = E \left(\int_0^t H_s^2 \, ds \right)$.

c) $E \left(\sup_{0 \le t \le T} \left| \int_0^t H_s \, dW_s \right|^2 \right) \le 4E \left(\int_0^T H_s^2 \, ds \right)$.

Proof.   a) For $t \in (t_k, t_{k+1}]$, we have

$$\int_0^t H_s \, dW_s = \sum_{i=0}^{k-1} H_i (W_{t_{i+1}} - W_{t_i}) + H_k (W_t - W_{t_k}).$$

Now $W_t(\cdot)$ is continuous a.s. in t; hence so $\int_0^t H_s \, dW_s$. Suppose that $0 \le s \le t \le T$. Recall that

$$\int_0^t H_s \, dW_s = \sum_{i=0}^{n} H_i (W_{t_{i+1} \wedge t} - W_{t_i \wedge t}),$$

where H_i is $\mathcal{F}_{t_i}$-measurable. Now if $s \le t_i$, then

$$\begin{aligned}
E \left(H_i \left[W_{t_{i+1} \wedge t} - W_{t_i \wedge t} \right] | \mathcal{F}_s \right) &= E \left(E \left(H_i \left[W_{t_{i+1} \wedge t} - W_{t_i \wedge t} \right] | \mathcal{F}_{t_i} \right) | \mathcal{F}_s \right) \\
&= E \left(H_i E \left(W_{t_{i+1} \wedge t} - W_{t_i \wedge t} | \mathcal{F}_{t_i} \right) | \mathcal{F}_s \right) \\
&= 0 = H_i \left(W_{t_{i+1} \wedge s} - W_{t_i \wedge s} \right)
\end{aligned}$$

because $t_{i+1} \wedge s = t_i \wedge s = s$. If $s \ge t_i$, then

$$\begin{aligned}
E \left(H_i \left[W_{t_{i+1} \wedge t} - W_{t_i \wedge t} \right] | \mathcal{F}_s \right) &= H_i E \left(W_{t_{i+1} \wedge t} - W_{t_i} | \mathcal{F}_s \right) \\
&= H_i \left(W_{t_{i+1} \wedge s} - W_{t_i \wedge s} \right).
\end{aligned}$$

Consequently, for $s \le t$,

$$E \left(\int_0^t H_s \, dW_s | \mathcal{F}_s \right) = \int_0^s H_u \, dW_u$$

and $\left(\int_0^t H \, dW \right)$ is a continuous martingale.

b) Now suppose $i < j$ so that $i + 1 \le j$. Then

$$E \left(H_i H_j \left(W_{t_{i+1} \wedge t} - W_{t_i \wedge t} \right) \left(W_{t_{j+1} \wedge t} - W_{t_j \wedge t} \right) \right)$$

$$= E \left(E \left(H_i H_j \left(W_{t_{i+1} \wedge t} - W_{t_i \wedge t} \right) \left(W_{t_{j+1} \wedge t} - W_{t_j \wedge t} \right) | \mathcal{F}_{t_j} \right) \right)$$
$$= E \left(H_i H_j \left(W_{t_{i+1} \wedge t} - W_{t_i \wedge t} \right) E \left(W_{t_{j+1} \wedge t} - W_{t_j \wedge t} | \mathcal{F}_{t_j} \right) \right)$$
$$= 0.$$

Also,

$$E \left(H_i^2 \left(W_{t_{i+1} \wedge t} - W_{t_i \wedge t} \right)^2 \right) = E \left(H_i^2 E \left(\left(W_{t_{i+1} \wedge t} - W_{t_i \wedge t} \right)^2 | \mathcal{F}_{t_i} \right) \right)$$
$$= E \left(H_i^2 \left(t_{i+1} \wedge t - t_i \wedge t \right) \right).$$

Consequently,

$$E \left(\left(\int_0^t H dW \right)^2 \right) = \sum_{i=0}^n E \left(H_i^2 \left(t_{i+1} \wedge t - t_i \wedge t \right) \right)$$
$$= E \left(\int_0^t H_s^2 ds \right) = \int_0^t E \left(H_s^2 \right) ds.$$

c) For the final part, apply Doob's maximal inequality, Corollary 6.2.17, to the martingale $\left(\int_0^t H_s dW_s \right)$.

$\square$

Notation 6.3.4. We write $\mathcal{H}$ for the space of processes adapted to $(\mathcal{F}_t)$ that satisfy $E \left(\int_0^T H_s^2 ds \right) < \infty$.

Lemma 6.3.5. *Suppose* $\{ H_s \} \in \mathcal{H}$. *Then there is a sequence* $\{ H_s^n \}$ *of simple processes such that*

$$\lim_{n \to \infty} E \left(\int_0^T |H_s - H_s^n|^2 ds \right) = 0.$$

Outline of the Proof. Fix $f \in \mathcal{H}$, and define a sequence of simple functions converging to f by setting

$$f_n(t, \omega) = n \int_{\frac{k-1}{n}}^{\frac{k}{n}} f(s, \omega) ds \text{ for } t \in \left[\frac{k}{n}, \frac{k+1}{n} \right].$$

If the integral diverges, replace it by 0. By Fubini's theorem, this only happens on a null set in Ω since f is integrable on $\mathbb{T} \times \Omega$.

Note that, using progressive measurability (recall Definition 6.1.16), as a random variable, the preceding integral is $\mathcal{F}_{\frac{k}{n}}$-measurable, so that f_n is a simple process as defined in Definition 6.3.1. We show in the following that $\int_0^T |f_n(t, \omega) - f(t, \omega)|^2 dt$ converges to 0 whenever $f(\cdot, \omega) \in L^2[0, T]$, and also that, for all such $\omega \in \Omega$,

$$\int_0^T |f_n(t, \omega)|^2 dt \leq \int_0^T |f(t, \omega)|^2 dt.$$

Thus the dominated convergence theorem allows us to conclude that

$$E\left(\int_0^T |f_n - f|\, dt\right) \to 0 \text{ as } n \to \infty.$$

We write $f_h = f_n$ when $h = \frac{1}{n}$. The proof now reduces to a problem in $L^2[0,T]$, namely to show that if $f \in L^2[0,T]$ is fixed, then as $h \downarrow 0$, the 'time averages' f_h defined for $t \in \left[(k-1)h, kh \wedge h^{-1}\right)$ by $f_h(t) = \frac{1}{h}\int_{(k-1)h}^{kh} f(s)ds$ and 0 outside $[h, h^{-1})$ remain L^2-dominated by f and converge to f in L^2-norm. To prove this, first consider the estimate

$$\int_0^T f_h^2(t)dt \le \sum_{k=1}^{[\frac{T}{h}]} \left|\int_{(k-1)h}^{kh} f(s)ds\right|^2,$$

which is exact if $\frac{T}{h} \in \mathbb{N}$ or $T = \infty$. Now the Schwarz inequality, applied to $1 \cdot f$, shows that each term in the latter sum is bounded above by $h \cdot \int_{(k-1)h}^{kh} f^2(s)ds$; hence the sum is bounded by $h \cdot \int_0^{[\frac{T}{h}]\cdot h} f^2(s)ds \le h \cdot \int_0^T f^2(s)ds$, which proves domination. To prove the convergence, consider $\varepsilon > 0$ and note that if f is a step function, then f_h will converge to f as $h \downarrow 0$. Since the step functions are dense in $L^2[0,T]$, choose a step function f^ε such that $\|f^\varepsilon - f\| < \varepsilon$ (with $\|\cdot\|$ denoting the norm in $L^2[0,T]$). Note that since f_h is also a step function, $f_h - f_h^\varepsilon = (f - f^\varepsilon)_h$. Moreover, by the definition of f_h, it is easy to verify that $\|f_h - f_h^\varepsilon\| \le \|f - f^\varepsilon\|$. Therefore we can write

$$\|f_h - f\| = \|f_h^\varepsilon - f^\varepsilon + (f - f^\varepsilon)_h - (f - f^\varepsilon)\| \le \|f_h^\varepsilon - f^\varepsilon\| + 2\|f - f^\varepsilon\|.$$

But the first term goes to 0 as $h \downarrow 0$ since f_h is a step function, while the second is less than 2ε. This proves the result.   □

The Integral as a Stochastic Process

Theorem 6.3.6. *Suppose* $(W_t)_{t \ge 0}$ *is a Brownian motion on the filtration* $(\mathcal{F}_t)$. *Then there is a unique linear map* I *from* $\mathcal{H}$ *into the space of continuous* $\mathcal{F}_t$-*martingales on* $[0,T]$ *such that:*

a) If H is a simple process in $\mathcal{H}$, then

$$I(H)_t = \int_0^t H_s dW_s.$$

b) If $t \le T$,

$$E\left((I(H)_t)^2\right) = E\left(\int_0^t H_s^2 ds\right).$$

The second identity is called the isometry property *of the integral.*

Proof. For H a simple process, one defines $I(H)_t = \int_0^t H_s dW_s$. Suppose $H \in \mathcal{H}$ and (H^n) is a sequence of simple processes converging to H. Then

$$I(H^{n+p} - H^n)_t = \int_0^t (H_s^{n+p} - H_s^n) dW_s$$

$$= \int_0^t H_s^{n+p} dW_s - \int_0^t H_s^n dW_s.$$

From Doob's inequality (Corollary 6.2.17),

$$E\left(\sup_{0 \le t \le T} |I(H^{n+p} - H^n)_t|^2\right) \le 4E\left(\int_0^T |H_s^{n+p} - H_s^n|^2 ds\right). \quad (6.4)$$

Consequently, there is a subsequence (H^{k_n}) such that

$$E\left(\sup_{t \le T} |I(H^{k_{n+1}})_t - I(H^{k_n})_t|^2\right) \le 2^{-n}.$$

Almost surely, the sequence of continuous functions $I(H^{k_n})_t$, $0 \le t \le T$ is uniformly convergent on $[0, T]$ to a function $I(H)_t$. Letting $p \to \infty$ in (6.4), we see that

$$E\left(\sup_{t \le T} |I(H)_t - I(H^n)_t|^2\right) \le 4E\left(\int_0^T |H_s - H_s^n|^2 ds\right).$$

This argument also implies that $I(H)$ is independent of the approximating sequence (H^n).

Now $E(I(H^n)_t | \mathcal{F}_s) = I(H^n)_s$ a.s. The integrals $\{I(H^n), I(H)\}$ belong to $L^2(\Omega, \mathcal{F}, P)$, so

$$\|E(I(H)_t | \mathcal{F}_s) - I(H)_s\|_2 \le \|E(I(H)_t | \mathcal{F}_s) - E(I(H^n)_t | \mathcal{F}_s)\|_2$$
$$+ \|E(I(H^n)_t | \mathcal{F}_s) - I(H^n)_s\|_2 + \|I(H^n)_s - I(H)_s\|_2.$$

The right-hand side can be made arbitrarily small, so $I(H)_t$ is an $\mathcal{F}_t$-martingale.

The remaining results follow by continuity and from the density in $\mathcal{H}$ of simple processes. $\qquad \square$

Notation 6.3.7. We write $I(H)_t = \int_0^t H_s dW_s$ for $H \in \mathcal{H}$.

Lemma 6.3.8. *For $H \in \mathcal{H}$,*

a) $E\left(\sup_{0 \le t \le T} \left|\int_0^t H_s dW_s\right|^2\right) \le 4E\left(\int_0^T |H|_s^2 ds\right).$

b) *If τ is an $(\mathcal{F}_t)$-stopping time such that $\tau \le T$, then*

$$\int_0^\tau H_s dW_s = \int_0^T 1_{\{s \le \tau\}} H_s dW_s.$$

Proof.   a) Let (H^n) be a sequence of simple processes approximating H. We know that

$$E\left(I(H^n)_T^2\right) = E\left(\int_0^T |H_s^n|^2\, ds\right)$$

so, taking limits, we have

$$E\left(I(H)_T^2\right) = E\left(\int_0^T |H_s|^2\, ds\right).$$

Also,

$$E\left(\sup_{t \le T} I(H^n)_t^2\right) \le 4E\left(\int_0^T |H_s^n|^2\, ds\right).$$

Taking limits, the result follows.

b) Suppose τ is a stopping time of the form

$$\tau = \sum_{1 \le i \le n} t_i \mathbf{1}_{A_i}, \tag{6.5}$$

where $A_i \cap A_j = \emptyset$ for $i \ne j$ and $A_i \in \mathcal{F}_{t_i}$. Then

$$\int_0^T \mathbf{1}_{\{s > \tau\}} H_s\, dW_s = \int_0^T \left(\sum_{1 \le i \le n} \mathbf{1}_{A_i} \mathbf{1}_{\{s > t_i\}}\right) H_s\, dW_s.$$

Now for each i the process $\mathbf{1}_{\{s > t_i\}} \mathbf{1}_{A_i} H_s$ is adapted and in $\mathcal{H}$; it is zero if $s \le t_i$ and equals $\mathbf{1}_{A_i} H_s$ otherwise. Therefore

$$\int_0^T \left(\sum_{1 \le i \le n} \mathbf{1}_{A_i} \mathbf{1}_{\{s > t_i\}}\right) H_s\, dW_s$$

$$= \sum_{1 \le i \le n} \mathbf{1}_{A_i} \int_{t_i}^T H_s\, dW_s = \int_\tau^T H_s\, dW_s.$$

Consequently, for τ of the form (6.5),

$$\int_0^T \mathbf{1}_{\{s \le \tau\}} H_s\, dW_s = \int_0^\tau H_s\, dW_s.$$

Now an arbitrary stopping time τ can be approximated by a decreasing sequence of stopping times τ_n where

$$\tau_n = \sum_{i=0}^{2^n} \frac{(k+1)T}{2^n} \mathbf{1}_A,$$

where $A = \left\{ \frac{kT}{2^n} \leq \tau < \frac{(k+1)T}{2^n} \right\}$, so that $\lim \tau_n = \tau$ a.s. Consequently, because $I(H)$ is almost surely continuous in t,

$$\lim_{n \to \infty} \int_0^{\tau_n} H_s dW_s = \int_0^\tau H_s dW_s \quad \text{a.s.}$$

Also,

$$E \left(\left| \int_0^T \mathbf{1}_{\{s \leq \tau\}} H_s dW_s - \int_0^T \mathbf{1}_{\{s \leq \tau_n\}} H_s dW_s \right|^2 \right)$$

$$= E \left(\int_0^T \mathbf{1}_{\{\tau < s \leq \tau_n\}} H_s^2 ds \right),$$

and this converges to zero by the dominated convergence theorem. Therefore

$$\lim_{n \to \infty} \int_0^T \mathbf{1}_{\{s \leq \tau_n\}} H_s dW_s = \int_0^T \mathbf{1}_{\{s \leq \tau\}} H_s dW_s$$

both a.s. and in $L^2(\Omega)$; the result follows.

$\square$

Notation 6.3.9. Write

$$\widehat{\mathcal{H}} = \left\{ \{H_s\} : H \text{ is } \mathcal{F}_t\text{-adapted and } \int_0^T H_s^2 ds < \infty \text{ a.s.} \right\}.$$

The preceding definition and results for the stochastic integral can be extended from $\mathcal{H}$ to $\widehat{\mathcal{H}}$.

Theorem 6.3.10. *There is a unique linear map $\widehat{I}$ of $\widehat{\mathcal{H}}$ into the space of continuous processes defined on $[0, T]$ such that:*

a) *If $\{H_t\}_{0 \leq t \leq T}$ is in $\mathcal{H}$, then for all $t \in [0, T]$ the processes $\widehat{I}(H)_t$ and $I(H)_t$ are indistinguishable.*

b) *If $\{H^n\}_{n \geq 0}$ is a sequence in $\widehat{\mathcal{H}}$ such that $\int_0^T (H_s^n)^2 ds$ converges to zero in probability, then*

$$\sup_{0 \leq t \leq T} \left| \widehat{I}(H^n)_t \right|$$

converges to zero in probability.

Remark 6.3.11. In fact, $\widehat{I}(H)_t$ is a local martingale, meaning that there is a non-decreasing sequence T_n of stopping times with limit T such that $T_n \leq T_{n+1} \leq T$ and for each n, $\widehat{I}(H)_{T_n \wedge t}$ is a martingale.

Notation 6.3.12. One writes $\widehat{I}(H)_t = \int_0^t H_s dW_s$.

Proof. From Theorem 6.3.6, we know that for $H \in \mathcal{H}$, $I(H)$ is defined. Suppose $H \in \widehat{\mathcal{H}}$. Define

$$T_n = \inf\left\{0 \le u \le T : \int_0^u H_s^2 ds \ge n\right\} \wedge T.$$

(Here and elsewhere we adopt the convention that $\inf(\emptyset) = \infty$.) Because H_s is adapted, $\int_0^t H_s^2 ds$ is adapted and T_n is an $(\mathcal{F}_t)$-stopping time.

Then write $H_s^n = \mathbf{1}_{\{s < T_n\}} H_s$. The processes H^n are therefore in $\mathcal{H}$ and

$$\int_0^t H_s^n dW_s = \int_0^t \mathbf{1}_{\{s \le T_n\}} H_s^{n+1} dW_s = \int_0^{T_n \wedge t} H_s^{n+1} dW_s.$$

Therefore, on the set $\left\{\int_0^T H_u^2 du < n\right\}$, for all $t \le T$,

$$I(H^n)_t = I(H^{n+1})_t.$$

Now

$$\bigcup_{n \ge 0}\left\{\int_0^T H_u^2 du < n\right\} = \left\{\int_0^T H_u^2 du < \infty\right\}.$$

Thus we define $\widehat{I}(H)_t$ by putting $\widehat{I}(H)_t = I(H^n)_t$ on $\left\{\int_0^T H_s^2 ds < n\right\}$. Clearly $\widehat{I}(H)_t = I(H)_t$ and is continuous a.s.

For the second assertion, write

$$B = \left\{\int_0^T H_u^2 du \ge \frac{1}{N}\right\}, \qquad A = \left\{\omega : \sup_{0 \le t \le T}\left|\widehat{I}(H)_t\right| \ge \varepsilon\right\}.$$

Then

$$P(A) = P(A \cap B) + P(A \cap B^c) \le P(B) + P(A \cap B^c).$$

Therefore, for any $\varepsilon > 0$,

$$P\left(\sup_{0 \le t \le T}\left|\widehat{I}(H)_t\right| \ge \varepsilon\right) \le P\left(\int_0^T H_u^2 du \ge \frac{1}{N}\right)$$
$$+ P\left(\left\{\int_0^T H_u^2 du < \frac{1}{N}\right\} \cap \left\{\sup_{0 \le t \le T}\left|\widehat{I}(H)_t\right| \ge \varepsilon\right\}\right). \qquad (6.6)$$

Write

$$\tau_N = \inf\left\{s \le T : \int_0^s H_u^2 du \ge \frac{1}{N}\right\} \wedge T.$$

On the set B^c, we have

$$\int_0^t H_s dW_s = \int_0^t \mathbf{1}_{\{s \le \tau_N\}} H_s dW_s.$$

Therefore, with $G_s = H_s 1_{\{s \leq \tau_N\}}$, $G_s = H_s$ on B^c, and from Doob's inequality, it follows that

$$E\left(\sup_{0 \leq t \leq T} \left|\int_0^t G_s dW_s\right|^2\right) \leq 4E\left(\int_0^t (G_s)^2 ds\right) \leq \frac{4}{N}. \qquad (6.7)$$

Using Chebychev's inequality, we also have, by (6.7),

$$P(A \cap B^c) = P\left(B^c \cap \left\{\sup_{0 \leq t \leq T} \left|\widehat{I}(H)_t\right| \geq \varepsilon\right\}\right)$$

$$\leq P\left(\sup_{0 \leq t \leq T} \left|\widehat{I}(G)_t\right| \geq \varepsilon\right)$$

$$\leq \frac{1}{\varepsilon^2} E\left(\sup_{0 \leq t \leq T} \left|\int_0^t G_s dW_s\right|^2\right) \leq \frac{4}{N\varepsilon^2}.$$

Consequently, from (6.6),

$$P\left(\sup_{0 \leq t \leq T} \left|\widehat{I}(H)_t\right| \geq \varepsilon\right) \leq P\left(\int_0^T H_u^2 du \geq \frac{1}{N}\right) + \frac{4}{N\varepsilon^2}.$$

Hence we see that if (H^n) is a sequence in $\widehat{\mathcal{H}}$ such that $\left(\int_0^T (H_u^n)^2 du\right)$ converges to zero in probability, then $\left(\sup_{0 \leq t \leq T} \left|\widehat{I}(H^n)_t\right|\right)$ converges to zero in probability. The continuity of the operator $\widehat{I}$ is therefore established.

If $H \in \widehat{\mathcal{H}}$, then, with $H_s^n = 1_{\{s < T_n\}} H_s$, we see that $\left(\int_0^T (H_s - H_s^n)^2 ds\right)$ converges to zero in probability. Using the continuity property, we see that the map $\widehat{I}$ is uniquely defined.

Similarly, for $H, K \in \widehat{\mathcal{H}}$, suppose there are the approximating sequences $H^n, K^n \in \mathcal{H}$. Now $\left(\int_0^T (H_s - H_s^n)^2 ds\right)$ and $\left(\int_0^T (K_s - K_s^n)^2 ds\right)$ converge to zero in probability as $n \to \infty$. Furthermore,

$$I(\alpha H^n + \beta K^n)_t = \alpha I(H^n)_t + \beta I(K^n)_t.$$

Letting $n \to \infty$, we see that $\widehat{I}$ is a linear map. $\qquad \square$

6.4 The Itô Calculus

If $f(t)$ is a real-valued, differentiable function for $t \geq 0$ and $f(0) = 0$, then

$$f(t)^2 = 2\int_0^t f(s)\dot{f}(s)ds = 2\int_0^t f(s)df(s).$$

However, if W is a Brownian motion we know that $E\left(W_t^2\right) = t$. Consequently, W_t^2 cannot be equal to $2\int_0^t W_s dW_s$ because this integral is a (local) martingale and $E\left(2\int_0^t W_s dW_s\right) = 0$.

The Itô calculus is described for a class of processes known as Itô processes, which we will now define.

Itô Processes and the Differentiation Rule

Definition 6.4.1. Suppose $(\Omega, \mathcal{F}, P)$ is a probability space with a filtration $(\mathcal{F}_t)_{t \geq 0}$, and (W_t) is a standard $(\mathcal{F}_t)$-Brownian motion. A real-valued *Itô process* $(X_t)_{t \geq 0}$ *is a process of the form*

$$X_t = X_0 + \int_0^t K_s ds + \int_0^t H_s dW_s,$$

where

(a) X_0 is $\mathcal{F}_0$-measurable,

(b) K and H are adapted to $\mathcal{F}_t$, and

(c) $\int_0^T |K_s| ds < \infty$ a.s. and $\int_0^T |H_s^2| ds < \infty$ a.s.

We can then obtain a uniqueness property that is a consequence of the following result.

Lemma 6.4.2. *Suppose the process $\int_0^t K_s ds = M_t$ is a continuous martingale, where $\int_0^T |K_s| ds < \infty$ a.s. Then $M_t = 0$ a.s. for all $t \leq T$, and there is a set $N \subset \Omega$ of measure zero such that, for $\omega \notin N$, $K_s(\omega) = 0$ for almost all s.*

Proof. Suppose initially that $\int_0^T |K_s| ds \leq C < \infty$ a.s. Then, with $t_i^n = \frac{iT}{N}$ for $0 \leq i \leq n$, we have

$$\sum_{i=1}^n (M_{t_i^n} - M_{t_{i-1}^n})^2 \leq \sup_i \left| M_{t_i^n} - M_{t_{i-1}^n} \right| \sum_{i=1}^n \left| M_{t_i^n} - M_{t_{i-1}^n} \right|$$

$$= \sup_i \left| M_{t_i^n} - M_{t_{i-1}^n} \right| \sum_{i=1}^n \left| \int_{t_{i-1}^n}^{t_i^n} K_s ds \right|$$

$$\leq \sup_i \left| M_{t_i^n} - M_{t_{i-1}^n} \right| \sum_{i=1}^n \int_{t_{i-1}^n}^{t_i^n} |K_s| ds$$

$$\leq C \sup_i \left| M_{t_i^n} - M_{t_{i-1}^n} \right|.$$

Consequently, $\lim_{n \to \infty} \sum_{i=1}^n (M_{t_i^n} - M_{t_{i-1}^n})^2 = 0$ a.s., so by the bounded convergence theorem, we have

$$\lim_{n \to \infty} E \left(\sum_{i=1}^n \left(M_{t_i^n} - M_{t_{i-1}^n} \right)^2 \right) = 0 = E \left(M_t^2 - M_0^2 \right),$$

as M is a martingale. By definition, $M_0 = 0$ a.s. Consequently, $M_t = 0$ a.s., and so $M_t = 0$ a.s. for $t \leq T$.

Now relax the assumption that $\int_0^T |K_s| \, ds$ is bounded. Write

$$T_n = \inf \left\{ 0 \leq s \leq T : \int_0^s |K_u| \, du \geq n \right\} \wedge T.$$

Then T_n is a stopping time because K is adapted, and $\lim_{n \to \infty} T_n = T$. The preceding result shows that $M_{t \wedge T_n} = 0$ a.s., and so $\lim_{n \to \infty} M_{t \wedge T_n} = 0 = M_t$ a.s. □

Corollary 6.4.3. *Suppose M is a martingale of the form $\int_0^t H_s dW_s + \int_0^t K_s ds$ with $\int_0^t H_s^2 ds < \infty$ a.s. and $\int_0^t |K_s| \, ds < \infty$ a.s. Then $\int_0^t K_s ds$ is a martingale that is zero a.s. and there is a P-null set $N \subset \Omega$ such that, for $\omega \notin N$, $K_s(\omega) = 0$ for almost all s.*

Corollary 6.4.4. *Suppose the Itô process X has representations*

$$X_t = X_0 + \int_0^t K_s ds + \int_0^t H_s dW_s,$$

$$X_t = X_0' + \int_0^t K_s' ds + \int_0^t H_s' dW_s.$$

Then $X_0 = X_0'$ a.s., $H_s = H_s'$ a.s. $(ds \times dP)$, and $K_s = K_s'$ a.s. $(ds \times dP)$. In particular, if X is a martingale, then $K = 0$.

Proof. Clearly $X_0 = X_0'$. Therefore

$$\int_0^t (K_s - K_s') ds = \int_0^t (H_s' - H_s) dW_s,$$

and $\int_0^t (K_s - K_s') ds$ is a martingale. The result follows from Lemma 6.4.2. □

Remark 6.4.5. Suppose $(W_t)_{t \geq 0}$ is a Brownian motion and

$$\pi = \{0 = t_0 \leq t_1 \leq \cdots \leq t_N = t\}$$

is a partition of $[0, t]$. Write

$$|\pi| = \max_i (t_{i+1} - t_i).$$

Then

$$E\left(\sum_{i=0}^{N-1} (W_{t_{i+1}} - W_{t_i})^2 \right) = E\left(\sum_{i=0}^{N-1} (W_{t_{i+1}}^2 - W_{t_i}^2) \right) = t. \tag{6.8}$$

In fact, we can show that $\sum_{i=0}^{N-1} (W_{t_{i+1}} - W_{t_i})^2$ converges to t almost surely as $|\pi| \to 0$.

Choose a sequence (π_n) of partitions with $|\pi_n| \to 0$ as $n \to \infty$. Write

$$Q_n = \sum_{\pi_n} (W_{t_{i+1}} - W_{t_i})^2;$$

then we have shown that $Q_n \to t$ in L^2 as $n \to \infty$. By Chebychev's inequality, we have, for any $\varepsilon > 0$, that

$$P(|Q_n - t| > \varepsilon) \leq \frac{E\left((Q_n - t)^2\right)}{\varepsilon^2}.$$

Set $E\left((Q_n - t)^2\right) = q_n$, so that $q_n \to 0$ as $n \to \infty$. Choosing a subsequence, we can assume that $q_n < \frac{1}{2^{2n}}$. Letting $\varepsilon_n = \frac{1}{2^n}$ and writing $A_n = \left\{|Q_n - t| > \frac{1}{2^n}\right\}$, we obtain $P(A_n) \leq \frac{1}{2^n}$, so that $\sum_{n=1}^{\infty} P(A_n) < \infty$. By the first Borel-Cantelli lemma, it follows that $P(\cap_{n \geq 1} A_n) = 0$, and hence that $Q_n \to t$ a.s. as $n \to \infty$.

For a general, continuous (local) martingale $(M_t)_{t \geq 0}$,

$$\lim_{|\pi| \to 0} \sum_{i=0}^{N} \left(M_{t_{i+1}} - M_{t_i}\right)^2$$

exists and is a predictable, continuous increasing process denoted by $\langle M \rangle_t$. From Jensen's inequality, M^2 is a submartingale and it turns out that $\langle M \rangle$ is the unique (continuous) increasing process in the Doob-Meyer decomposition of M^2. This decomposition is entirely analogous to the Doob decomposition described in Section 5.3, but the technical complexities involved are substantially greater in continuous time. For details, see the development in [109, Chapter 10] or [199, Chapter 3]. $\langle M \rangle$ is called the (predictable) quadratic variation of M. Consequently, (6.8) states that for a Brownian motion W,

$$\langle W \rangle_t = t.$$

For $H \in \widehat{\mathcal{H}}$ we have seen that $M_t = \int_0^t H_s dW_s$ is a local martingale. It is shown in [109] that in this case

$$\langle M \rangle_t = \int_0^t H_s^2 ds \quad \text{a.s.}$$

In some sense, (6.8) indicates that, very formally, $(dW)^2 \simeq dt$, or $(dW) \simeq \sqrt{dt}$.

Suppose X is an Itô process on $0 \leq t \leq T$,

$$X_t = X_0 + \int_0^t K_s ds + \int_0^t H_s dW_s, \tag{6.9}$$

where $\int_0^T |K_s| ds < \infty$ a.s. and $\int_0^T |H_s|^2 ds < \infty$ a.s. Considering partitions $\pi = \{0 = t_0 \leq t_1 \leq \cdots \leq t_N = t\}$ of $[0, t]$, it can be shown that

$$\lim_{|\pi| \to 0} \sum_{i=0}^{N} \left(X_{t_{i+1}} - X_{t_i}\right)^2$$

converges a.s. to

$$\int_0^t |H_s|^2 \, ds.$$

That is, $\langle X \rangle_t = \langle M \rangle_t$, where $M_t = \int_0^t H_s dW_s$ is the martingale term in the representation (6.9) of X.

Again, if X is a differentiable process (that is, if $H_s = 0$ in (6.9)), then the usual chain rule states that, for a differentiable function f,

$$f(X_t) = f(X_0) + \int_0^t f'(X_s) dX_s.$$

However, if X is an Itô process, the differentiation rule (commonly known as the Itô formula) has the following form.

Theorem 6.4.6. *Suppose* $\{X_t\}_{t \geq 0}$ *is an Itô process of the form*

$$X_t = X_0 + \int_0^t K_s ds + \int_0^t H_s dW_s.$$

Suppose f is twice differentiable. Then

$$f(X_t) = f(X_0) + \int_0^t f'(X_s) dX_s + \frac{1}{2} \int_0^t f''(X_s) d \langle X \rangle_s.$$

Here, by definition, $\langle X \rangle_t = \int_0^t H_s^2 ds$; *that is, the (predictable) quadratic variation of X is the quadratic variation of its martingale component*

$$\int_0^t H_s dW_s.$$

Also,

$$\int_0^t f'(X_s) dX_s = \int_0^t f'(X_s) K_s ds + \int_0^t f'(X_s) H_s dW_s.$$

For a proof see [109]. More generally, the differentiation rule can be proved in the following form.

Theorem 6.4.7. *If $F : [0, \infty) \times \mathbb{R} \to \mathbb{R}$ is continuously differentiable in the first component and twice continuously differentiable in the second, then*

$$F(t, X_t) = F(0, X_0) + \int_0^t \frac{\partial F}{\partial s}(s, X_s) ds$$

$$+ \int_0^t \frac{\partial F}{\partial x}(s, X_s) dX_s + \frac{1}{2} \int_0^t \frac{\partial^2 F}{\partial x^2}(s, X_s) d \langle X \rangle_s.$$

Example 6.4.8. (i) Let us consider the case when $K_s = 0, H_s = 1$. Then

$$X_t = X_0 + W_t,$$

where W_t is standard Brownian motion. Taking $f(x) = x^2$, we have $\langle X \rangle_t = \langle W \rangle_t = t$ so

$$X_t^2 = X_0^2 + 2 \int_0^t W_s dW_s + \frac{1}{2} \int_0^t 2 ds.$$

That is,

$$X_t^2 - X_0^2 - t = 2 \int_0^t W_s dW_s.$$

For any $T < \infty$, we have $E\left(\int_0^T W_s^2 ds\right) < \infty$, so from Theorem 6.3.6, $\int_0^t W_s dW_s$ is a martingale. If $X_0 = 0$, then $X_t = W_t$ and we see that $W_t^2 - t$ is a martingale.

(ii) An often-used model for a price process is the so-called 'log-normal' model. In this case, it is supposed the price process S_t evolves according to the stochastic dynamics

$$\frac{dS_t}{S_t} = \mu dt + \sigma dW_t, \tag{6.10}$$

where μ and σ are real constants and $S_0 = X_0$. This means that

$$S_t = X_0 + \int_0^t S_s \mu ds + \int_0^t S_s \sigma dW_s.$$

Assuming such a process S exists, it is therefore an Itô process with

$$K_s = \mu S_s, \qquad\qquad H_s = \sigma S_s.$$

Then $\langle X \rangle_t = \int_0^t \sigma^2 S_s^2 ds$. Assuming $S_t > 0$ and applying Itô's formula with $f(x) = \log x$ (formally, because the logarithmic function is not twice continuously differentiable everywhere),

$$\log S_t = \log X_0 + \int_0^t \mu \frac{dS_s}{S_s} + \frac{1}{2} \int_0^t \left(-\frac{1}{S_s^2}\right) \sigma^2 S_s^2 ds$$

$$= \log X_0 + \int_0^t \left(\mu - \frac{\sigma^2}{2}\right) ds + \int_0^t \sigma dW_s$$

$$= \log X_0 + \left(\mu - \frac{\sigma^2}{2}\right) t + \sigma W_t.$$

Consequently,

$$S_t = X_0 \exp\left\{\left(\mu - \frac{\sigma^2}{2}\right) t + \sigma W_t\right\}.$$

Exercise 6.4.9. Consider the function

$$F(t, x) = X_0 \exp\left\{\left(\mu - \frac{\sigma^2}{2}\right)t + \sigma x\right\}.$$

Apply the Itô formula of Theorem 6.4.7 to $S_t = F(t, W_t)$ to show that S_t does satisfy the log-normal equation (6.10). This 'justifies' our formal application of the Itô formula.

Exercise 6.4.10. Let B be a Brownian motion, and suppose that the processes X, Y have dynamics given by

$$\begin{aligned}
dX_t &= X_t(\mu_X dt + \sigma_X dB_t), \\
dY_t &= Y_t(\mu_Y dt + \sigma_Y dB_t).
\end{aligned}$$

Define Z by $Z_t = \frac{Y_t}{X_t}$. Show that Z is also log-normal, with dynamics

$$dZ_t = Z_t(\mu_Z dt + \sigma_Z dB_t),$$

and determine the coefficients μ_Z and σ_Z in terms of the coefficients of X and Y.

Multidimensional Itô Processes

Definition 6.4.11. Suppose we have a probability space $(\Omega, \mathcal{F}, P)$ with a filtration $(\mathcal{F}_t)_{t \geq 0}$. An m-dimensional $(\mathcal{F}_t)$-Brownian motion is a process $W_t = (W_t^1, W_t^2, \ldots, W_t^m)$ whose components W_t^i are standard, independent $(\mathcal{F}_t)$-Brownian motions.

We can extend our definition of an Itô process to the situation where the (scalar) stochastic integral involves an m-dimensional Brownian motion.

Definition 6.4.12. $(X_t)_{0 \leq t \leq T}$ is an Itô process if

$$X_t = X_0 + \int_0^t K_s ds + \sum_{i=1}^m \int_0^t H_s^i dW_s^i,$$

where the K and H^i are adapted to $(\mathcal{F}_t)$, $\int_0^T |K_s| ds < \infty$ a.s., and

$$\int_0^T |H_s^i|^2 ds < \infty \text{ a.s. for all } i = 1, 2, \ldots, m.$$

An n-dimensional Itô process is then a process $X_t = (X_t^1, \ldots, X_t^N)$, each component of which is an Itô process in the sense of Definition 6.4.12. The differentiation rule takes the following form.

Theorem 6.4.13. *Suppose* $X_t = (X_t^1, \ldots, X_t^N)$ *is an n-dimensional Itô process with*

$$X_t^i = X_0^i + \int_0^t K_s^i ds + \sum_{j=1}^m \int_0^t H_s^{ij} dW_s^j,$$

and suppose $f : [0, T] \times \mathbb{R}^n \to \mathbb{R}$ is in $C^{1,2}$ (the space of functions once continuously differentiable in t and twice continuously differentiable in $x \in \mathbb{R}^n$). Then

$$f(t, X_t^1, \ldots, X_t^n) = f(0, X_0^1, \ldots, X_0^n) + \int_0^t \frac{\partial f}{\partial s}(s, X_s^1, \ldots, X_s^n)ds$$

$$+ \sum_{i=1}^n \int_0^t \frac{\partial f}{\partial x_i}(s, X_s^1, \ldots, X_s^n)dX_s^i$$

$$+ \frac{1}{2} \sum_{i,j=1}^n \int_0^t \frac{\partial^2 f}{\partial x_i \partial x_j}(s, X_s^1, \ldots, X_s^n)d\langle X^i, X^j \rangle_s .$$

Here

$$dX_s^i = K_s^i ds + \sum_{j=1}^m H_s^{i,j} dW_s^j, \qquad d\langle X^i, X^j \rangle_s = \sum_{r=1}^m H_s^{i,r} H_s^{j,r} ds.$$

Remark 6.4.14. For components

$$X_t^p = X_0^p + \int_0^t K_s^p ds + \sum_{j=1}^m \int_0^t H_s^{pj} dW_s^j,$$

$$X_t^q = X_0^q + \int_0^t K_s^q ds + \sum_{j=1}^m \int_0^t H_s^{qj} dW_s^j,$$

it is shown in [227] that for partitions $\pi = \{0 = t_0 \leq t_1 \leq \cdots \leq t_N = t\}$,

$$\lim_{|\pi| \to 0} \sum_i \left(X_{t_{i+1}}^p - X_{t_i}^p \right) \left(X_{t_{i+1}}^q - X_{t_i}^q \right)$$

converges in probability to

$$\int_0^t \sum_{r=1}^m H_s^{pr} H_s^{qr} ds.$$

This process is the predictable covariation of X^p and X^q and is denoted by

$$\langle X^p X^q \rangle_t = \sum_{r=1}^m \int_0^t H_s^{pr} H_s^{qr} ds. \tag{6.11}$$

We note that $\langle X^p X^q \rangle$ is symmetric and bilinear as a function on Itô processes.

Taking

$$Y_t = Y_0 + \int_0^t K_s' ds, \qquad X_t = X_0 + \int_0^t K_s ds + \sum_{j=1}^m H_s^j dW_s^j,$$

we see that $\langle X, Y \rangle_t = 0$. Furthermore, formula (6.11) gives

$$\left\langle \int_0^t H_s^{pi} dW_s^i, \int_0^t H_s^{qj} dW_s^j \right\rangle = \begin{cases} \int_0^t H_s^{pi} H_s^{qi} ds & \text{if } i = j, \\ 0 & \text{if } i \neq j. \end{cases}$$

Remark 6.4.15. We noted in 6.4.5 that if $(M_t)_{t \geq 0}$ is a continuous local martingale, then $\langle M \rangle_t$ is the unique continuous increasing process in the Doob-Meyer decomposition of the submartingale M_t^2. If

$$X_t = X_0 + \int_0^t K_s ds + \int_0^t H_s dM_s,$$

where H and K are adapted, $\int_0^T |K_s| \, ds < \infty$ a.s., and $\int_0^T H_s^2 ds < \infty$ a.s., the differentiation formula has the form

$$f(X_t) = f(X_0) + \int_0^t \frac{\partial f}{\partial x}(X_s) K_s ds$$
$$+ \int_0^t \frac{\partial f}{\partial x}(X_s) H_s dM_s + \frac{1}{2} \int_0^t \frac{\partial^2 f}{\partial x^2}(X_s) H_s^2 d \langle M \rangle_s.$$

Using without proof the analogue of the Itô rule (Theorem 6.4.6) for general square integrable martingales M (see [109, p. 138]), we can prove the converse of Theorem 6.2.5.

Theorem 6.4.16. Suppose $(W_t)_{t \geq 0}$ is a continuous (scalar) local martingale on the filtered probability space $(\Omega, \mathcal{F}, P, \mathcal{F}_t)$, such that $(W_t^2 - t)_{t \geq 0}$ is a local martingale. Then (W_t) is a Brownian motion.

Proof. We must show that, for $0 \leq s \leq t$, the random variable $W_t - W_s$ is independent of $\mathcal{F}_s$ and is normally distributed with mean 0 and covariance $t - s$.

In terms of characteristic functions, this means we must show that

$$E\left(e^{iu(W_t - W_s)} | \mathcal{F}_s\right) = E\left(e^{iu(W_t - W_s)}\right) = \exp\left\{-\frac{u^2(t - s)}{2}\right\} \quad \text{for all } u \in \mathbb{R}.$$

To this end, consider the (complex-valued) function

$$f(x) = e^{iux}.$$

Applying the differentiation rule to the real and imaginary parts of $f(x)$, we have

$$f(W_t) = e^{iuW_t} = f(W_s) + \int_s^t iue^{iuW_r} dW_r - \frac{1}{2} \int_s^t u^2 e^{iuW_r} dr \qquad (6.12)$$

because $d \langle W \rangle_r = dr$ by hypothesis. Furthermore, the real and imaginary parts of $iu \int_s^t e^{iuW_r} dW_r$ are in fact square integrable martingales because the integrands are bounded by 1. Consequently, $E\left(iu\int_s^t e^{iuW_r}dW_r \,|\mathcal{F}_s\right) = 0$ a.s. For any $A \in \mathcal{F}_s$, we may multiply (6.12) by $1_A e^{-iuW_s}$ and take expectations to deduce that

$$E\left(e^{iu(W_t-W_s)}1_A\right) = P(A) - \frac{1}{2}u^2 \int_0^t E\left(e^{iu(W_r-W_s)}1_A\right)dr.$$

Solving this equation, we have

$$E\left(e^{iu(W_t-W_s)}1_A\right) = P(A)\exp\left\{-\frac{u^2(t-s)}{2}\right\}.$$

$\square$

6.5   Stochastic Differential Equations

We first establish a useful result known as Gronwall's lemma.

Lemma 6.5.1. *Suppose α and β are integrable functions on $[a,b]$. If there is a constant H such that*

$$\alpha(t) \leq \beta(t) + H\int_a^t \alpha(s)ds \text{ for } t \in [a,b], \tag{6.13}$$

then

$$\alpha(t) \leq \beta(t) + H\int_a^t e^{H(t-s)}\beta(s)ds.$$

Note that if $\beta(t) = B$, a constant, then

$$\alpha(t) \leq Be^{H(t-a)}. \tag{6.14}$$

Proof. Write

$$A(t) = \int_a^t \alpha(s)ds, \qquad\qquad g(t) = A(t)e^{-Ht}.$$

Then

$$g'(t) = \alpha(t)e^{-Ht} - HA(t)e^{-Ht} \leq \beta(t)e^{-Ht}$$

from (6.13). Integrating, we obtain

$$g(t) - g(a) \leq \int_a^t \beta(s)e^{-Hs}ds.$$

That is,

$$A(t) \leq e^{Ht}\int_a^t \beta(s)e^{-Hs}ds.$$

Using (6.13) again, we have

$$\alpha(t) \leq \beta(t) + HA(t) = \beta(t) + H \int_a^t \beta(s) e^{H(t-s)} ds.$$

$\square$

Definition 6.5.2. Suppose $(\Omega, \mathcal{F}, P)$ is a probability space with a filtration $(\mathcal{F}_t)_{0 \leq t \leq T}$. Let $(W_t) = ((W_t^1, \ldots, W_t^m))$ be an m-dimensional $(\mathcal{F}_t)$-Brownian motion and $f(x, t)$ and $\sigma(x, t)$ be measurable functions of $x \in \mathbb{R}^n$ and $t \in [0, T]$ with values in $\mathbb{R}^n$ and $L(\mathbb{R}^m, \mathbb{R}^n)$, the space of $m \times n$ matrices, respectively. We take ξ to be an $\mathbb{R}^n$-valued, $\mathcal{F}_0$-measurable random variable.

A process X_t, $0 \leq t \leq T$ is a *solution* of the stochastic differential equation

$$dX_t = f(X_t, t)dt + \sigma(X_t, t)dW_t$$

with initial condition $X_0 = \xi$ if for all t the integrals

$$\int_0^t f(X_s, s)ds \text{ and } \int_0^t \sigma(X_s, s)dW_s$$

are well-defined and

$$X_t = \xi + \int_0^t f(X_s, s)ds + \int_0^t \sigma(X_s, s)dW_s \text{ a.s.} \qquad (6.15)$$

Theorem 6.5.3. *Suppose the assumptions of Definition 6.5.2 apply. In addition, assume that ξ, f, and σ satisfy*

$$|f(x, t) - f(x', t)| + |\sigma(x, t) - \sigma(x', t)| \leq K |x - x'|, \qquad (6.16)$$

$$|f(x, t)|^2 + |\sigma(x, t)|^2 \leq K_0^2 \left(1 + |x|^2\right), \qquad (6.17)$$

$$E\left(|\xi|^2\right) < \infty.$$

Then there is a solution X of (6.15) such that

$$E\left(\sup_{0 \leq t \leq T} |X_t|^2\right) < C\left(1 + E\left(|\xi|^2\right)\right).$$

Note, for the matrix σ, that $|\sigma|^2 = \text{Tr}(\sigma\sigma^*)$. This solution is unique in the sense that, if X_t' is also a solution, then they are indistinguishable in the sense of Definition 6.1.12.

Proof. Uniqueness: Suppose that X and X' are solutions of (6.15). Then, for all $t \in [0, T]$,

$$X_t - X_t' = \int_0^t \left(f(X_s, s) - f(X_s', s)\right) ds + \int_0^t \left(\sigma(X_s, s) - \sigma(X_s', s)\right) dW_s.$$

Therefore

$$|X_t - X_t'|^2 \leq 2 \left(\int_0^t (f(X_s, s) - f(X_s', s)) \, ds \right)^2$$

$$+ 2 \left(\int_0^t (\sigma(X_s, s) - \sigma(X_s', s)) \, dW_s \right)^2.$$

Taking expectations, we obtain

$$E\left(|X_t - X_t'|^2\right) \leq 2 \int_0^t E\left((f(X_s, s) - f(X_s', s))^2\right) ds$$

$$+ 2 \int_0^t E\left(|\sigma(X_s, s) - \sigma(X_s', s)|^2\right) ds.$$

Write $\phi(t) = E\left(|X_t - X_t'|^2\right)$ and use the Lipschitz conditions (6.16) to deduce that

$$\phi(t) \leq 2(T + 1)K^2 \int_0^t \phi(s) ds.$$

Gronwall's inequality (Lemma 6.5.1) therefore implies that $\phi(t) = 0$ for all $t \in [0, T]$. Consequently,

$$|X_t - X_t'| = 0 \text{ a.s.}$$

The process $|X_t - X_t'|$ is continuous, so there is a set $N \in \mathcal{F}_0$ of measure zero such that if $\omega \notin N, X_t(\omega) = X_t'(\omega)$ for all $t \in [0, T]$. That is, X' is a modification of X.

   Existence: Write $X_t^0 = \xi$ for $0 \leq t \leq T$. Define a sequence of processes $\left(X_t^N\right)$ by

$$X_t^N = \xi + \int_0^t f(X_s^{n-1}, s) ds + \int_0^t \sigma(X_s^{n-1}, s) dW_s. \qquad (6.18)$$

It can be shown that $\sigma(X_s^{n-1}, s) \in \mathcal{H}$, so the stochastic integrals are defined.

   Using arguments similar to those in the uniqueness proof, we can show that

$$E\left(|X_t^{n+1} - X_t^N|^2\right) \leq L \int_0^t E\left(|X_s^n - X_s^{n-1}|^2\right) ds, \qquad (6.19)$$

where $L = 2(1 + T)K^2$. Iterating (6.19), we see that

$$E\left(|X_t^{n+1} - X_t^N|^2\right) \leq L^n \int_0^t \frac{(t - s)^{n-1}}{(n-1)!} E\left(|X_s^1 - \xi|^2\right) ds$$

and

$$E\left(\left|X_s^1 - \xi\right|^2\right) \leq LTK^2\left(1 + E\left(|\xi|^2\right)\right).$$

Therefore

$$E\left(\left|X_t^{n+1} - X_t^N\right|^2\right) \leq C\frac{T^n}{n!}. \tag{6.20}$$

Also,

$$\sup_{0 \leq t \leq T} \left|X_t^{n+1} - X_t^N\right| \leq \int_0^T \left|f\left(X_s^n, s\right) - f\left(X_s^{n-1}, s\right)\right| ds$$

$$+ \sup_{0 \leq t \leq T} \left|\int_0^t \left(\sigma(X_s^n, s) - \sigma(X_s^{n-1}, s)\right) dW_s\right|;$$

so, using the vector form of Doob's inequality (Corollary 6.2.17), we have

$$E\left(\sup_{0 \leq t \leq T} \left|X_t^{n+1} - X_t^n\right|^2\right) \leq 2TK^2 \int_0^T E\left(\left|X_s^n - X_s^{n-1}\right|^2\right) ds$$

$$+ CE\left(\int_0^T \left|X_s^n - X_s^{n-1}\right|^2 ds\right)$$

$$\leq C_1 \frac{T^{n-1}}{(n-1)!}$$

using (6.20). Consequently,

$$\sum_{n=1}^\infty P\left(\sup_{0 \leq t \leq T} \left|X_t^{n+1} - X_t^n\right| > \frac{1}{n^2}\right) \leq \sum_{n=1}^\infty n^4 C_1 \frac{T^{n-1}}{(n-1)!}.$$

The series on the right converges. Therefore, almost surely, the series $\xi + \sum_{n=0}^\infty (X_t^{n+1} - X_t^n)$ converges uniformly in t, and so X_t^n converges to some X_t uniformly in t.

Each X^n is a continuous process, so X is a continuous process. Now

$$E\left(\left|X_t^n\right|^2\right) \leq 3\left[E\left(|\xi|^2\right) + K_0^2 T \int_0^t \left(1 + E\left(\left|X_s^{n-1}\right|^2\right)\right) ds\right.$$

$$\left. + K_0^2 \int_0^t \left(1 + E\left(\left|X_s^{n-1}\right|^2\right)\right) ds\right],$$

so

$$E\left(\left|X_t^n\right|^2\right) \leq C\left(1 + E\left(|\xi|^2\right)\right) + C\int_0^t E\left(\left|X_s^{n-1}\right|^2\right) ds.$$

By recurrence, taking $C > 1$,

$$
\begin{aligned}
E\left(|X_t^n|^2\right) &\leq \left(1 + E\left(|\xi|^2\right)\right)\left(C + C^2 t + \cdots + C^{n-1}\frac{t^n}{n!}\right) \\
&\leq C\left(1 + E\left(|\xi|^2\right)\right)e^{Ct}.
\end{aligned}
$$

Using the bounded convergence theorem, we can take the limit in (6.18) to deduce that

$$
X_t = \xi + \int_0^t f(X_s, s)ds + \int_0^t \sigma(X_s, s)dW_s \text{ a.s.}
$$

Therefore, X is the unique solution of the equation (6.15). $\square$

6.6   Markov Property of Solutions of SDEs

Definition 6.6.1. Let $(\Omega, \mathcal{F}, P)$ be a probability space with filtration $(\mathcal{F}_t)_{t \geq 0}$. An adapted process (X_t) is said to be a *Markov process* with respect to the filtration $(\mathcal{F}_t)$ if

$$
E\left(f(X_t)\,|\,\mathcal{F}_s\right) = E\left(f(X_t)\,|\,X_s\right) \text{ a.s. for all } t \geq s \geq 0
$$

for every bounded real-valued Borel function f defined on $\mathbb{R}^d$.

Consider a stochastic differential equation as in (6.15) with coefficients satisfying the conditions of Theorem 6.5.3 so the solution exists. Consider a point $x \in \mathbb{R}^n$ and for $s \leq t$ write $X_s(x, t)$ for the solution process of the equation

$$
X_s(x, t) = x + \int_s^t f\left(X_s(x, u), u\right) du + \int_s^t \sigma\left(X_s(x, u), u\right) dW_u. \quad (6.21)
$$

We quote the following results.

Theorem 6.6.2. $X_s(x, t)$ *is a continuous function of its arguments, and if the coefficients f and σ are C^1 functions of their first argument, the solution $X_s(x, t)$ is C^1 in x.*

Proof. See Kunita [204]. $\square$

Write $X_s(x, t, \omega)$ for the solution of (6.21), so $X_s(x, t, \omega) : \mathbb{R}^d \times [s, T] \times \Omega \to \mathbb{R}^d$, and $\mathcal{F}^W(s, t)$ for the completion of the σ-field generated by $W_{s+u} - W_s$, $0 \leq u \leq t - s$.

Theorem 6.6.3. *For $t \in [s, T]$, the restriction of $X_s(x, u, \omega)$ to $\mathbb{R}^d \times [s, t] \times \Omega$ is $\mathcal{B}(\mathbb{R}^d) \times \mathcal{B}([s, t]) \times \mathcal{F}^W(s, t)$-measurable.*

Proof. [109, Lemma 14.23]. $\square$

We next prove the 'flow' property of solutions of equation (6.21).

Lemma 6.6.4. *If $X_s(x,t)$ is the solution of (6.21) and $X_r(x,t)$ is the solution of (6.21) starting at time r with $r \le s \le t$, then $X_r(x,t) = X_s(X_r(x,s),t)$ in the sense that one is a modification of the other.*

Proof. By definition,

$$X_r(x,t) = x + \int_r^t f(X_r(x,u),u)\,du + \int_r^t \sigma(X_r(x,u),u)\,dW_u$$

$$= X_r(x,s) + \int_s^t f(X_r(x,u),u)\,du + \int_s^t \sigma(X_r(x,u),u)\,dW_u.$$

$$(6.22)$$

However, for any $y \in \mathbb{R}^n$,

$$X_s(y,t) = y + \int_s^t f(X_s(y,u),u)\,du + \int_s^t \sigma(X_s(y,u),u)\,dW_u.$$

Therefore, using the continuity of the solution,

$$X_s(X_r(x,s),t) = X_r(x,s) + \int_s^t f(X_s(X_r(x,s),u),u)\,du$$

$$+ \int_s^t \sigma(X_s(X_r(x,s),u),u)\,dW_u. \quad (6.23)$$

Using the uniqueness of the solution, we see from (6.22) and (6.23) that $X_r(x,s)$ is a modification of $X_s(X_r(x,s),t)$. $\qquad \square$

Before establishing the Markov property of solutions of (6.21), we prove a general result on conditional expectations.

Lemma 6.6.5. *Given a probability space $(\Omega, \mathcal{G}, P)$ and measurable spaces $(E, \mathcal{E})$, $(F, \mathcal{F})$, suppose that $\mathcal{A} \subset \mathcal{G}$ and $X : \Omega \to E$ and $Y : \Omega \to F$ are random variables such that X is $\mathcal{A}$-measurable and Y is independent of $\mathcal{A}$.*

For any bounded real-valued Borel function Φ defined on $(E \times F, \mathcal{E} \times \mathcal{F})$, consider the function ϕ defined for all $x \in E$ by

$$\phi(x) = E(\Phi(x,Y)).$$

Then ϕ is a Borel function on $(E, \mathcal{E})$ and

$$E(\Phi(X,Y)\,|\,\mathcal{A}) = \phi(X) \quad a.s.$$

Proof. Write P_Y for the probability law of Y. Then

$$\phi(x) = \int_F \Phi(x,y)\,dP_Y(y).$$

The measurability of Φ follows from Fubini's theorem.

Suppose Z is any $\mathcal{A}$-measurable random variable. Write $P_{X,Z}$ for the probability law of (X, Z). Then, because Y is independent of (X, Z),

$$
\begin{aligned}
E\left(\Phi(X, Y)Z\right) &= \iint \Phi(x, y)z\, dP_{X,Z}(x, z)\, dP_Y(y) \\
&= \int \left(\int \Phi(x, y)\, dP_Y(y) \right) z\, dP_{X,Z}(x, z) \\
&= \int \phi(x)z\, dP_{X,Z}(x, y) \\
&= E\left(\phi(X)Z\right).
\end{aligned}
$$

This identity is true for all such Z; the result follows. □

Lemma 6.6.6. *Suppose $X_s(x, t, \omega)$ is the solution of* (6.21) *and $g : \mathbb{R}^d \to \mathbb{R}$ is a bounded Borel-measurable function. Then*

$$
f(x, \omega) = g\left(X_s(x, t, \omega)\right)
$$

is $\mathcal{B}(\mathbb{R}^d) \times \mathcal{F}^W(s, t)$-measurable.

Proof. Write $\mathcal{A}$ for the collection of sets $A \in \mathcal{B}(\mathbb{R}^d)$ for which the lemma is true with $g = \mathbf{1}_A$. If $f(x, \omega) = \mathbf{1}_A\left(X_s(x, t, \omega)\right)$, then

$$
\{(x, \omega) : f(x, \omega) = 1\} = \{(x, \omega) : X_s(x, t, \omega) \in A\} \in \mathcal{B}(\mathbb{R}^d) \times \mathcal{F}^W(s, t).
$$

The lemma is therefore true for all $A \in \mathcal{B}(\mathbb{R}^d)$, and the result follows for general g by approximation with simple functions. □

We now show that solutions of stochastic differential equations of the form (6.21) are Markov processes with respect to the right-continuous (and completed) filtration $(\mathcal{F}_t)$ generated by the Brownian motion $(W_t)_{t \geq 0}$ and the initial value $x \in \mathbb{R}^d$.

Theorem 6.6.7. *Suppose $X_0(x, t)$ is the solution of* (6.21) *such that $X_0(x, 0) = x \in \mathbb{R}^d$. For any bounded real-valued Borel function g defined on $\mathbb{R}^d$, we have*

$$
E\left(g(X_t) \,|\, \mathcal{F}_s\right) = E\left(g(X_t) \,|\, X_s\right) \quad \text{for all } 0 \leq s \leq t.
$$

More precisely, if

$$
\phi(z) = E\left(g\left(X_s(z, t)\right)\right),
$$

then

$$
E\left(g(X_t) \,|\, \mathcal{F}_s\right) = \phi\left(X_0(x, s)\right) \quad a.s.
$$

Proof. Suppose $g : \mathbb{R}^d \to \mathbb{R}$ is any bounded Borel-measurable function. As in Lemma 6.6.6, write $f(x, \omega) = g\left(X_s(x, t, \omega)\right)$. Then, for each $x \in \mathbb{R}^d$, $f(x, \cdot)$ is $\mathcal{F}^W(s, t)$-measurable and thus independent of $\mathcal{F}_s$.

Write, as in Lemma 6.6.5,

$$\phi(x) = E\left(g\left(X_s(x,t,\omega)\right)\right).$$

If Z is any $\mathcal{F}_s$-measurable random variable,

$$E\left(g\left(X_s(Z,t,\omega)\right)|\mathcal{F}_s\right) = \phi(Z). \tag{6.24}$$

From the flow property of the solutions, Lemma 6.6.4, it follows that

$$X_t = X_0(x,t) = X_s\left(X_0(x,s),t\right)$$

and $X_0(x,s)$ is $\mathcal{F}_s$-measurable. Substituting $Z = X_0(x,s)$ in (6.24), therefore,

$$E\left(g\left(X_0(x,t)\right)|\mathcal{F}_s\right) = E\left(g(X_t)|\mathcal{F}_s\right) = \phi\left(X_0(x,s)\right) = \phi(X_s).$$

Consequently, $E\left(g(X_t)|\mathcal{F}_s\right) = E\left(g(X_t)|X_s\right)$ and the result follows. □

Theorem 6.6.8. *Suppose $X_0(x,s) = X_s \in \mathbb{R}^d$ is the solution of (6.21), and consider the process*

$$\beta_s(1,t) = \beta_t = e^{-\int_s^t r(u,X_u)du},$$

where $r(s,x)$ is a positive measurable function. Then

$$d\beta_t = -r(t,X_t)\beta_t dt, \qquad\qquad \beta_s = 1,$$

and the augmented process $(\beta_t, X_t) \in \mathbb{R}^{d+1}$ is given by an equation similar to (6.21). Consequently, the augmented process is Markov and, for any bounded Borel function $f : \mathbb{R}^d \to \mathbb{R}$,

$$E\left(e^{-\int_s^t r(u,X_u)du}f(X_t)|\mathcal{F}_s\right) = \phi(X_s),$$

where

$$\phi(x) = E\left(e^{-\int_s^t r(u,X_s(x,u))du}f\left(X_s(x,t)\right)\right).$$

Chapter 7

Continuous-Time European Options

In this chapter, we shall develop a continuous-time theory that is the analogue of that in Chapters 1 to 3. The simple model will consist of a riskless bond and a risky asset, which can be thought of as a stock. The dynamics of our model are described in Section 7.1. The following two sections present the fundamental results of Girsanov and martingale representation. These are then applied to discuss the hedging and pricing of European options. In particular, we establish the famous results of Black and Scholes, results that are applied widely in the finance industry in spite of the simplified nature of the model. Recall that the Black-Scholes pricing formula for a European call was derived in Section 2.7 as the limit of a sequence of prices in binomial models.

7.1 Dynamics

We describe the dynamics of the Black-Scholes option pricing model. Our processes will be defined on a complete probability space $(\Omega, \mathcal{F}, P)$. The time parameter t will take values in the intervals $[0, \infty)$ or $[0, T]$. We suppose the market contains a riskless asset, or bond, whose price at time t is S_t^0, and a risky asset, or stock, whose price at time t is S_t^1.

Let r be a non-negative constant that represents the instantaneous interest rate on the bond. (This instantaneous interest rate should not be confused with the interest rate over a period of time in discrete models.) We then suppose that the evolution in the price of the bond S_t^0 is described by the ordinary differential equation

$$dS_t^0 = rS_t^0 dt. \tag{7.1}$$

If the initial value at time 0 of the bond is $S_0^0 = 1$, then (7.1) can be solved to give

$$S_t^0 = e^{rt} \text{ for } t \geq 0. \tag{7.2}$$

Let μ and $\sigma > 0$ be constants and $(B_t)_{t \geq 0}$ be a standard Brownian motion on $(\Omega, \mathcal{F}, P)$. We suppose that the evolution in the price of the risky asset S_t^1 is described by the stochastic differential equation

$$dS_t^1 = S_t^1(\mu dt + \sigma dB_t). \tag{7.3}$$

If the initial price at time 0 of the risky asset is S_0^1, then (7.3) can be solved to give

$$S_t^1 = S_0^1 \exp\left\{\mu t - \frac{\sigma^2}{2}t + \sigma B_t\right\}. \tag{7.4}$$

Taking logarithms, we have

$$\log S_t^1 = \log S_0^1 + \left(\mu - \frac{\sigma^2}{2}\right)t + \sigma B_t, \tag{7.5}$$

and we see that $\log S_t^1$ evolves like a Brownian motion with drift $(\mu - \frac{\sigma^2}{2})t$ and volatility σ. In particular, $\log S_t^1$ is a normal random variable, which is often expressed by saying S_t^1 is 'log-normal'. It is immediate from (7.4) and (7.5) that (S_t^1) has continuous trajectories, and $\log S_t^1$ has independent stationary increments (so $\frac{S_t^1 - S_v^1}{S_v^1}$ is independent of the σ-field $\sigma(S_u^1 : u \leq v)$ and $\frac{S_t^1 - S_v^1}{S_v^1}$ is identically distributed to $\frac{S_{t-v}^1 - S_0^1}{S_0^1}$).

7.2   Girsanov's Theorem

Girsanov's theorem shows how martingales, in particular Brownian motion, transform under a different probability measure. We first define certain spaces of martingales.

The set of martingales for which convergence results hold is the set of *uniformly integrable* martingales. As we noted in Chapters 5 and 6, this is not a significant restriction if the time horizon is finite (i.e., $T < \infty$). Recall Definition 5.3.1 applied to a martingale: if (M_t) is a martingale, for $0 \leq t < \infty$ or $0 \leq t \leq T$, (M_t) is uniformly integrable if

$$\int_{\{|M_t(\omega)| \geq K\}} |M_t(\omega)| dP(\omega)$$

converges to 0 uniformly in t as $K \to \infty$.

If $(X_t)_{t \geq 0}$ is any real, measurable process, we shall write

$$X_t^* = \sup_{s \leq t} |X_s|.$$

We shall write $\mathcal{M}$ for the space of right-continuous, uniformly integrable martingales. Consistent with Notation 6.2.9, $\mathcal{M}_{\ell oc}$ will denote the set of processes that are locally in $\mathcal{M}$ (i.e., we say that $M \in \mathcal{M}_{\ell oc}$ if there exists an increasing sequence of stopping times (T_n) such that $M_t^{T_n} = M_{t \wedge T_n} \in \mathcal{M}$). We call $\mathcal{M}_{\ell oc}$ the space of local martingales. Let $\mathcal{L}$ be the subset of $\mathcal{M}_{\ell oc}$ consisting of those local martingales for which $M_0 = 0$ a.s.

For $M \in \mathcal{M}$ and $p \in [1, \infty]$, write

$$\|M\|_{\mathcal{H}^p} = \|M_\infty^*\|_p.$$

Here $\|\cdot\|_p$ denotes the norm on $L^p(\Omega, \mathcal{F}, \mathcal{P})$. Then $\mathcal{H}^p$ is the space of martingales in $\mathcal{M}$ such that

$$\|M\|_{\mathcal{H}^p} < \infty.$$

In particular, $\mathcal{H}^2$ is the space of square integrable martingales.

Suppose $(\Omega, \mathcal{F}, P)$ is a probability space with a filtration $(\mathcal{F}_t)_{t \geq 0}$. Let Q be a second probability measure on $(\Omega, \mathcal{F})$ that is absolutely continuous with respect to P. Write

$$M_t = \begin{cases} \frac{dQ}{dP} & \text{if } t = \infty, \\ E(M_\infty | \mathcal{F}_t) & \text{if } t < \infty. \end{cases}$$

Remark 7.2.1. In continuous time, versions of martingales are considered that are right-continuous and have left limits. There is a right-continuous version of M with left limits if the filtration $(\mathcal{F}_t)$ satisfies the usual conditions (see [109, Theorem 4.11]).

Lemma 7.2.2. *$(X_t M_t)$ is a local martingale under P if and only if (X_t) is a local martingale under Q.*

Proof. We prove the result for martingales. The extension to local martingales can be found in [168, Proposition 3.3.8]. Suppose $s \leq t$ and $A \in \mathcal{F}_s$. Then

$$\int_A X_t dQ = \int_A X_t M_t dP = \int_A X_s M_s dP = \int_A X_s dQ,$$

and the result follows. $\square$

Suppose $(\Omega, \mathcal{F}, P)$ is a probability space. Recall from Theorem 6.4.16 that a real process $(B_t)_{t \geq 0}$ is a standard Brownian motion if:

a) $t \to B_t(\omega)$ is continuous a.s.,

b) B_t is a (local) martingale, and

c) $B_t^2 - t$ is a (local) martingale.

This characterisation of Brownian motion using properties a)-c) is due to Lévy, and it is shown in Theorem 6.4.16 that these properties imply the other well-known properties of Brownian motion, including, for example, that B is a Gaussian process with independent increments.

Write $\mathcal{F}_t^0 = \sigma(B_s : s \leq t)$ for the σ-field on Ω generated by the history of the Brownian motion up to time t. Then $(\mathcal{F}_t)_{t \geq 0}$ will denote the right-continuous complete filtration generated by the $\mathcal{F}_t^0$.

We show how (B_t) behaves under a change of measure.

Theorem 7.2.3 (Girsanov). *Suppose* $(\theta_t)_{0 \leq t \leq T}$ *is an adapted, measurable process such that* $\int_0^T \theta_s^2 ds < \infty$ *a.s. and also so that the process*

$$\Lambda_t = \exp\left\{ -\int_0^t \theta_s dB_s - \frac{1}{2} \int_0^t \theta_s^2 ds \right\}$$

is an $(\mathcal{F}_t, P)$ *martingale. Define a new measure* Q_θ *on* $\mathcal{F}_T$ *by putting*

$$\left. \frac{dQ_\theta}{dP} \right|_{\mathcal{F}_T} = \Lambda_T.$$

Then the process

$$W_t = B_t + \int_0^t \theta_s ds$$

is a standard Brownian motion on $(\mathcal{F}_t, Q_\theta)$.

Remark 7.2.4. A sufficient condition, widely known as Novikov's condition, for Λ to be a martingale is that

$$E\left(\exp\left\{ \frac{1}{2} \int_0^T \theta_s^2 ds \right\} \right) < \infty$$

(see [109]).

Proof. Using the Itô rule and definition of Λ, we see, as in Exercise 6.4.9, that

$$\Lambda_t = 1 - \int_0^t \Lambda_s \theta_s dB_s. \tag{7.6}$$

Clearly $\Lambda_t > 0$ a.s. and as Λ is a martingale

$$E\left(\Lambda_t \right) = 1.$$

Now for $A \in \mathcal{F}_T$, $Q_\theta(A) = \int_A \Lambda_T dP \geq 0$ and $Q_\theta(\Omega) = \int_\Omega \Lambda_T dP = E\left(\Lambda_t \right) = 1$, so Q_θ is a probability measure.

To show that (W_t) is a standard Brownian motion, we verify that it satisfies the conditions a)-c) above, which are required for the application of Theorem 6.4.16. By definition, (W_t) is a continuous process almost surely, as (B_t) is continuous a.s. and an indefinite integral is a continuous process. For the second condition, we must show that (W_t) is a local $(\mathcal{F}_t)$-martingale under the measure Q_θ. Equivalently, from Lemma 7.2.2 we must show that $(\Lambda_t W_t)$ is a local martingale under P. Applying the Itô rule to (7.6) and (W_t), we have

$$\Lambda_t W_t = W_0 + \int_0^t \Lambda_s dW_s + \int_0^t W_s d\Lambda_s + \int_0^t d \langle \Lambda, W \rangle_s$$

$$= W_0 + \int_0^t \Lambda_s dB_s + \int_0^t \Lambda_s \theta_s ds - \int_0^t W_s \Lambda_s \theta_s dB_s - \int_0^t \Lambda_s \theta_s ds$$

$$= W_0 + \int_0^t \Lambda_s (1 - W_s \theta_s) dB_s$$

and, as a stochastic integral with respect to B, $(\Lambda_t W_t)$ is a (local) martingale under P.

The third condition is established similarly since

$$W_t^2 = 2 \int_0^t W_s dW_s + \langle W \rangle_t = 2 \int_0^t W_s dW_s + t.$$

We must prove that $W_t^2 - t$ is a local $(\mathcal{F}_t, Q_\theta)$-martingale. However,

$$W_t^2 - t = 2 \int_0^t W_s dW_s,$$

and we have established that W_s is a (local) martingale under Q_θ. Consequently, the stochastic integral is a (local) martingale under Q_θ and the result follows. □

Hitting Times of Brownian Motion

We shall need the following results on hitting times of Brownian motion. Their proofs involve an exponential martingale M of a form similar to Λ. Suppose $(B_t)_{t \geq 0}$ is a standard Brownian motion with $B_0 = 0$ adapted to the filtration $(\mathcal{F}_t)$. Write

$$T_a = \inf \{ s \geq 0 : B_s = a \} \text{ for } a \in \mathbb{R}. \tag{7.7}$$

As usual, we take $\inf \{\emptyset\} = \infty$.

Theorem 7.2.5. T_a in (7.7) is a stopping time that is almost surely finite and

$$E \left(e^{-\lambda T_a} \right) = e^{-\sqrt{2\lambda}|a|} \text{ for } \lambda \geq 0.$$

Proof. Suppose $a \geq 0$. Because B is continuous, we have, with $\mathbb{Q}^+$ denoting the positive rationals,

$$\{ T_a \leq t \} = \bigcap_{\varepsilon \in \mathbb{Q}^+} \left\{ \sup_{r \leq t} B_r > a - \varepsilon \right\} = \bigcap_{\varepsilon \in \mathbb{Q}^+} \bigcap_{\substack{r \in \mathbb{Q}^+ \\ r \leq t}} \{ B_r > a - \varepsilon \} \in \mathcal{F}_t.$$

Consequently, T_a is a stopping time.

For any $\sigma \geq 0$, the process

$$M_t = \exp \left\{ \sigma B_t - \frac{\sigma^2}{2} t \right\}$$

is an $(\mathcal{F}_t)$-martingale by Theorem 6.2.5. For $n \in \mathbb{Z}^+$, consider the stopping time $T_a \wedge n$. Then, from the optional stopping theorem (Theorem 6.2.10), we have

$$E\left(M_{T_a \wedge n}\right) = E\left(M_0\right) = 1.$$

However,

$$M_{T_a \wedge n} = \exp\left\{\sigma B_{T_a \wedge n} - \frac{\sigma^2}{2}(T_a \wedge n)\right\} \le \exp\left\{\sigma a\right\}.$$

Now if $T_a < \infty$, then $\lim_{n \to \infty} M_{T_a \wedge n} = M_{T_a}$. If $T_a = \infty$, then $B_t \le a$ for all $t \ge 0$, so that $\lim_{n \to \infty} M_{T_a \wedge n} = 0$.

Using Lebesgue's dominated convergence theorem, we have

$$E\left(\mathbf{1}_{\{T_a < \infty\}} M_{T_a}\right) = 1. \tag{7.8}$$

Now $B_{T_a} = a$ if $T_a < \infty$. Therefore,

$$E\left(\mathbf{1}_{\{T_a < \infty\}} e^{\sigma a} e^{-\frac{\sigma^2}{2} T_a}\right) = 1;$$

consequently,

$$E\left(\mathbf{1}_{\{T_a < \infty\}} e^{\frac{\sigma^2}{2} T_a}\right) = e^{-\sigma a}. \tag{7.9}$$

Letting $\sigma \to 0$ in (7.9), we see that

$$E\left(\mathbf{1}_{\{T_a < \infty\}}\right) = P(T_a < \infty) = 1.$$

Hence almost every sample path of the Brownian motion reaches the value a, and

$$E\left(e^{\frac{-\sigma^2}{2} T_a}\right) = e^{-\sigma a}.$$

Now $(-B_t)$ is also an $(\mathcal{F}_t)$-Brownian motion, so the case $a < 0$ can be deduced by noting that

$$T_a = \inf_{s \ge 0}\left\{s \ge 0 : -B_s = -a\right\}.$$

$\square$

An application of Girsanov's theorem enables us to deduce the following extension.

Corollary 7.2.6. *Suppose μ, a are real numbers. Write*

$$T_a(\mu) = \inf\left\{t \ge 0 : \mu t + B_t = a\right\}.$$

Then

$$E\left(e^{-\alpha T_a(\mu)}\right) = \exp\left\{\mu a - |a|\sqrt{\mu^2 + 2\alpha}\right\} \text{ for all } \alpha > 0.$$

Proof. Introduce the probability measure Q by setting

$$\frac{dQ}{dP}\bigg|_{\mathcal{F}_t} = \exp\left\{\mu B_t - \frac{\mu^2}{2}t\right\}.$$

From Girsanov's theorem, the process $\widetilde{B}$ is a standard Brownian motion under Q, where

$$\widetilde{B}_t = B_t - \mu t.$$

Clearly, the hitting time $T_a(\mu)$ of $\widetilde{B}_t + \mu t$ is the same as the hitting time $T_a(0)$ of B_t. Therefore, for all $\alpha > 0$ and $t > 0$, we have

$$E\left(\exp\left\{-\alpha(T_a(\mu) \wedge t)\right\}\right)$$

$$= E\left(\exp\left\{-\alpha(T_a(0) \wedge t)\right\}\exp\left\{\mu B_{T_a(0)\wedge t} - \frac{\mu^2}{2}(T_a(0) \wedge t)\right\}\right).$$

Now $\exp\left\{-\alpha(T_a(0) \wedge t)\right\} \le e^{-\alpha t}$. Noting that $T_a(0) = T_a$, we have for $t < T_a$ that $t < \infty$. Therefore

$$\exp\left\{\mu B_{T_a \wedge t} - \frac{\mu^2}{2}(T_a \wedge t)\right\}\mathbf{1}_{\{t<T_a\}} \le \exp\left\{\mu B_t - \frac{\mu^2}{2}t\right\},$$

which has expected value 1, and

$$E\left(\exp\left\{-\alpha(T_a \wedge t)\right\}\exp\left\{\mu B_{T_a \wedge t} - \frac{\mu^2}{2}(T_a \wedge t)\right\}\mathbf{1}_{\{t<T_a\}}\right) \le e^{-\alpha t}.$$

Suppose initially that $a \ge 0$, and write

$$\widetilde{M}_t = \exp\left\{-\alpha(T_a \wedge t)\right\}\exp\left\{\mu B_{T_a \wedge t} - \frac{\mu^2}{2}(T_a \wedge t)\right\}.$$

Then $\widetilde{M}_t \le \exp\left\{\mu a\right\}$ and, again by the dominated convergence theorem,

$$E\left(\lim_{t\to\infty} \widetilde{M}_t\right) = E\left(\mathbf{1}_{\{t<T_a\}}e^{-\alpha T_a}e^{\mu B_{T_a} - \frac{\mu^2}{2}T_a}\right)$$

$$= e^{\mu a}E\left(\mathbf{1}_{\{T_a<\infty\}}e^{-(\alpha+\frac{\mu^2}{2})T_a}\right)$$

$$= e^{\mu a}e^{-\sqrt{2\alpha+\mu^2}|a|}.$$

Again the case when $a < 0$ can be discussed by considering $-B$.

We have therefore established that

$$E\left(\mathbf{1}_{\{T_a<\infty\}}e^{-\alpha T_a}e^{\mu B_{T_a} - \frac{\mu^2}{2}T_a}\right) = E\left(\mathbf{1}_{\{T_a(\mu)<\infty\}}e^{-\alpha T_a(\mu)}\right)$$

$$= e^{\mu a - \sqrt{2\alpha+\mu^2}|a|}.$$

Letting $\alpha \to 0$, we see that

$$P(T_a < \infty) = e^{\mu a - |\mu a|},$$

this probability being equal to 1 if μ and a have the same sign. Furthermore, as $e^{-\alpha T_a} = 0$ on $\{T_a = \infty\}$, we have

$$E\left(e^{-\alpha T_a(\mu)}\right) = e^{\mu a - \sqrt{2\alpha + \mu^2}|a|}.$$

$\square$

7.3   Martingale Representation

We first recall concepts related to martingales and stable subspaces of martingales.

Definition 7.3.1. Two local martingales $M, N \in \mathcal{M}_{\ell oc}$ are *orthogonal* if their product MN is in $\mathcal{L}$.

Remark 7.3.2. We then write $M \perp N$. Note that orthogonality implies that $M_0 N_0 = 0$ a.s.

We now have the following result.

Lemma 7.3.3. *Suppose $M, N \in \mathcal{H}^2$ are orthogonal. For every stopping time τ the random variables M_τ, N_τ are orthogonal in $L^2(\Omega, \mathcal{F}, P)$ and $MN \in \mathcal{H}^1$.*

Conversely, if $M_0 N_0 = 0$ a.s. and the random variables M_τ, N_τ are orthogonal in L^2 for every stopping time τ, then $M \perp N$.

Proof. Because M_∞^* and N_∞^* are in L^2, the product $M_\infty^* N_\infty^*$ is in L^1. Furthermore,

$$(MN)_\infty^* = \sup_t |M_t N_t| \leq M_\infty^* N_\infty^*$$

so $MN \in \mathcal{H}^1$ if $M \perp N$, and $M_0 N_0 = 0$ a.s. Consequently,

$$E\left(M_\tau N_\tau\right) = E\left(M_0 N_0\right) = 0.$$

Conversely, suppose that, for any stopping time τ, we have $M_\tau \in L^2$ and $N_\tau \in L^2$. Then $M_\tau N_\tau \in L^1$, so $E\left(|M_\tau N_\tau|\right) < \infty$ and $E\left(M_\tau N_\tau\right) = 0$. From [109, Lemma 4.18] this condition is sufficient for MN to be a uniformly integrable martingale, and the result follows. $\square$

Notation 7.3.4. If $(X_t)_{t \geq 0}$ is a stochastic process and τ is a stopping time, X^τ will denote the process X stopped at time τ (i.e., $X_t^\tau = X_{t \wedge \tau}$).

Definition 7.3.5. A linear subspace $\mathcal{K} \subset \mathcal{H}^2$ is called *stable* if

(a) It is closed in the L^2 norm.

(b) If $M \in \mathcal{K}$ and τ is a stopping time, then $M^\tau \in \mathcal{K}$.

(c) If $M \in \mathcal{K}$ and $A \in \mathcal{F}_0$, then $\mathbf{1}_A M \in \mathcal{K}$.

Theorem 7.3.6. *Suppose $\mathcal{K}$ is a stable subspace of $\mathcal{H}^2$. Write $\mathcal{K}^\perp$ for the set of martingales $N \in \mathcal{H}^2$ such that $E\left(M_\infty N_\infty\right) = 0$ for all $M \in \mathcal{K}$. Then $\mathcal{K}^\perp$ is a stable subspace, and $M \perp N$ for all $M \in \mathcal{K}, N \in \mathcal{K}^\perp$.*

Proof. Suppose $M \in \mathcal{K}, N \in \mathcal{K}^\perp$, and τ is a stopping time. Then $E\left(L_\infty N_\infty\right) = 0$ for all $L \in \mathcal{K}$. Now $M^\tau \in \mathcal{K}$, so

$$E\left(M_\infty^\tau N_\infty\right) = E\left(M_\tau N_\infty\right) = 0.$$

Therefore

$$E\left(E\left(M_\tau N_\infty \mid \mathcal{F}_\tau\right)\right) = E\left(M_\tau E\left(N_\infty \mid \mathcal{F}_\tau\right)\right) = E\left(M_\tau N_\tau\right) = 0.$$

Taking $\tau = 0$, we have $\mathbf{1}_A M \in \mathcal{K}$ for any $A \in \mathcal{F}_0$, so $E\left(\mathbf{1}_A M_0 N_0\right) = 0$. Therefore, $M_0 N_0 = 0$ a.s. and M and N are orthogonal. We also have

$$E\left((\mathbf{1}_A M_\tau) N_\tau\right) = E\left(M_\infty (\mathbf{1}_A N^\tau)_\infty\right) = 0,$$

so $\mathbf{1}_A N^\tau \in \mathcal{K}^\perp$ for any $N \in \mathcal{K}^\perp$, any stopping time τ, and any $A \in \mathcal{F}_0$. Consequently, $\mathcal{K}^\perp$ is a stable subspace. $\qquad\square$

Corollary 7.3.7. *Suppose $\mathcal{K} \subset \mathcal{H}^2$ is a stable subspace. Then every element $M \in \mathcal{H}^2$ has a unique decomposition*

$$M = N + N^1,$$

where $N \in \mathcal{K}$ and $N^1 \in \mathcal{K}^\perp$.

Proof. Suppose $\mathcal{K}_\infty$ is the closed subspace of $L^2(\Omega, \mathcal{F}_\infty)$ generated by the random variables M_∞ for $M \in \mathcal{K}$. $\mathcal{K}_\infty^\perp$ is defined analogously. Then $\mathcal{K}_\infty$ and $\mathcal{K}_\infty^\perp$ give an orthogonal decomposition of $L^2(\Omega, \mathcal{F}_\infty)$ and, for any $M \in \mathcal{H}^2$, M_∞ has a unique decomposition,

$$M_\infty = N_\infty + N_\infty^1,$$

where $N_\infty \in \mathcal{K}_\infty$ and $N_\infty^1 \in \mathcal{K}_\infty^\perp$. Then define N (resp. N^1) to be the right-continuous version, with left limits, of the martingale

$$N_t = E\left(N_\infty \mid \mathcal{F}_t\right), \qquad (\text{resp. } N_t^1 = E\left(N_\infty^1 \mid \mathcal{F}_t\right)).$$

$$\square$$

Remark 7.3.8. From the isometry properties of the stochastic integral, it can be shown that the stable subspace generated by $M \in \mathcal{H}^2$ is the set of all stochastic integrals with respect to M. See [199, page 140].

We now prove the basic representation theorem for Brownian martingales; the proof is adapted from [109]. The stochastic process $(B_t)_{t\geq 0}$ is a Brownian motion on the probability space $(\Omega, \mathcal{F}, P)$ with respect to the filtration $\mathcal{F}_t^0 = \sigma\{B_s : s \leq t\}$. For $t \geq 0$, $\mathcal{F}_t$ is the completion of $\mathcal{F}_t^0$, so that $(\mathcal{F}_t)_{t\geq 0}$ is the filtration generated by B, which satisfies the 'usual conditions' of right-continuity and completeness. We have seen that if $(H_t)_{0\leq t\leq T}$ is a measurable adapted process on $[0,T]$ such that $E\left(\int_0^T H_s^2 ds\right) < \infty$, then $\left(\int_0^t H_s dB_s\right)$ is a square integrable martingale. The representation result tells us that all square integrable martingales on $(\mathcal{F}_t)_{0\leq t\leq T}$ are of this form.

Theorem 7.3.9. *Suppose $(M_t, \mathcal{F}_t)_{0\leq t\leq T}$ is a square integrable martingale, where $\mathcal{F}_t$ is the completion of the Brownian filtration $\mathcal{F}_t^0$. Then there is a measurable, adapted process $(H_t)_{0\leq t\leq T}$ such that $E\left(\int_0^T H_s^2 ds\right) < \infty$ and*

$$M_t = M_0 + \int_0^t H_s dB_s \text{ a.s. for all } t \in [0,T]. \tag{7.10}$$

Proof. First note that, by subtracting $M_0 = E(M_t)$ from each side of (7.10), we can assume $M_0 = 0$. Second, M_T is $\mathcal{F}_T$-measurable and square integrable, so all we have to establish is that any square integrable, $\mathcal{F}_T$-measurable, zero mean random variable M_T has a representation

$$M_T = \int_0^T H_s dB_s \text{ a.s.}$$

Write $\mathcal{H}_T^2$ for the space of square integrable $(\mathcal{F}_t)$-martingales on $[0,T]$. We can consider the stable subspace of $\mathcal{H}_T^2$ generated by stochastic integrals with respect to (B_t); this is closed in the norm of $L^2(\Omega, \mathcal{F}_T)$. Consequently, the martingale (M_t) has a projection on this stable subspace, which we denote by $(Y_t)_{0\leq t\leq T}$. From Remark 7.3.8, (Y_t) is a stochastic integral with respect to (B_t), so there is a measurable adapted integrand $(H_t)_{0\leq t\leq T}$ such that

$$Y_t = \int_0^t H_s dB_s \text{ a.s. for } t \in [0,T].$$

By construction, $M_t - Y_t$ is orthogonal to the stable subspace $\mathcal{H}_T^2(B)$ of $\mathcal{H}_T^2$ generated by stochastic integrals with respect to B. We can therefore suppose that the martingale (M_t) is orthogonal to $\mathcal{H}_T^2(B)$ and show that this implies $M_T = 0$ a.s. Write

$$\sigma_n = \inf\{t : |M_t| \geq n\}, \qquad M_t^N = \frac{1}{2n} M_{t\wedge\sigma_n}.$$

Then $|M_t^N| \leq \frac{1}{2}$ and (M_t) is orthogonal to both the martingales (B_t) and $(B_t^2 - t) = \left(2\int_0^t B_s dB_s\right)$.

A new probability measure Q can be defined on $\mathcal{F}_T$ by putting $\frac{dQ}{dP} = \Lambda = 1 + M_t^N$. Now (B_t) and $(B_t^2 - t)$ are continuous martingales on $(\Omega, \mathcal{F}, Q)$. Consequently, (B_t) is a Brownian motion under Q as well as P, so P and Q coincide on $\mathcal{F}_T$. This implies that $M_t^N = 0$ a.s. Letting $n \to \infty$, we see that $M_T = 0$ a.s. and the theorem is proved. $\qquad\square$

We now extend Theorem 7.3.9 to the situation where the filtration is generated by the weak solution of a stochastic differential equation. Suppose we have a probability space $(\Omega, \mathcal{F}, P)$ and an $\mathbb{R}^n$-valued stochastic process $(x_t)_{0 \le t \le T}$. Denote by $(\mathcal{F}_t)$ the filtration generated by (x_t), and let (B_t) be an $(\mathcal{F}_t)$-Brownian motion, such that

$$x_t \doteq x_0 + \int_0^t f(s, x) ds + \int_0^t \sigma(s, x) dB_s \text{ a.s. } (P).$$

Here, f and σ satisfy measurability and growth conditions, as in Theorem 6.5.3.

In the continuous-time setting, the *predictable* σ-field on $\Omega \times [0, T]$ is the σ-field generated by the left-continuous processes. A process is called predictable if it is measurable with respect to this σ-field. (Compare this with the discrete-time definition given in Section 2.2.)

Theorem 7.3.10. *Suppose* $(N_t)_{0 \le t \le T}$, *with* $N_0 = 0$, *is a square integrable* P-martingale with respect to the filtration $(\mathcal{F}_t)_{0 \le t \le T}$. *Then there is an* $\mathcal{F}_t$-predictable process (γ_t) *such that*

$$\int_0^t E\left(|\gamma_s|^2\right) ds < \infty, \qquad N_t = \int_0^t \gamma_s dB_s \text{ a.s.}$$

Proof. For $n \in \mathbb{Z}^+$, define

$$T_n = \min\left\{T, \inf\left\{t : \int_0^t \left|\sigma_s^{-1} f_s\right|^2 ds \ge n\right\}\right\}.$$

Then T_n is an $(\mathcal{F}_t)$-stopping time and $\lim T_n = T$. Write

$$\Lambda_t^* = \exp\left\{-\int_0^t \sigma_s^{-1} f_s dB_s - \frac{1}{2} \int_0^t \left|\sigma_s^{-1} f_s\right|^2 ds\right\}, \qquad (7.11)$$

and define a new measure P_n^* by setting

$$\left.\frac{dP^*}{dP}\right|_{\mathcal{F}_t} = \Lambda_{t \wedge T_n}^*.$$

For each n, P_n^* is a probability measure. From Girsanov's theorem (Theorem 7.2.3), the process

$$z_t^n = B_t + \int_0^{t \wedge T_n} \sigma_s^{-1} f_s ds$$

is a Brownian motion under P_n^*. Write

$$\mathcal{Z}_t^n = \sigma \{z_s^n : 0 \le s \le t\}.$$

From Theorem 7.3.9, if $\left(\widetilde{N}_t\right)$ is a square integrable, zero mean martingale under P_n^* with respect to the filtration $(\mathcal{Z}_t^n)$, then there is a process (ϕ_t^n), adapted to $(\mathcal{Z}_t^n)$, such that $\widetilde{N}_t = \int_0^t \phi_s^n dz_s^n$ a.s. Now

$$z_t^n = \int_0^t \sigma_s^{-1}(\sigma_s dB_s + f_s ds) = \int_0^t \sigma_s^{-1} dx_s \text{ for } t < T_n,$$

so $(\mathcal{Z}_t^n) = (\mathcal{F}_{t \wedge T_n})$. We have shown that if $\left(\widetilde{N}_t\right)$ is a square integrable, zero-mean P_n^*-martingale with respect to the filtration $(\mathcal{Z}_t^n)$, then

$$\widetilde{N}_{t \wedge T_n} = \int_0^{t \wedge T_n} \phi_s^n \sigma_s^{-1} dx_s \text{ a.s.} \tag{7.12}$$

Now, from Lemma 7.3.3, if (B_t) is a square integrable P-martingale with respect to the filtration $(\mathcal{F}_t)$, then $\left(\widetilde{N}_t\right)$ is a square integrable martingale with respect to the filtration $(\mathcal{F}_{t \wedge T_n})$, where

$$\widetilde{N}_t = (\Lambda_{t \wedge T_n}^*)^{-1} N_{t \wedge T_n}.$$

In this situation, we certainly have $\widetilde{N}_t = \widetilde{N}_{t \wedge T_n}$, so, from (7.10),

$$\widetilde{N}_t = \int_0^{t \wedge T_n} \phi_s^n \sigma_s^{-1} dx_s = \int_0^{t \wedge T_n} \phi_s^n dB_s + \int_0^{t \wedge T_n} \phi_s^n \sigma_s^{-1} f_s ds.$$

From (7.11), we have

$$\Lambda_{t \wedge T_n}^* = 1 - \int_0^{t \wedge T_n} \Lambda_s^* \sigma_s^{-1} f_s dB_s.$$

Therefore, using the Itô rule,

$$
\begin{aligned}
N_{t \wedge T_n} &= \widetilde{N}_t \Lambda_{t \wedge T_n}^* \\
&= \int_0^{t \wedge T_n} \widetilde{N}_s d\Lambda_s^* + \int_0^{t \wedge T_n} \Lambda_s^* d\widetilde{N}_s + \left\langle \widetilde{N}, \Lambda^* \right\rangle_{t \wedge T_n} \\
&= -\int_0^{t \wedge T_n} \widetilde{N}_s \Lambda_s^* \sigma_s^{-1} f_s dB_s + \int_0^{t \wedge T_n} \Lambda_s^* \phi_s^n \sigma_s^{-1} dx_s - \int_0^t \Lambda_s^* \phi_s^n \sigma_s^{-1} ds \\
&= \int_0^{t \wedge T_n} \gamma_s^n dB_s,
\end{aligned}
$$

where $\gamma_s^n = \Lambda_s(\phi_s^n - \widetilde{N}_s \sigma_s^{-1} f_s)$. Furthermore,

$$E\left(N_{t \wedge T_n}^2\right) = \int_0^t E\left(\left(\mathbf{1}_{\{0 \le s \le T_n\}} \gamma_s^n\right)^2\right) ds \le E\left(N_T^2\right) < \infty.$$

The representation is unique, so that

$$\gamma_s^n = \gamma_s^m \text{ for } s \le T_n \text{ for all } m \ge n.$$

Define γ_s to be the process such that

$$\gamma_s = \gamma_s^n \text{ for } 0 \le s \le T_n.$$

Then

$$N_{t \wedge T_n} = \int_0^{t \wedge T_n} \gamma_s dB_s, \qquad N_t = \int_0^t \gamma_s dB_s \text{ for } t < T_n.$$

However, $\lim T_n = T$ a.s. so that $\int_0^1 E\left(|\gamma_s|^2\right) ds < \infty$, and the result follows. $\qquad\qquad\square$

Corollary 7.3.11. *Suppose B is a P-Brownian motion and $(\mathcal{F}_t)_{t \ge 0}$ is the completed filtration generated by B. Suppose further that $(\theta_t)_{t \ge 0}$ is a predictable process such that if Λ is given by*

$$d\Lambda_t = \Lambda_t \theta_t dB_t, \qquad\qquad \Lambda_0 = 1,$$

then Λ is a (positive) martingale under $(P, (\mathcal{F}_t)_{t \ge 0})$. Define a probability measure Q by

$$\left.\frac{dQ}{dP}\right|_{\mathcal{F}_t} = \Lambda_t,$$

so that, by Girsanov's theorem, W is a Brownian motion under Q, where $dW_t = dB_t - \theta_t dt$. Then, if M is an $(\mathcal{F}_t, Q)$-martingale, there is a predictable process ψ such that

$$M_t = M_0 + \int_0^t \psi_s dW_s.$$

Corollary 7.3.12. *By considering stopping times and pasting, the representation results apply to locally square integrable martingales.*

Description of the Integrand in a Markov Model

In the Markov case, when the coefficients are sufficiently differentiable, the form of the integrand in the martingale representation can be made more explicit.

Suppose again that $B = \left(B^1, B^2, \dots, B^m\right)$ is an m-dimensional Brownian motion defined for $t \ge 0$ on $(\Omega, \mathcal{F}, P)$. Consider the stochastic differential equation

$$dx_t = f(t, x_t)dt + \sigma(t, x_t)dB_t \text{ for } t \ge 0, \qquad (7.13)$$

where $f : [0, \infty) \times \mathbb{R}^n \to \mathbb{R}^n$ and $\sigma : [0, \infty) \times \mathbb{R}^n \to \mathbb{R}^n \times \mathbb{R}^n$ are measurable functions that are three times differentiable in x, and, together with their derivatives, have linear growth in x.

Write $\xi_{s,t}(x)$ for the solution of (7.13) for $t \geq s$ under the initial condition

$$\xi_{s,s}(x) = x \in \mathbb{R}^n.$$

From results of Bismut [25] or Kunita [204], we know there is a set $N \subset \Omega$ of measure zero such that, for $\omega \notin N$, there is a version of $\xi_{s,t}(x)$ that is twice differentiable in x and continuous in t and s. Write

$$D_{s,t}(k) = \frac{\partial \xi_{s,t}(x)}{\partial x}$$

for the Jacobian of the map $x \to \xi_{s,t}(x)$. Then D is the solution of the linearised equation

$$dD_{s,t}(x) = f_x(t, x_t)D_{s,t}dt + \sigma_x(t, x_t)D_{s,t}dB_t$$

with initial condition $D_{s,s}(x) = I$, the $n \times n$ identity matrix. It is known that the inverse $D_{s,t}^{-1}(x)$ exists; see [25], [204].

Suppose $g : [0, \infty) \times \mathbb{R}^n \to \mathbb{R}^m$ satisfies conditions similar to those of f and define the exponential $M_{s,t}(x)$ by

$$M_{s,t}(x) = 1 + \int_0^t M_{s,r}(x)g(r, \xi_{s,r}(x))dB_r.$$

Write $(\mathcal{F}_t)$ for the right-continuous complete family of σ-fields generated by B. As g satisfies a linear growth condition, a new probability measure $\overline{P}$ can be defined by putting

$$\left. \frac{d\overline{P}}{dP} \right|_{\mathcal{F}_t} = M_{0,t}(x_0).$$

From Girsanov's theorem, W is an $(\mathcal{F}_t)$-Brownian motion under $\overline{P}$ if

$$dW_t = dB_t - g(t, \xi_{0,t}(x_0))dt. \qquad (7.14)$$

Suppose $c : \mathbb{R}^n \to \mathbb{R}$ is a C^2 function, that, together with its derivatives has linear growth, and consider the $\overline{P}$-martingale

$$N_t = E_{\overline{P}}\left(c(\xi_{0,T}(x_0)) \,|\, \mathcal{F}_t\right) \text{ for } 0 \leq t \leq T.$$

Then, from Theorem 7.3.10, (N_t) has a representation

$$N_t = N_0 + \int_0^t \gamma_s dW_s \text{ for } 0 \leq t \leq T, \qquad (7.15)$$

where $\int_0^T E\left(|\gamma_s|^2\right) ds < \infty$. We can now describe γ.

Theorem 7.3.13.

$$
\gamma_t = E\bigg\{ \int_t^T dW_r^* g_\xi\left(r, \xi_{0,r}(x_0)\right) D_{0,r}(x_0) c\left(\xi_{0,T}(x_0)\right)
$$

$$
+ c_\xi\left(\xi_{0,T}(x_0)\right) D_{0,T}(x_0) \,|\mathcal{F}_t \bigg\} D_{0,t}^{-1}(x_0)\sigma\left(t, \xi_{0,t}(x_0)\right).
$$

Here the asterisk denotes the transpose.

Proof. For $0 \le t \le T$, write $x = \xi_{0,t}(x_0)$. From the semigroup property of the solution of stochastic differential equations, Lemma 6.6.4, it follows that

$$
\xi_{0,T}(x_0) = \xi_{t,T}\left(\xi_{0,t}(x_0)\right) = \xi_{t,T}(x). \tag{7.16}
$$

Differentiating (7.16), we have

$$
D_{0,T}(x_0) = D_{t,T}(x) D_{0,t}(x_0).
$$

Furthermore, the exponential M satisfies

$$
M_{0,T}(x_0) = M_{0,t}(x_0) M_{t,T}(x).
$$

For $y \in \mathbb{R}^n$, define

$$
V(t,y) = E\left(M_{t,T}(y) c(\xi_{t,T}(y))\right)
$$

and consider the martingale

$$
\begin{aligned}
N_t &= E_{\overline{P}}\left(c(\xi_{0,t}(x_0)) \,|\mathcal{F}_t\right) \\
&= \frac{E\left(M_{0,T}(x_0) c(\xi_{0,T}(x_0)) \,|\mathcal{F}_t\right)}{E\left(M_{0,T}(x_0) \,|\mathcal{F}_t\right)} \\
&= E\left(M_{t,T}(x) c(\xi_{t,T}(x)) \,|\mathcal{F}_t\right) \\
&= E\left(M_{t,T}(x) c(\xi_{t,T}(x))\right) \\
&= V(t,x),
\end{aligned}
$$

the last two equalities being due to the Markov property and Theorem 6.6.7, respectively. The differentiability of $\xi_{t,T}(x)$ in x and t was established by Kunita [204].

Under $\overline{P}$, we have

$$
\xi_{0,t}(x_0)
$$

$$
= x_0 + \int_0^t \left(f(x, \xi_{0,s}(x_0)) + \sigma g(s, \xi_{0,s}(x_0))\right) ds + \int_0^t \sigma(s, \xi_{0,s}(x_0)) dW_s.
$$

Expand $V(t,x) = V(t, \xi_{0,t}(x_0))$ by the Itô rule to get

$$
V(t, \xi_{0,t}(x_0)) = N_t = V(0, x_0) + \int_0^t \left(\frac{\partial V}{\partial t} + LV\right)(s, \xi_{0,s}(x_0)) ds
$$

$$+ \int_0^t \frac{\partial V}{\partial x}(s, \xi_{0,s}(x_0))\sigma(s, \xi_{0,s}(x_0))dW_s, \quad (7.17)$$

where

$$L = \sum_{i=1}^n \left(f^i + \sum_{j=1}^m \sigma_{ij}g^j \right) \frac{\partial}{\partial x_i} + \frac{1}{2} \sum_{i,j=1}^n a_{ij} \frac{\partial^2}{\partial x_i \partial x_j},$$

and (a_{ij}) is the matrix $\sigma\sigma^*$.

A *special semimartingale* is a semimartingale that is the sum of a (local) martingale and a predictable process of (locally) integrable variation. The decomposition is unique when it exists (see [109, Theorem 12.38]).

Now N is a special semimartingale, so the decompositions given in (7.15) and (7.17) must be the same. As there is no bounded variation term in (7.15), we must have

$$\frac{\partial V}{\partial t} + LV = 0 \text{ with } V(T, x) = c(x).$$

In addition,

$$\gamma_s = \frac{\partial V}{\partial x}(x, \xi_{0,s}(x_0))\,\sigma(s, \xi_{0,s}(x_0)).$$

However, $\xi_{t,T}(x) = \xi_{0,T}(x_0)$ so, from the differentiability and linear growth of g, we have

$$\frac{\partial V(t, x)}{\partial x} = E\left(\frac{\partial M_{t,T}(x)}{\partial x}c(\xi_{0,T}(X_0)) + M_{t,T}(x)\frac{\partial c}{\partial x}(\xi_{t,T}(x)) \right).$$

Now, using the existence of solutions of stochastic differential equations that are differentiable in their initial conditions, we have,

$$\frac{\partial M_{t,T}(x)}{\partial x} = \int_t^T g(r, \xi_{t,r}(x))\frac{\partial M_{t,r}(x)}{\partial x}dB_r$$

$$+ \int_t^T dB_r g_\xi(r, \xi_{t,r}(x))\frac{\partial \xi_{t,r}(x)}{\partial x}M_{t,r}(x). \quad (7.18)$$

Equation (7.18) can be solved by variation of constants to obtain

$$\frac{\partial M_{t,T}(x)}{\partial x} = M_{t,T}(x)\int_t^T dW_r^* g(r, \xi_{t,r}(x))D_{t,r}(x).$$

Therefore, with $x = \xi_{0,t}(x_0)$,

$$\frac{\partial V(t, x)}{\partial x} = E\left(M_{t,T}(x)\left[\int_t^T dW_r^* g_\xi(r, \xi_{t,r}(x))D_{t,r}(x)c(\xi_{0,T}(x_0)) \right.\right.$$

$$+ c_\xi(\xi_{t,T}(x))D_{t,T}(x)])$$

$$= E_{\overline{P}}\left(\int_t^T dW_r^* g_\xi(r, \xi_{0,r}(x_0))D_{0,r}(x_0)c(\xi_{0,T}(x_0)) \right.$$

$$+ c_\xi(\xi_{0,T}(x_0))D_{0,T}(x_0)\,|\mathcal{F}_t)\,D_{0,t}^{-1}(x_0)$$

and the result follows.   □

7.4 Self-Financing Strategies

One-factor Model

In a pricing model with one risky stock, a hedging strategy is a measurable process $(\phi_t) = (H_t^0, H_t^1)$, with values in $\mathbb{R}^2$ that is adapted to the filtration $(\mathcal{F}_t)_{t \geq 0}$, where $\mathcal{F}_t = \sigma\{B_u : u \leq t\} = \sigma\{S_u^1 : u \leq t\}$. The quantity H_t^0 (resp. H_t^1) denotes the amount of the bond S_t^0 (resp. the risky asset S_t^1) that is held at time t. Consequently, the value, or wealth, of the portfolio at time t is

$$V_t(\phi) = H_t^0 S_t^0 + H_t^1 S_t^1. \tag{7.19}$$

In discrete time, we have established (see (2.2)) that a self-financing strategy should satisfy the identity

$$V_{n+1}(\phi) - V_n(\phi) = \phi_{n+1}(S_{n+1} - S_n)$$
$$= H_{n+1}^0(S_{n+1}^0 - S_n^0) + H_{n+1}^1(S_{n+1}^1 - S_n^1).$$

The natural continuous-time analogue of this condition therefore appears to be

$$dV_t(\phi) = H_t^0 dS_t^0 + H_t^1 dS_t^1. \tag{7.20}$$

Indeed, if H^0 and H^1 are of bounded variation, then

$$dV_t(\phi) = H_t^0 dS_t^0 + H_t^1 dS_t^1 + S_t^0 dH_t^0 + S_t^1 dH_t^1 \tag{7.21}$$

and equation (7.20) is equivalent to saying that

$$S_t^0 dH_t^0 + S_t^1 dH_t^1 = 0, \tag{7.22}$$

which is the analogue of (2.5). The intuitive meaning of (7.22) is that changes in the holdings of the bond, $S_t^0 dH_t^0$, can only take place due to corresponding changes in holding of the stock $S_t^1 dH_t^1$; that is, there is no net inflow or outflow of capital.

We consider European claims with an expiration time T. Consequently, for (7.20) to make sense, we require

$$\int_0^T |H_t^0|\, dt < \infty \text{ a.s.}, \qquad \int_0^T \left(H_t^1 S_t^1\right)^2 dt < \infty \text{ a.s.} \tag{7.23}$$

From the dynamics for S^1, namely

$$dS_t^1 = S_t^1(\mu dt + \sigma dB_t),$$

we have

$$\left. \begin{aligned} \int_0^T H_t^0 dS_t^0 &= \int_0^T H_t^0 re^{rt} dt, \\ \int_0^T H_t^1 dS_t^1 &= \int_0^T (H_t^1 S_t^1 \mu) dt + \int_0^T (H_t^1 S_t^1 \sigma) dB_t. \end{aligned} \right\} \tag{7.24}$$

We can therefore give the following definition.

Definition 7.4.1. A *self-financing strategy* $\phi = (\phi_t)_{0 \leq t \leq T}$ is given by two measurable adapted processes $(H_t^0), (H_t^1)$ satisfying (7.20) and (7.23). The corresponding wealth process is given for all $t \in [0, T]$ by

$$V_t(\phi) = H_t^0 S_t^0 + H_t^1 S_t^1 = H_0^0 S_0^0 + H_0^1 S_0^1 + \int_0^t H_u^0 dS_u^0 + \int_0^t H_u^1 dS_u^1 \text{ a.s.}$$

Notation 7.4.2. Write $\widetilde{S}_t^1 = e^{-rt} S_t^1$ for the discounted price of the risky asset and $\widetilde{V}_t(\phi) = e^{-rt} V_t(\phi)$ for the discounted wealth process.

We can then establish the following result, whose discrete-time analogue was discussed in Section 2.2.

Theorem 7.4.3. *Suppose* $\phi = (\phi_t) = \left(H_t^0, H_t^1\right)_{0 \leq t \leq T}$ *is a pair of measurable adapted processes satisfying* (7.23). *Then* ϕ *is a self-financing strategy if and only if*

$$\widetilde{V}_t(\phi) = V_0(\phi) + \int_0^t H_u^1 d\widetilde{S}_u^1 \text{ a.s. for all } t \in [0, T]. \qquad (7.25)$$

Proof. Suppose $\phi = \left(H_t^0, H_t^1\right)$ is self-financing, so (7.20) holds. Then

$$
\begin{aligned}
d\widetilde{V}_t(\phi) &= d\left(e^{-rt} V_t(\phi)\right) \\
&= -r\widetilde{V}_t(\phi)dt + e^{-rt} dV_t(\phi) \\
&= -re^{-rt}(H_t^0 e^{rt} + H_t^1 S_t^1)dt + e^{-rt} H_t^0 d(e^{rt}) + e^{-rt} H_t^1 dS_t^1 \\
&= H_t^1(-re^{-rt} S_t^1 dt + e^{-rt} dS_t^1) \\
&= H_t^1 d\widetilde{S}_t,
\end{aligned}
$$

and (7.25) follows.

The converse follows by considering $V_t(\phi) = e^{rt}\widetilde{V}_t(\phi)$, reversing the steps above and using (7.25). $\qquad\square$

Remark 7.4.4. Although in general the continuous-time stochastic integral requires predictable integrands (see [109]), we have only required the trading strategies $\phi = \left(H_t^0, H_t^1\right)$ to be measurable and adapted. This is possible here because the filtration $(\mathcal{F}_t)_{t \geq 0}$ is generated by the continuous Brownian motion (B_t) (or, equivalently, by the continuous process (S_t^1)).

Let us write SF for the set of self-financing strategies $\phi = (\phi_t)_{0 \leq t \leq T}$ so that $\phi_t = (H_t^0, H_t^1)$, where H^0 and H^1 satisfy (7.23). If there are no contributions or withdrawals, the corresponding wealth process is given in (7.19) as

$$V_t(\phi) = H_t^0 S_t^0 + H_t^1 S_t^1.$$

Suppose more generally (as we did for discrete-time models in Section 5.6) that the model allows contributions to the wealth process (say, from dividends) or withdrawals (consumption). Let these be modelled by the adapted right-continuous, increasing processes D_t (for contributions) and C_t (for accumulated consumption). Then

$$V_t(\phi) = H_t^0 S_t^0 + H_t^1 S_t^1 + D_t - C_t$$
$$= V_0(\phi) + \int_0^t H_u^0 dS_u^0 + \int_0^t H_u^1 dS_u^1 + D_t - C_t.$$

The self-financing condition (7.20) now becomes

$$S_t^0 dH_t^0 + S_t^1 dH_t^1 = dD_t - dC_t.$$

7.5 An Equivalent Martingale Measure

Consider the situation of Section 7.1 where the bond is described by a price process (S_t^0) satisfying

$$dS_t^0 = rS_t^0 dt \tag{7.26}$$

and the risky asset has a price process (S_t^1) satisfying

$$dS_t^1 = S_t^1(\mu dt + \sigma dB_t). \tag{7.27}$$

The discounted price of the risky asset is

$$\widetilde{S}_t = e^{-rt} S_t^1$$

with dynamics

$$d\widetilde{S}_t = -re^{-rt} S_t^1 + e^{-rt} dS_t^1 = \widetilde{S}_t \left((\mu - r)dt + \sigma dB_t\right).$$

If we apply Girsanov's theorem Theorem 7.2.3 with $\theta_t = \frac{\mu - r}{\sigma}$, we see there is a probability measure P^μ defined on $\mathcal{F}_T$ by putting

$$\frac{dP^\mu}{dP} = \Lambda_T = \exp\left\{-\int_0^t \theta_s dB_s - \frac{1}{2}\int_0^t \theta_s^2 ds\right\}$$

such that if

$$W_t^\mu = \left(\frac{\mu - r}{\sigma}\right)t + B_t, \tag{7.28}$$

then $(W_t^\mu)_{0 \leq t \leq T}$ is a standard Brownian motion under P^μ. Under P^μ, we then have

$$d\widetilde{S}_t = \widetilde{S}_t \sigma dW_t^\mu, \qquad \widetilde{S}_t = S_0 \exp\left\{\sigma W_t^\mu - \frac{\sigma^2}{2}t\right\}.$$

Definition 7.5.1. A strategy $\phi = (H_t^0, H_t^1)_{0 \leq t \leq T}$ is *admissible* if it is self-financing and the discounted value process

$$\widetilde{V}_t(\phi) = H_t^0 + H_t^1 \widetilde{S}_t$$

is non-negative and square integrable under P.

Definition 7.5.2. A *European contingent claim* is a positive random variable h, measurable with respect to $\mathcal{F}_T$.

If $h = f(S_T)$ and $f(S_T) = (S_T - K)^+$ (resp. $f(S_T) = (K - S_T)^+$), then the option is a *European call option* (resp. *European put option*).

Recall that a claim is attainable (sometimes also called *replicable*) if its value at time T (the exercise time) is equal to the value $V_T(\phi) = H_T^0 S_t^0 + H_T^1 S_T^1$ of an admissible strategy ϕ.

Suppose that $\phi \in SF$, and $(\phi_t) = (H_t^0, H_t^1)$. The corresponding wealth process is

$$V_t(\phi) = H_t^0 S_t^0 + H_t^1 S_t^1 = V_0(\phi) + \int_0^t H_u^0 dS_u^0 + \int_0^t H_u^1 dS_u^1.$$

From (7.26) and (7.27), it follows that

$$V_t(\phi) = V_0(\phi) + \int_0^t r H_u^0 S_u^0 du + \int_0^t H_u^1 S_u^1 (\mu du + \sigma dB_u)$$

$$= V_0(\phi) + \int_0^t r V_u(\phi_u) du + \int_0^t \sigma H_u^1 S_u^1 dW_u^\mu,$$

where W^μ is defined in (7.28). Consider the discounted wealth

$$\widetilde{V}_t(\phi) = \left(S_t^0\right)^{-1} V_t(\phi_t).$$

From the differentiation rule, we have

$$\widetilde{V}_t(\phi) = V_0(\phi_0) + \int_0^t \sigma H_u^1 \left(S_u^0\right)^{-1} \left(S_u^1\right) dW_u^\mu.$$

We have seen above that, under the measure P^μ, the process W^μ is a standard Brownian motion. Consequently, under P^μ, we see that

$$M_t = \int_0^t \sigma H_u^1 \left(S_u^0\right)^{-1} S_u^1 dW_u^\mu$$

is a local martingale. In fact, consider the stopping times

$$T_n = \inf\left\{ t \geq 0 : \int_0^t \left(\sigma H_u^1 \left(S_u^0\right)^{-1} S_u^1\right)^2 du \geq n \right\}.$$

Then the (T_n) are increasing and $\lim_n T_n = T$. Furthermore, $(M_{t \wedge T_n})$ is a uniformly integrable martingale under measure P^μ for each n. Consequently, $\left(\widetilde{V}_t(\phi_t)\right)$ is a local martingale under P^μ.

Suppose $\xi = \xi(\omega)$ is a non-negative $\mathcal{F}$-measurable random variable with $E^\mu(\xi) < \infty$, where $E^\mu(\cdot)$ denotes expectation with respect to the measure P^μ. A strategy $\phi \in SF$ will belong to $SF(\xi)$ if

$$\widetilde{V}_t(\phi_t) \geq -E^\mu(\xi \,|\, \mathcal{F}_t) \text{ a.s. for } t \geq 0.$$

A strategy $\phi \in SF(\xi)$ will provide a hedge against a maximum loss of ξ. Under this condition, Fatou's lemma ([279, Chap. II, §6]) can be applied, and the local martingale $\tilde{V}_t(\phi)$ is a supermartingale under measure P^μ.

Consequently, if τ_1 and τ_2 are two stopping times, with $\tau_i \leq T$ and $\tau_1 \leq \tau_2$ a.s., then

$$E^\mu(\tilde{V}_{\tau_2}(\phi)\,|\mathcal{F}_{\tau_1}) \leq \tilde{V}_{\tau_1}(\phi_{\tau_1})$$

from the optimal stopping theorem. In particular, if $V_0(\phi_0) = x \geq 0$ and $\phi \in SF(\xi)$, then

$$E^\mu\left(\tilde{V}_\tau(\phi)\right) = E^\mu\left(e^{-rt}V_\tau(\phi_\tau)\right) \leq x. \tag{7.29}$$

Let us summarise these observations in the following result.

Lemma 7.5.3. *a) If $\phi \in SF$, then $\tilde{V}(\phi)$ is a local martingale.*

b) If $\phi \in SF(\xi)$, then $\tilde{V}(\phi)$ is also a supermartingale.

c) If $\phi \in SF(0)$, then $\tilde{V}(\phi)$ is also a non-negative supermartingale.

Note that if $\phi \in SF(0)$, then simultaneous borrowing from the bank and stocks is not permitted.

Definition 7.5.4. A strategy $\phi \in SF$ is said to provide an *arbitrage opportunity* if, with $V_0(\phi) = x \leq 0$, we have $V_T(\phi) \geq 0$ a.s. and

$$P(V_T(\phi) > 0) > 0.$$

We can then establish the following lemma.

Lemma 7.5.5. *If ξ is a non-negative $\mathcal{F}$-measurable random variable with*

$$E^\mu(\xi) < \infty,$$

then any $\phi \in SF(\xi)$ does not provide an arbitrage opportunity.

Proof. Equation (7.29) rules out the possibility of ϕ providing an arbitrage opportunity. $\qquad\square$

Superhedging

We turn to the continuous-time analogue of the brief discussion outlined in Section 2.4.

Definition 7.5.6. Suppose $T > 0$ and f_T is an $\mathcal{F}_T$-measurable, non-negative random variable. A strategy $\phi \in SF$ is a *hedge* for the European claim f_T with initial investment x if

$$V_0(\phi) = x, \qquad\qquad V_T(\phi) \geq f_T \text{ a.s.}$$

We call $\phi \in SF$ an (x, f_T)-hedge.

Although the next definition can be given for strategies in SF, we shall restrict ourselves to strategies in $SF(0)$.

Definition 7.5.7. The *investment price $C(T, f_T)$ for the European claim f_T at time $T > 0$ is the smallest initial investment with which the investor can attain an amount f_T at time T using strategies from $SF(0)$.*

More precisely, write $\Sigma(T, x, f_T)$ for the set of (x, f_T)-hedges that belong to $SF(0)$. Then

$$C(T, f_T) = \inf \left\{ x \geq 0 : \Sigma(T, x, f_T) \neq \emptyset \right\}.$$

For any European claim f_T, we must therefore do the following:

a) Determine the investment price $C(T, f_T)$.

b) Determine the (x, f_T)-hedging strategy $\phi \in SF(0)$ for $x = C(T, f_T)$.

Remark 7.5.8. A European option with exercise time $T > 0$ and payment f_T gives the buyer of the contract the right to obtain an amount f_T at time T.

Clearly, if the seller of the contract can start with an amount $x = C(T, f_T)$ and obtain $V_T(\phi_T) \geq f_T$ at time T, then $C(T, f_T)$ is the fair, or rational, price for the option from the seller's point of view. The discussion in Section 2.4 also applies here, in the continuous-time setting, and shows that (in a complete market, which is the case here) $C(T, f_T)$ is also the fair price from the buyer's standpoint.

Recall that the price at time t of the European call option on S^1 with an exercise time T and a strike price K corresponds to taking $f_T = (S_T^1 - K)^+$.

Pricing

Suppose that $\phi \in SF(0)$. Then, since $\widetilde{V}(\phi)$ is a supermartingale, we have

$$x = V_0(\phi_0) \geq E^\mu \left(e^{-rT} V_T(\phi_T) \right).$$

If, further, ϕ is an (x, f_T)-hedge, then

$$x \geq E^\mu \left(e^{-rT} f_T \right).$$

Consequently, the rational investment price $C(T, f_T)$ satisfies

$$C(T, f_T) \geq E^\mu \left(e^{-rT} f_T \right).$$

From (7.27),

$$dS_t^1 = S_t^1 (\mu dt + \sigma dB_t),$$

where B is a standard Brownian motion under P. Write $S^1(\mu)$ for the solution of (7.27). Then, from (7.28), under the measure P^μ, the process $S^1(\mu)$ satisfies

$$dS_t^1 = S_t^1 \left(rdt + \sigma dW_t^\mu \right),$$

where W^μ is a standard Brownian motion.

Let us write $S^1(r)$ for the solution of

$$dS_t^1(r) = S_t^1(r)(rdt + \sigma dB_t).$$

Then

$$\text{Law}\left(f_T\left(S^1(\mu)\right)|P^\mu\right) = \text{Law}\left(f_T\left(S^1(r)\right)|P\right)$$

and

$$E^\mu\left(e^{-rT}f_T\right) = E^\mu\left(e^{-rT}f_T\left(S^1(\mu)\right)\right) = E\left(e^{-rT}f_T\left(S^1(r)\right)\right).$$

This quantity has the unexpected property that it does not depend on μ.

Suppose that f_T is a non-negative $\mathcal{F}_T$-measurable random variable such that

$$E^\mu\left(f_T^2\right) < \infty. \tag{7.30}$$

Recall that $E^\mu\left(f_T^2\right) = E\left(\Lambda_T f_T^2\right)$, where

$$\Lambda_T = \exp\left\{-\int_0^T \theta dB_s - \frac{1}{2}\int_0^T \theta^2 ds\right\}, \qquad \theta = \frac{\mu - r}{\sigma}.$$

A sufficient condition for (7.30) is that $E\left(f_T^{2+\delta}\right) < \infty$ for some $\delta > 0$.

Consider the square integrable $(P^\mu, \mathcal{F}_t)$-martingale

$$N_t = E^\mu\left(e^{-rT}f_T\,|\mathcal{F}_t\right) \text{ for } 0 \le t \le T.$$

From the martingale representation result, Theorem 7.3.9, there is a predictable process γ such that

$$E\left(\int_0^T \gamma_s^2 ds\right) < \infty, \qquad N_t = N_0 + \int_0^t \gamma_s dW_s^\mu \text{ a.s.} \tag{7.31}$$

Here, $N_0 = E^\mu\left(e^{-rT}f_T\right)$.

Now take

$$H_t^1 = \gamma_t e^{rt}\sigma^{-1}\left(S_t^1\right)^{-1} \qquad\qquad H_t^0 = N_t - \sigma^{-1}\gamma_t;$$

consider the trading strategy $\phi_t^* = (H_t^0, H_t^1)$.

Lemma 7.5.9. *a) The strategy ϕ_t^* is self-financing and*

b)

$$N_t = \widetilde{V}_t(\phi^*) = e^{-rt}V_t(\phi^*).$$

Proof. By definition,

$$V_t(\phi^*) = H_t^0 S_t^0 + H_t^1 S_t^1 = N_t S_t^0. \tag{7.32}$$

Therefore

$$
\begin{aligned}
dV_t(\phi^*) &= N_t dS_t^0 + S_t^0 dN_t \\
&= r N_t S_t^0 dt + S_t^0 \gamma_t dW_t^\mu \\
&= \left(N_t - \sigma^{-1} \gamma_t \right) dS_t^0 + \sigma^{-1} \gamma_t S_t^0 \left(r dt + \sigma dW_t^\mu \right) \\
&= H_t^0 dS_t^0 + H_t^1 dS_t^1.
\end{aligned}
$$

Consequently, the strategy $\phi_t^* = \left(H_t^0, H_t^1 \right)$ is self-financing. (The condition (7.22) is satisfied because of (7.31) and the path-continuity of N.)

From (7.32) we see that

$$
N_t = \left(S_t^0 \right)^{-1} V_t(\phi^*),
$$

that is, the P^μ-martingale N is the discounted wealth process of the strategy ϕ_t^*. Also,

$$
V_t(\phi^*) = N_t S_t^0 = E^\mu \left(S_t^0 \left(S_t^0 \right)^{-1} f_T \,|\, \mathcal{F}_t \right) = E^\mu \left(e^{-r(T-t)} f_T \,|\, \mathcal{F}_t \right).
$$

In particular,

$$
V_0(\phi^*) = E^\mu \left(e^{-rT} f_T \right), \qquad\qquad V_T(\phi^*) = f_T.
$$

These equations mean that ϕ^* is an (x, f_T)-hedge with initial capital

$$
x = E^\mu \left(e^{-rT} f_T \right).
$$

Clearly, if $\phi \in SF$ is any hedge for f_T with initial capital x, then

$$
V_T(\phi) \geq V_T(\phi_T^*) = f_T.
$$

Consequently, the rational price for the European option f_T is

$$
C(T, f_T) = E^\mu \left(e^{-rT} f_T \right) = E \left(e^{-rT} f_T \left(S^1(r) \right) \right).
$$

From (7.5) this is $E \left[e^{-rT} f_T \left(S^1(r) \right) \right]$ and so $C(T, f_T)$ does not depend on μ. $\square$

In summary, we have shown that the following results hold.

Theorem 7.5.10. *Suppose that f_T represents a European claim, which can be exercised at time T. That is, f_T is an $\mathcal{F}_T$-measurable random variable and*

$$
E^\mu \left(e^{-rT} f_T \right) < \infty.
$$

Then the rational price for f_T is

$$
C(T, f_T) = E^\mu \left(e^{-rT} f_T \left(S^1(\mu) \right) \right) = E \left(e^{-rT} f_T \left(S^1(r) \right) \right).
$$

There is a minimal hedge $\phi_t^* = \left(H_t^0, H_t^1 \right)$ given by

$$
H_t^1 = \sigma^{-1} \gamma_t e^{rt} \left(S_t^1 \right)^{-1}, \qquad\qquad H_t^0 = N_t - e^{rt} S_t^1 H_t^1.
$$

Here (N_t) is the martingale $\left(E^\mu \left(e^{-rT} f_T \,|\, \mathcal{F}_t \right) \right)$ and (γ_t) is the integrand in its martingale representation.

Multifactor Models

Definition 7.5.11. In the setting of a probability space $(\Omega, \mathcal{F}, P)$, an equivalent measure $\tilde{P}$ is called a *martingale measure* if, under $\tilde{P}$, all discounted asset prices are martingales. $\tilde{P}$ is sometimes called a *risk-neutral measure*.

Remark 7.5.12. We have seen that, in the case of one risky asset, $\tilde{P} = P^\mu$ is a martingale measure. We now extend these ideas to multifactor Brownian pricing models.

Suppose $(B_t) = \left(B_t^1, \ldots, B_t^m\right)_{0 \le t \le T}$ is an m-dimensional Brownian motion on $(\Omega, \mathcal{F}, P)$ and let $(\mathcal{F}_t)$ be the filtration generated by B. Suppose now we have a bond with price S_t^0, or bank account, whose instantaneous interest rate is r_t, and n risky assets with time t prices $S_t^1, \ldots, S_t^n$.

With $S_0^0 = 1$, we have $S_t^0 = \exp\left\{\int_0^t r_u du\right\}$. The dynamics of the risky assets are described by the equations

$$dS_t^i = \mu_i(t)S_t^i dt + S_t^i \left(\sum_{j=1}^n \sigma_{ij}(t)dB_t^j\right).$$

Here μ_i, σ_{ij}, and r are adapted processes. The prices $\frac{S_t^1}{S_t^0}, \ldots, \frac{S_t^n}{S_t^0}$ are the discounted prices. The differentiation rule gives

$$d\left(\frac{S_t^i}{S_t^0}\right) = (\mu_i(t) - r_t)\frac{S_t^i}{S_t^0}dt + \frac{S_t^i}{S_t^0}\sum_{j=1}^m \sigma_{ij}(t)dB_t^j. \tag{7.33}$$

Definition 7.5.13. $(\mu_i(t) - r_t)$ is called the *risk premium*.

Definition 7.5.14. If we can find processes $\theta_1(t), \ldots, \theta_n(t)$ such that

$$\mu_i(t) - r_t = \sum_{j=1}^m \sigma_{ij}(t)\theta_j(t) \text{ for } i = 1, 2, \ldots, n, \tag{7.34}$$

then the adapted process

$$\theta(t) = (\theta_1(t), \ldots, \theta_n(t))$$

is called the *market price of risk*. Equation (7.33) then becomes

$$d\left(\frac{S_t^i}{S_t^0}\right) = \frac{S_t^i}{S_t^0}\left(\sum_{j=1}^m \sigma_{ij}(t)\left(\theta_j(t)dt + dB_t^j\right)\right).$$

Now consider the linear system (7.34). Three cases can arise:

a) It has a unique solution $(\theta(t)) = (\theta_1(t), \ldots, \theta_n(t))$.

b) It has no solution.

c) It has more than one solution.

In the first and third cases, we have a solution process $\theta(t)$. Consider the process

$$\Lambda_t = \exp\left\{-\int_0^t \theta(u)dB(u) - \frac{1}{2}\int_0^t |\theta(u)|^2\,du\right\},$$

and define a new measure P^θ by setting

$$\left.\frac{dP^\theta}{dP}\right|_{\mathcal{F}_T} = \Lambda_T.$$

The vector form of Girsanov's theorem states that, under P^θ, $W^\theta = (W^{\theta_i})_{1\le i\le m}$ is an m-dimensional martingale with dynamics

$$dW_t^\theta = \theta(t)dt + dB_t.$$

A hedging strategy is a measurable adapted process $\phi_t = (H_t^j)_{1\le j\le N}$, where H_t^i represents the number of units of asset i held at time t. Its corresponding wealth process is

$$V_t(\phi_t) = H_t^0 S_t^0 + H_t^1 S_t^1 + \cdots + H_t^n S_t^n.$$

The strategy ϕ is said to be self-financing if

$$dV_t(\phi_t) = \sum_{i=0}^n H_t^i dS_t^i$$

so that

$$V_t(\phi_t) = V_0(\phi_0) + \int_0^t \sum_{i=0}^n H_u^i dS_u^i$$

$$= V_0(\phi_0) + \int_0^t rH_u^0 S_u^0\,du + \sum_{i=1}^n \int_0^t H_u^i S_u^i\left(\mu_i(u) + \sum_{j=1}^m \sigma_{ij}(u)dB^j(u)\right)$$

$$= V_0(\phi_0) + \int_0^t rV_u(\phi_u)\,du + \sum_{i=1}^n \sum_{j=1}^m \int_0^t H_u^i S_u^i \sigma_{ij}\left(\theta_j(u)du + dB^j(u)\right)$$

$$= V_0(\phi_0) + \int_0^t rV_u(\phi_u)\,du + \sum_{i=1}^n \sum_{j=1}^m \int_0^t H_u^i S_u^i \sigma_{ij}(u)dW_u^{\theta_j}.$$

Therefore, the discounted wealth

$$\widetilde{V}_t(\phi_t) = \left(S_t^0\right)^{-1} V_t(\phi_t) = V_0(\phi_0) + \sum_{i=1}^n \sum_{j=1}^m \int_0^t H_u^i S_u^i \sigma_{ij}(u)dW_u^{\theta_j}$$

is a local martingale under P^θ. For $\phi \in SF(\xi)$, the proof of the first part of this section shows that $V_t(\phi_t)$ is a supermartingale.

Consequently, there is no arbitrage if (7.34) has at least one solution. When the solution is unique, we must have $m = n$ and the matrix $\sigma = (\sigma_{ij}(t))$ is non-singular.

If f_T is a European claim to be exercised at time T, we can consider the martingale

$$N_t = E^\theta \left(e^{-rT} f_T \,|\, \mathcal{F}_t \right).$$

By the martingale representation result, this can be written as

$$N_t = N_0 + \int_0^t \gamma_u dW_u^\theta,$$

where $\gamma_u = \left(\gamma_u^1, \gamma_u^2, \ldots, \gamma_u^n \right)$ is a measurable, adapted process such that

$$E \left(\int_0^T |\gamma_u|^2 \, du \right) < \infty.$$

Write Δ_t for the matrix $\operatorname{diag} \left(\left(S_t^1 \right)^{-1}, \ldots, \left(S_t^N \right)^{-1} \right)$, and let

$$\left(H_t^1, H_t^2, \ldots, H_t^N \right) = S_t^0 \Delta_t \sigma'(t)^{-1} \gamma_t, \qquad H_t^0 = N_t - \left(\sigma'(t)^{-1} \gamma_t \right) \cdot \mathbf{1},$$

where $\mathbf{1} = (1, \ldots, 1)$. Then $\left(H_t^0, H_t^1, \ldots, H_t^N \right)$ is a self-financing strategy that hedges the claim f_T and the market is complete.

In the case where (7.34) has multiple solutions, although there are no arbitrage opportunities, there are claims that cannot be hedged, and the market is incomplete. If (7.34) has no solution, there is no martingale measure and the market may allow arbitrage.

In Chapters 3 and 4, we derived the two fundamental theorems of asset pricing in discrete time models. The extensive literature on this topic began with two papers by Harrison and Pliska ([149],[150]). In continuous time, the technical issues surrounding the 'equivalence' of the no-arbitrage condition and the existence of an EMM for the model are more intricate still, and we do not pursue this here. The reader is referred especially to the papers by Stricker [287] and Delbaen and Schachermayer [76].

7.6 Black-Scholes Prices

In this section, we suppose the European option has the form $f\left(S_T^1\right)$. We require some integrability properties of $f : (0, \infty) \to [0, \infty)$, so we assume that for some non-negative c, k_1, k_2

$$f(s) \le c(1 + s^{k_1})s^{-k_2}. \tag{7.35}$$

From Theorem 7.5.10, the rational price for the option f is independent of μ and is given by

$$C(T, f) = E\left(e^{-rT} f\left(S_T^1(r)\right)\right), \tag{7.36}$$

where S^1 is the solution of

$$dS_t^1 = S_t^1(r\,dt + \sigma\,dW_t).$$

Here W is a standard Brownian motion on $(\Omega, \mathcal{F}, P)$.

The wealth process of the corresponding minimal hedge is

$$V_t(\phi_t^*) = E\left(e^{-r(T-t)} f\left(S_T^1(r)\right) \mid \mathcal{F}_t\right).$$

Now, from (7.36),

$$S_0^1 = S_0^1 \exp\left\{\left(r - \frac{\sigma^2}{2}\right) t + \sigma W_t\right\}.$$

From the Markov property, we have

$$
\begin{aligned}
V_t(\phi^*) &= E\left(e^{-r(T-t)} f\left(S_T^1(r)\right) \mid \mathcal{F}_t\right) \\
&= E\left(e^{-r(T-t)} f\left(S_T^1\right) \mid S_t^1\right) \\
&= e^{-r(T-t)} F\left(T - t, S_t^1\right).
\end{aligned}
\tag{7.37}
$$

Here

$$
\begin{aligned}
F(T - t, s) &= \\
\frac{1}{\sqrt{2\pi}} \int_{-\infty}^{\infty} & f\left(s \exp\left\{\sigma y\sqrt{T-t} + \left(r - \frac{\sigma^2}{2}\right)(T-t)\right\}\right) e^{-\frac{1}{2}y^2}\,dy \\
&= \frac{1}{s} \int_{-\infty}^{\infty} f(y) g\left(T - t, \frac{y}{s}, r - \frac{\sigma^2}{2}, \sigma\right)\,dy,
\end{aligned}
$$

where

$$g(t, z, \alpha, \beta) = \frac{1}{\beta z \sqrt{2\pi t}} \exp\left\{-\frac{(\log z - \alpha t)^2}{2\beta^2 t}\right\}.$$

From the integrability condition, the function $F(T - t, s)$ is differentiable in t and s. Furthermore, $E\left(F\left(T - t, S_t^1\right)\right) < \infty$. Write $G(t, x) = F(T - t, e^{rt}x)$. Then, from (7.37),

$$V_t(\phi^*)e^{-rt} = e^{-rT} G(t, e^{-rt}S_t^1).$$

Using the Itô differentiation rule, we obtain

$$d\left(V_t(\phi^*)e^{-rt}\right) = e^{-rT} d\left(G(t, e^{-rt}S_t^1)\right)$$

$$= e^{-rT} \left(\frac{\partial G}{\partial x}(t, e^{-rt} S_t^1) d(e^{-rt} S_t^1) + \left(\frac{\partial G}{\partial t} + \frac{1}{2} \frac{\partial^2 G}{\partial x^2} \sigma^2 \left(S_t^1\right)^2 e^{-2t} \right) dt \right).$$

That is,

$$V_t(\phi^*)e^{-rt} = N_t = E\left(e^{-rT} f\left(S_T^1(r)\right) | \mathcal{F}_t\right)$$

$$= e^{-rT} E\left(f\left(S_T^1(r)\right)\right) + e^{-rT} \int_0^t \frac{\partial G}{\partial x} \cdot d(e^{-ru} S_u^1)$$

$$+ e^{-rT} \int_0^t \left(\frac{\partial G}{\partial t} + \frac{1}{2} \frac{\partial^2 G}{\partial x^2} \cdot \sigma^2 \left(S_u^1\right)^2 e^{-2u} \right) du. \qquad (7.38)$$

Now $d(e^{-ru} S_u^1) = \sigma e^{-ru} S_u^1 dW_u$. Consequently, the first integral in (7.38) is a local martingale. As the left-hand side of (7.38) is a martingale, the bounded variation process in (7.38) must be identically zero, as in (7.17). This implies that

$$\frac{\partial G}{\partial t} + \frac{1}{2} \frac{\partial^2 G}{\partial x^2} \sigma^2 \left(S_t^1\right)^2 e^{-2t} = 0,$$

with $G(T, x) = f(e^{rT} x)$. Noting that $\frac{\partial G}{\partial t} = e^{rt}(\frac{\partial(}{\partial F} x)x)$, we have therefore proved the following theorem.

Theorem 7.6.1. *Consider a European option with exercise time $T > 0$ and payment function $f\left(S_T^1\right)$, where f satisfies the integrability condition (7.35). Then the rational price for the option is*

$$C(T, f_T) = e^{-rT} F(T, S_0^1),$$

where

$$F(T, S_0^1) = \frac{1}{\sqrt{2\pi}} \int_{-\infty}^{\infty} f\left(S_0^1 \exp\left\{ \left(r - \frac{\sigma^2}{2}\right) T + \sigma y \sqrt{T} \right\} \right) e^{-\frac{1}{2} y^2} dy.$$

The minimal hedge $\phi_t^* = (H_t^0, H_t^1)$ is

$$H_t^1 = e^{-r(T-t)} \frac{\partial F}{\partial s} \left(T - t, S_t^1\right),$$

$$H_t^0 = e^{-rT} \left(F\left(T - t, S_t^1\right) - S_t^1 \frac{\partial F}{\partial s} \left(T - t, S_t^1\right) \right).$$

The corresponding wealth process is

$$V_t(\phi^*) = e^{-r(T-t)} F\left(T - t, S_t^1\right).$$

This is also the rational price for the option at time t.

The Black-Scholes Formula for a European Call

For the standard European call option we have $f\left(S_T^1\right) = \left(K - S_T^1\right)^+$. Specialising the above results, we recover the Black-Scholes pricing formula (2.32) as well as identifying the minimal hedge portfolio.

Theorem 7.6.2 (Black-Scholes). *The rational price of a standard European call option is*

$$C(T, \left(K - S_T^1\right)^+) = S_0^1 \Phi(d_+) - Ke^{-rT} \Phi(d_-).$$

Here $\Phi(y) = \frac{1}{\sqrt{2\pi}} \int_{-\infty}^y e^{-\frac{1}{2}z^2} dz$ *is the standard normal cumulative distribution function, and*

$$d_+ = \frac{\log\left(\frac{S_0^1}{K}\right) + T\left(r + \frac{\sigma^2}{2}\right)}{\sigma\sqrt{T}}, \qquad d_- = \frac{\log\left(\frac{S_0^1}{K}\right) + T\left(r - \frac{\sigma^2}{2}\right)}{\sigma\sqrt{T}}.$$

(Note that $d_- = d_+ - \sigma\sqrt{T}$.)
   The minimal hedge $\phi_t^* = (H_t^0, H_t^1)$ *has*

$$H_t^1 = \Phi\left(\frac{\log\left(\frac{S_t^1}{K}\right) + (T - t)\left(r + \frac{\sigma^2}{2}\right)}{\sigma\sqrt{T - t}}\right),$$

$$H_t^0 = -e^{-rT} K \Phi\left(\frac{\log\left(\frac{S_t^1}{K}\right) + (T - t)\left(r - \frac{\sigma^2}{2}\right)}{\sigma\sqrt{T - t}}\right).$$

The corresponding wealth process is

$$V_t(\phi^*) = H_t^0 S_t^0 + H_t^1 S_t^1.$$

Proof. With $f(s) = (s - K)^+$, we have, from Theorem 7.6.1,

$$F(t, s) = \frac{1}{\sqrt{2\pi}} \int_{-\infty}^\infty f\left(s \exp\left\{\sigma y\sqrt{t} + \left(r - \frac{\sigma^2}{2}\right)t\right\}\right) e^{-\frac{1}{2}y^2} dy$$

$$= \frac{1}{\sqrt{2\pi}} \int_{y(t,s)}^\infty \left(s \exp\left\{\sigma y\sqrt{t} + \left(r - \frac{\sigma^2}{2}\right)t\right\} - K\right) e^{-\frac{1}{2}y^2} dy,$$

where $y(t, s)$ is the solution of

$$s \exp\left\{\sigma y\sqrt{t} + \left(r - \frac{\sigma^2}{2}\right)t\right\} = K, \qquad (7.39)$$

so

$$y(t, s) = \sigma^{-1} t^{-\frac{1}{2}} \left(\log\left(\frac{K}{s}\right) - \left(r - \frac{\sigma^2}{2}\right)t\right).$$

Consequently,

$$F(t, s) = \frac{e^{rt}}{\sqrt{2\pi}} \int_{y(t,s)}^{\infty} s \exp \left\{ \sigma y \sqrt{y} - \sigma^2 \frac{t}{2} - \frac{1}{2} y^2 \right\} dy - K \left[1 - \Phi(y(t, s)) \right]$$

$$= \frac{se^{rt}}{\sqrt{2\pi}} \int_{y(t,s)-\sigma\sqrt{t}}^{\infty} e^{-\frac{1}{2}x^2} dx - K \left[1 - \Phi(y(t, s)) \right]$$

$$= se^{rt} \left[1 - \Phi(y(t, s) - \sigma\sqrt{t}) \right] - K \left[1 - \Phi(y(t, s)) \right].$$

From Theorem 7.5.10, the rational price for the standard European call option is

$$C(T, (S_T - K)^+) = e^{-rT} F(T, S_0)$$

$$= S_0 \Phi \left(\sigma\sqrt{T} - y(T, S_0) \right) - Ke^{-rT} \Phi(-y(T, S_0))$$

$$= S_0 \Phi(d_+) - Ke^{-rT} \Phi(d_-).$$

Now, from Theorem 7.6.1, the minimal hedge is $H_t^1 = e^{-r(T-t)} \frac{\partial F}{\partial s} (T-t, S_t)$ so, after some cancellations when performing the differentiation (see also Section 7.10), we obtain

$$H_t^1 = \Phi(\sigma\sqrt{T-t} - y(T-t, S_t))$$

$$= \Phi \left(\sigma\sqrt{T-t} - \sigma^{-1}(T-t)^{-\frac{1}{2}} \left(\log \left(\frac{K}{S_t} \right) - \left(r - \frac{\sigma^2}{2} \right) (T-t) \right) \right)$$

$$= \Phi \left(\frac{\log \left(\frac{S_t}{K} \right) + (T-t) \left(r + \frac{\sigma^2}{2} \right)}{\sigma\sqrt{T-t}} \right).$$

Now

$$V_t(\phi^*) = e^{-r(T-t)} F(T-t, S_t)$$

$$= S_t \Phi \left(\frac{\log \left(\frac{S_t}{K} \right) + (T-t) \left(r + \frac{\sigma^2}{2} \right)}{\sigma\sqrt{T-t}} \right)$$

$$- Ke^{-r(T-t)} \Phi \left(\frac{\log \left(\frac{S_t}{K} \right) + (T-t) \left(r - \frac{\sigma^2}{2} \right)}{\sigma\sqrt{T-t}} \right).$$

Then

$$H_t^0 = e^{-rt} V_t(\phi^*) - e^{-rt} H_t^1 S_t$$

$$= -Ke^{-rT} \Phi \left(\frac{\log \left(\frac{S_t}{K} \right) + \left(r - \frac{\sigma^2}{2} \right) (T-t)}{\sigma\sqrt{T-t}} \right),$$

and the result follows. □

Call-Put Parity

The simple relation between call and put prices was established in Chapter 1 by a basic no-arbitrage argument that was independent of the particular pricing model used. Recalling that the European put option with strike K and expiry T has the value

$$C\left(T, [K - S_T^1(r)]^+\right) = E\left(e^{-rT} [K - S_T^1(r)]^+\right),$$

we can give a simple 'model-dependent' version of the call-put parity formula in the Black-Scholes model as follows. Since $(K - S)^+ = (S - K)^+ - S + K$, we have

$$
\begin{aligned}
E&\left(e^{-rT} [K - S_T^1(r)]^+\right) \\
&= E\left(e^{-rT} \left(S_T^1(r) - K\right)^+ - e^{-rT} S_T^1(r) + e^{-rT} K\right) \\
&= E\left(e^{-rT} \left(S_T^1(r) - K\right)^+\right) - E\left(e^{-rT} S_T^1(r)\right) + E\left(e^{-rT} K\right) \\
&= C\left(T, \left(S_T^1(r) - K\right)^+\right) - S_0^1(r) + K e^{-rT}.
\end{aligned}
$$

Thus we can again relate the European put price P_T and European call price C_T by the formula (1.2) derived in Chapter 1 as

$$P_T = C_T - S_0 + K e^{-rT}.$$

Exercise 7.6.3. Show that in the one-factor Black-Scholes model the time t call value $C(S, K, T - t)$ and put value $P(S, K, T - t)$ for European options with strike K and expiry T are positive-homogeneous in the stock price S and the strike price K.
    Verify that $C(K e^{-r(T-t)}, S, T - t) = P(S e^{-r(T-t)}, K, T - t)$.

7.7   Pricing in a Multifactor Model

In Section 7.5, we considered a riskless bond $S_t^0 = e^{rt}$ and a single risky asset S_t^1. Suppose now that we have a vector of risky assets

$$S_t = \left(S_t^1, \ldots, S_t^d\right)$$

whose dynamics are described by stochastic differential equations of the form

$$dS_t^i = S_t^i \left(\mu^i(t, S_t) dt + \sum_{j=1}^d \lambda_{ij}(t, S_t) dW_t^i \right) \quad \text{for } i = 1, 2, \ldots, d.$$

When the μ^i and λ_{ij} are constant, we have the familiar log-normal stock price. To ensure the claim is attainable, the number of sources of noise - that is, the dimension of the Brownian motion w - is taken equal to the number of stocks. $\Lambda_t = \Lambda(t,S) = (\lambda_{ij}(t,S))$ is therefore a $d \times d$ matrix. We suppose Λ is non-singular, three times differentiable in S, and that $\Lambda^{-1}(t,S)$ and all derivatives of Λ are bounded. Writing $\mu(t,S) = (\mu^1(t,S), \ldots, \mu^d(t,S))'$, we also suppose μ is three times differentiable in S with all derivatives bounded.

Again suppose there is a bond S_t^0 with a fixed interest rate r, so $S_t^0 = e^{rt}$. The discounted stock price vector $\xi_t = (\xi_t^1, \ldots, \xi_t^d)'$ is then $\xi_t = e^{-rt}S_t$, so

$$d\xi_t^i = \xi_t^i \left(\left(\mu^i \left(t, e^{rt}\xi_t\right) - r \right) dt + \sum_{j=1}^d \lambda_{ij} \left(t, e^{rt}\xi_t\right) dW_t^j \right). \tag{7.40}$$

Writing

$$\Delta_t = \Delta(t,\xi_t) = \begin{pmatrix} \xi_t^1 & & 0 \\ & \ddots & \\ 0 & & \xi_t^d \end{pmatrix}$$

and $\rho = (r, r, \ldots, r)'$, equation (7.40) can be written as

$$d\xi_t = \Delta_t((\mu - \rho)dt + \Lambda_t dW_t). \tag{7.41}$$

As in Section 7.4, there is a flow of diffeomorphisms $x \to \xi_{s,t}(x)$ associated with this system, together with their non-singular Jacobians $D_{s,t}$.

In the terminology of Harrison and Pliska [150], the return process $Y_t = (Y_t^1, \ldots, Y_t^d)$ is here given by

$$dY_t = (\mu - \rho)dt + \Lambda dW_t. \tag{7.42}$$

The drift term in (7.42) can be removed by applying the Girsanov change of measure. Write

$$\eta(t,S) = \Lambda(t,S)^{-1}(\mu(t,S) - \rho),$$

and define the martingale M by

$$M_t = 1 - \int_0^t M_s \eta(s, S_s)' dW_s.$$

Then

$$M_t = \exp \left\{ -\int_0^t \eta_s' dW_s - \frac{1}{2} \int_0^t |\eta_s|^2 ds \right\}$$

is the Radon-Nikodym derivative of a probability measure P^μ. Furthermore, under P^μ,

$$\widetilde{W}_t = W_t + \int_0^t \eta(s, S_s)' ds$$

is a standard Brownian motion. Consequently, under P^μ, we have

$$dY_t = \Lambda_t d\widetilde{W}_t, \qquad\qquad d\xi_t = \Delta_t \Lambda_t d\widetilde{W}_t.$$

Therefore the discounted stock price process ξ is a martingale under P^μ.

Consider a function $\overline{\psi} : \mathbb{R}^d \to \mathbb{R}$, where $\overline{\psi}$ is twice differentiable and $\overline{\psi}$ and $\overline{\psi}_x$ are of at most linear growth in x. For some future time $T > t$, we shall be interested in finding the current price (i.e., the current valuation at time t) of a contingent claim of the form $\overline{\psi}(S_T)$. It is convenient to work with the discounted claim as a function of the discounted stock price, so we consider equivalently the current value of

$$\psi(\xi_T) = e^{-rT}\overline{\psi}(e^{rT}\xi_T) = e^{-rT}\overline{\psi}(S_T).$$

The function ψ has linear growth, so we may define the square integrable P^μ-martingale N by

$$N_t = E^\mu\left(\psi(\xi_T)\,|\mathcal{F}_t\right) \text{ for } 0 \le t \le T.$$

As in Section 7.5, the rational price for $\overline{\psi}$ is $E^\mu(\psi)$. Furthermore, if we can express N in the form

$$N_t = \widetilde{E}\left(\psi(\xi_T)\right) + \int_0^t \phi(s)' d\xi_s,$$

then the vector $H_t^1 = \left(\phi^1, \phi^2, \ldots, \phi^d\right)'$ is a hedge portfolio that generates the contingent claim. Then $H_t^0 = N_t - H_t^1 \cdot e^{-rt}S_t$. Applying Theorem 7.3.13, we immediately obtain the following.

Theorem 7.7.1. *We have*

$$N_t = \widetilde{E}\left(\psi(\xi_T)\right) + \int_0^t \phi(s)' d\xi_s,$$

where

$$\phi(s) = E^\mu\Big[\int_s^T \eta_\xi(u, e^{ru}\xi_{0,x}(x_0))D_{0,u}(x_0)d\widetilde{W}_u \cdot \psi\xi_{0,T}(x_0)$$
$$+ \psi_\xi(\xi_{0,T}(x_0))D_{0,T}(x_0)\,|\mathcal{F}_s\,\Big]D_{0,s}^{-1}(x_0).$$

Proof. From Theorem 7.3.13, under the measure P^μ, we have

$$N_t = \widetilde{E}\left(\psi(\xi_T)\right) + \int_0^t \gamma_s d\widetilde{W}_s,$$

where

$$\gamma_s = E^\mu\Big[\int_s^T \eta_\xi D_{0,u}(x_0)d\widetilde{W}_u \cdot \psi(\xi_{0,T}(x_0))$$
$$+ \psi_\xi(\xi_{0,T}(x_0))D_{0,T}(x_0)\,|\mathcal{F}_s\,\Big]D_{0,s}^{-1}(x_0)\Delta(\xi_{0,s}(x_0))\Lambda_s$$

since $d\xi_t = \Delta_t \Lambda_t d\widetilde{W}_t$, $\phi(s)$ has the stated form. $\square$

Remark 7.7.2. Note that if η is not a function of ξ (which is certainly the situation in the usual log-normal case where μ and Λ are constant), η_ξ is zero and the first term in ϕ vanishes.

The bond component H_t^0 in the portfolio is given by

$$H_t^0 = N_t - \sum_{i=1}^{d} \phi_t^i \xi_t^i, 0 \le t \le T,$$

and N_t is the price associated with the contingent claim at time t.

Examples

Stock price dynamics for which the hedging policy ϕ can be evaluated in closed form appear hard to find. However, if we consider a vector of log-normal stock prices, we can re-derive a vector form of the Black-Scholes results. Suppose, therefore, that the vector of stock prices $S = \left(S^1, S^2, \ldots, S^d\right)'$ evolves according to the equations

$$dS_t^i = S_t^i \left(\mu^i dt + \sum_{j=1}^{d} \lambda_{ij} dW_t^j \right), \tag{7.43}$$

where $\mu = \left(\mu^1, \mu^2, \ldots, \mu^d\right)$ and $\Lambda = (\lambda_{ij})$ are constant. The discounted stock price ξ is then given by (7.40).

Consider a contingent claim that consists of d European call options with expiry dates $T_1 \le T_2 \le \cdots \le T_d$ and exercise prices $c_1, c_2, \ldots, c_d$, respectively. Then

$$\psi\left(T_1, T_2, \ldots, T_d\right) = \sum_{k=1}^{d} \psi^k\left(\xi_{0,T_k}(x_0)\right) = \sum_{k=1}^{d} \left(\xi_{0,T_k}^k(x_0) - c_k e^{-rT_k}\right)^+.$$

From (7.43) we see that, with $a = (a_{ij}$ denoting the matrix $\Lambda\Lambda^*$, the Jacobian $D_{0,t}$ is just the diagonal matrix

$$D_{0,t} = \begin{bmatrix} e^{\sum_{j=1}^{d} \lambda_{1j} \widetilde{W}_t^j - \frac{1}{2}a_{11}t} & \cdots & 0 \\ \vdots & & \vdots \\ 0 & \cdots & e^{\sum_{j=1}^{d} \lambda_{dj} \widetilde{W}_t^j - \frac{1}{2}a_{dd}t} \end{bmatrix}$$

and its inverse is

$$D_{0,t}^{-1} = \begin{bmatrix} e^{-\left(\sum_{j=1}^{d} \lambda_{1j} \widetilde{W}_t^j - \frac{1}{2}a_{11}t\right)} & \cdots & 0 \\ \vdots & & \vdots \\ 0 & \cdots & e^{-\left(\sum_{j=1}^{d} \lambda_{dj} \widetilde{W}_t^j - \frac{1}{2}a_{dd}t\right)} \end{bmatrix}.$$

(The explicit, exponential form of the solution shows that $D_{0,t}$ is independent of x_0). Thus, the trading strategy ϕ_k that generates the contingent claim $\psi^k(\xi_{T_k})$ is

$$
\phi_k(s)' = E^\mu \left(\psi^k_\xi(\xi_{0,T_k}(x_0)) D_{0,T_k} \,|\, \mathcal{F}_s \right) D_{0,s}^{-1}
$$

$$
= \left(0, \ldots, 0, \widetilde{E} \left(1_{\{\xi_{0,T_k} > c_k e^{-rT_k}\}} \exp\left\{ \sum_{j=1}^d \lambda_{kj}(\widetilde{W}^j_{T_k} - \widetilde{W}^j_s) \right. \right. \right.
$$

$$
\left. \left. \left. -\frac{1}{2} a_{kk}(T_k - s) \right\} \,|\, \mathcal{F}_s \right), 0, \ldots, 0 \right),
$$

for $0 \le s \le T_k$. Note that $\phi_k(s) = 0$ for $s > T_k$ (i.e., $\phi_k(s)$ stops at T_k). However, from (7.43), it follows that

$$
\xi^k_{0,T_k}(x_0) = x^k_0 \exp\left\{ \sum_{j=1}^d \lambda_{kj} \widetilde{W}^j_{T_k} - \frac{1}{2} a_{kk} T_k \right\} > c_k e^{-rT_k} \qquad (7.44)
$$

if and only if

$$
\sum_{j=1}^d \lambda_{kj} \widetilde{W}^j_{T_k} > \log\left(\frac{c_k}{x^k_0} \right) + \left(\frac{1}{2} a_{kk} - r \right) T_k = \alpha_k,
$$

say; that is, if and only if

$$
\sum_{j=1}^d \lambda_{kj}(\widetilde{W}^j_{T_k} - \widetilde{W}^j_s) > \alpha_k - \sum_{j=1}^d \lambda_{kj} \widetilde{W}^j_s.
$$

Now, under $\widetilde{P}$, $\sum_{j=1}^d \lambda_{kj}(\widetilde{W}^j_{T_k} - \widetilde{W}^j_s)$ is normally distributed with mean zero and variance $a_{kk}(T_k - s)$ and is independent of $\mathcal{F}_s$. Therefore, the non-zero component of $\phi_k(s)$ is

$$
\int_{\alpha_k - \sum_{j=1}^d \lambda_{kj} \widetilde{W}^j_s}^\infty \exp\left\{ x - \frac{1}{2} a_{kk}(T_k - s) \right\}
$$

$$
\times \exp\left\{ \frac{-x^2}{2a_{kk}(T_k - s)} \right\} \frac{dx}{\sqrt{2\pi a_{kk}(T_k - s)}}
$$

$$
= \int_{\alpha_k - \sum_{j=1}^d \lambda_{kj} \widetilde{W}^j_s}^\infty \exp\left\{ \frac{-[x - a_{kk}(T_k - s)]^2}{2a_{kk}(T_k - s)} \right\} \frac{dx}{\sqrt{2\pi a_{kk}(T_k - s)}}
$$

$$
= \int_{\frac{\alpha_k - \sum \lambda_{kj} \widetilde{W}^j_s - a_{kk}(T_k - s)}{\sqrt{a_{kk}(T_k - s)}}}^\infty e^{-\frac{1}{2} y^2} \frac{dy}{\sqrt{2\pi}}
$$

$$
= \Phi\left(\frac{-\alpha_k + \sum \lambda_{kj} \widetilde{W}^j_s - a_{kk}(T_k - s)}{\sqrt{a_{kk}(T_k - s)}} \right).
$$

Again from (7.44) we have

$$\sum_{j=1}^{d} \lambda_{kj} \widetilde{W}_s^j = \log \left(\frac{\xi_{0,s}^k}{x_0^k} \right) x_0^k + \frac{1}{2} a_{kk} s,$$

which together with (7.44) gives

$$\phi_k(x) = \left(0, \ldots, 0, \Phi \left(\frac{\log \left(\frac{\xi_{0,s}^k (X_0)}{c_k} \right) - \frac{1}{2} a_{kk} (T_k - s) + r T_k}{\sqrt{a_{kk}(T_k - s)}} \right), 0, \ldots, 0 \right)'$$

or, in terms of the (non-discounted) price S_s^k,

$$\phi_k(s) = \left(0, \ldots, 0, \Phi \left(\frac{\log \left(\frac{S_s^k}{c_k} \right) - \frac{1}{2}(a_{kk} - r)(T_k - s)}{\sqrt{a_{kk}(T_k - s)}} \right), 0, \ldots, 0 \right)'$$

(7.45)

for $0 \leq s \leq T_k$. Therefore, the trading strategy ϕ generating

$$\psi(T_1, T_2, \ldots, T_k) = \sum_{k=1}^{d} \psi^k(\xi_{T_k})$$

can be written, by a minor abuse of notation, as $\phi(s) = (\phi_1(s), \ldots, \phi_d(s))'$, where

$$\phi_k(s) = \mathbf{1}_{\{s \leq T_k\}} \Phi \left(\frac{\log \left(\frac{S_t^k}{c_k} \right) - \left(\frac{1}{2} a_{kk} - r \right) (T_k - s)}{\sqrt{a_{kk}(T_k - s)}} \right).$$
(7.46)

Finally, we calculate the price of the claim

$$E^\mu \left(\psi \left(T_1, T_2, \ldots, T_d \right) \right) = \sum_{k=1}^{d} E^\mu \left(\psi^k(\xi_{T_k}) \right)$$

similarly. Indeed,

$$\sum_{k=1}^{d} E^\mu \left(\psi^k(\xi_{T_k}) \right) = \sum_{k=1}^{d} E^\mu \left(\xi_{T_k}^k - c_k e^{-r T_k} \right)^+$$

$$= \sum_{k=1}^{d} E\mu \left[\mathbf{1}_{\left\{ \sum_{j=1}^{d} \lambda_{kj} \widetilde{W}_{T_k}^j > \alpha_k \right\}} \right.$$

$$\left. \times \left(Z_0 \exp \left\{ \sum_{j=1}^{d} \lambda_{kj} \widetilde{W}_{T_k}^j - \frac{1}{2} a_{kk} T_k \right\} - c_k e^{-r T_k} \right) \right]$$

$$= \sum_{k=1}^{d} S_0^k \Phi \left(\frac{\log \left(\frac{S_0^k}{c_k} \right) + \left(\frac{1}{2}a_{kk} + r \right)T_k}{\sqrt{a_{kk}T_k}} \right)$$

$$- c_k e^{-rT_k \Phi} \left(\frac{\log \left(\frac{S_0^k}{c_k} \right) + \left(\frac{1}{2}a_{kk} + r \right)T_k}{\sqrt{a_{kk}T_k}} - \sqrt{a_{kk}T_k} \right)$$

(where we have used $\xi_0^k = S_0^k$ for $k = 1, 2, \ldots, d$). When $d = 1$, the above result reduces to the well-known Black-Scholes formula.

The following two exercises serve to introduce two further options closely related to the call and put.

Exercise 7.7.3. A binary call option with strike K pays \$1 if $S_T > K$ and 0 otherwise. Show that, under Black-Scholes dynamics for the stock price S, the value of the binary call at time $t \leq T$ is given by

$$B_C(S, K, T - t) = e^{-r(T-t)} \Phi \left(d_2 \left(\frac{e^{r(T-t)}}{K} \right) \right).$$

Hence verify that

$$\frac{\partial C}{\partial S}(S, K, T - t) = \frac{1}{S}[C(S, K, T - t) + K B_C(S, K, T - t)].$$

Explain how this provides a hedge for the call.

Exercise 7.7.4. Write $C_{t,T}(K), P_{t,T}(K)$ for the Black-Scholes prices at time $t \leq T$ of European call and put options with expiry T and strike K. Calculate $\max[C_{t,T}(K), P_{t,T}(K)]$.

A *chooser option* gives the holder the right to choose either the call or the put at time t. What is the rational price (at time 0) of such a chooser option for the above call and put?

7.8   Barrier Options

Consider a standard Brownian motion $(B_t)_{t \geq 0}$ defined on $(\Omega, \mathcal{F}, P)$. The filtration $(\mathcal{F}_t)$ is that generated by B. Recall that B_t is normally distributed, and

$$P(B_t < x) = \Phi \left(\frac{x}{\sqrt{t}} \right).$$

Therefore

$$P(B_t \geq x) = 1 - \Phi \left(\frac{x}{\sqrt{t}} \right) = \Phi \left(-\frac{x}{\sqrt{t}} \right).$$

For a real-valued process X, we shall write

$$M_t^X = \max_{0 \leq s \leq t} X_s, \qquad m_t^X = \min_{0 \leq s \leq t} X_s.$$

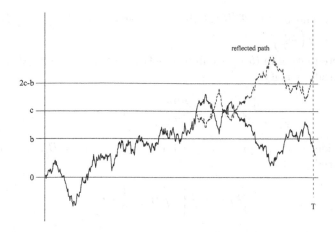

Figure 7.1: Reflection principle

If X is defined by

$$X_t = \mu t + \sigma B_t, \tag{7.47}$$

then

$$P(X_t < x) = \Phi\left(\frac{x - \mu t}{\sigma\sqrt{t}}\right)$$

and

$$-X_t = (-\mu)t + \sigma(-B_t).$$

The process $(-B_t)$ is also a standard Brownian motion, so $-X$ has the same form as X but with μ replaced by $-\mu$. Since $m^X = -M^{-X}$, we shall consider only M^X.

Consider the event

$$\{B_T < b, M_T^B > c\} \text{ for } T > 0.$$

For each path that hits level c before time T and ends up below b at time T there is, by the 'reflection principle' (see Figure 7.1), an equally probable path that hits level c and ends up above $2c - b$ at time T. Therefore

$$P\left(B_T < b, M_T^B > c\right) = P(B_T > 2c - b) = \Phi\left(\frac{b - 2c}{\sqrt{T}}\right).$$

Let us calculate the joint distribution function of B_T and M_T^B,

$$\begin{aligned} F^B(T, b, c) &= P\left(B_T < b, M_T^B < c\right) \\ &= P\left(B_T < b\right) - P\left(B_T < b, M_T^B > c\right) \\ &= \Phi\left(\frac{b}{\sqrt{T}}\right) - \Phi\left(\frac{b - 2c}{\sqrt{T}}\right). \end{aligned}$$

For $c < 0$ and $b < 0$, $F^B(T, b, c) = 0$. For $c > 0$, $B \geq c$, $F^B(T, b, c) = \Phi\left(\frac{c}{\sqrt{T}}\right) - \Phi\left(\frac{-c}{\sqrt{T}}\right)$.

Differentiating in (b, c), we find that the random variable (B_T, M_T^B) has the bivariate density

$$f^B(T, b, c) = \frac{2(2c - b)}{T\sqrt{T}} \phi\left(\frac{b - 2c}{\sqrt{T}}\right). \tag{7.48}$$

Consider now the process X defined by $X_t = \mu t + B_t$. Introduce the exponential process

$$\Lambda_t = \exp\left\{-\mu B_t - \frac{1}{2}\mu^2 t\right\}$$

and define a new measure P^μ by setting

$$\left.\frac{dP^\mu}{dP}\right|_{\mathcal{F}_t} = \Lambda_t.$$

Suppose that $c \geq 0$ and $b \leq c$. Then, from Girsanov's theorem, under P^μ, X_t is a standard Brownian motion and (X_T, M_T^X) has the same distribution under P^μ as (B_T, M_T^B) has under P. Then, writing $E^\mu(\cdot)$ for expectation with respect to P^μ and writing $A = \left\{X_t < b, M_t^X < c\right\}$, we obtain

$$\begin{aligned}
F^X(T, b, c) &= E(\mathbf{1}_A) \\
&= E^\mu(\Lambda_T^{-1}\mathbf{1}_A) \\
&= E^\mu\left(\exp\left\{\mu X_t - \frac{1}{2}\mu^2 T\right\}\mathbf{1}_A\right).
\end{aligned}$$

Under P^μ, the process X is a standard Brownian motion, so, if f is given by (7.48), then

$$\begin{aligned}
F^X(T, b, c) &= \int_0^c \int_{-\infty}^b \exp\left\{\mu z - \frac{1}{2}\mu^2 T\right\} f(T, z, y)\, dz\, dy \\
&= \int_{-\infty}^b \exp\left\{\mu z - \frac{1}{2}\mu^2 T\right\} \frac{1}{\sqrt{T}}\left[\phi\left(\frac{z}{\sqrt{T}}\right) - \phi\left(\frac{z - 2c}{\sqrt{T}}\right)\right] dz \\
&= \int_{-\infty}^0 \exp\left\{\mu(b + z) - \frac{1}{2}\mu^2 T\right\} \frac{1}{\sqrt{T}}\left[\phi\left(\frac{b + z}{\sqrt{T}}\right) - \phi\left(\frac{b + z - 2c}{\sqrt{T}}\right)\right] dz \\
&= \exp\left\{\mu b - \frac{1}{2}\mu^2 T\right\} \cdot (\Psi(b) - \Psi(b - 2c)), \tag{7.49}
\end{aligned}$$

where

$$\Psi(b) = \frac{1}{\sqrt{T}} \int_{-\infty}^0 \exp\{\mu z\} \cdot \phi\left(\frac{b + z}{\sqrt{T}}\right) dz$$

$$= \frac{1}{\sqrt{2\pi T}} \int_{-\infty}^{0} \exp\left\{\mu z - \left(\frac{b+z}{T}\right)^2\right\} dz$$

$$= \exp\left\{-\mu b + \frac{1}{2}\mu^2 T\right\} \int_{-\infty}^{0} \frac{1}{\sqrt{T}} \phi\left(\frac{b+z-\mu T}{\sqrt{T}}\right) dz$$

$$= \exp\left\{-\mu b + \frac{1}{2}\mu^2 T\right\} \Phi\left(\frac{b-\mu T}{\sqrt{T}}\right). \tag{7.50}$$

Substituting (7.50) into (7.49), we see that

$$F^X(T,b,c) = \Phi\left(\frac{b-\mu T}{\sqrt{T}}\right) - e^{2\mu c}\Phi\left(\frac{b-2c-\mu T}{\sqrt{T}}\right). \tag{7.51}$$

Once again, differentiating in (b,c), we find that the random variable (X_T, M_T^X) has the bivariate density

$$f^X(T,b,c) = \frac{2(2c-b)}{T\sqrt{T}} \phi\left(\frac{2c-b}{\sqrt{T}}\right) \cdot e^{\mu b - \frac{1}{2}\mu^2 T}.$$

Note that the processes $(\mu t + \sigma B_t)$ and $(\mu t - \sigma B_t)$ have the same law. Hence we consider the process

$$Y_t = \mu t + \sigma B_t \text{ for } \sigma > 0.$$

Write $F^Y(T,b,c) = P\left(Y_T < b, M_T^Y < c\right)$. Consider

$$\widehat{X}_t = \sigma^{-1} Y_t = \frac{\mu}{\sigma} t + B_t. \tag{7.52}$$

Then

$$P\left(Y_T < b, M_T^Y < c\right) = P\left(\widehat{X}(T) < \frac{b}{\sigma}, M^{\widehat{X}}(T) < \frac{c}{\sigma}\right)$$

$$= \Phi\left(\frac{b-\mu T}{\sigma\sqrt{T}}\right) - e^{2\mu c\sigma^{-2}}\Phi\left(\frac{b-2c-\mu T}{\sigma\sqrt{T}}\right) \tag{7.53}$$

from (7.51). Furthermore, (Y_T, M_T^Y) has bivariate density

$$f^Y(T,b,c) = \frac{2(2c-b)}{\sigma T\sqrt{T}} \phi\left(\frac{2c-b}{\sigma\sqrt{T}}\right) \exp\left\{\left(\mu b - \frac{1}{2}\mu^2 T\right)\sigma^{-2}\right\}. \tag{7.54}$$

These formulas enable us to derive the distribution of the first hitting time of level $y > 0$.

Lemma 7.8.1. If $\tau(y) = \inf_{t \geq 0}\{t : Y_t \geq y\}$, then

$$P(\tau(y) > T) = \Phi\left(\frac{y-\mu T}{\sigma\sqrt{T}}\right) - \exp\left\{\frac{2\mu y}{\sigma^2}\right\}\Phi\left(\frac{-y-\mu T}{\sigma\sqrt{T}}\right).$$

Proof. Clearly

$$\{\omega : \tau(y)(\omega) > t\} = \{\omega : M_t^Y(\omega) < y\},$$

so that

$$\begin{aligned}
P(\tau(y) > T) &= P(\omega : M_t^Y(\omega) < y) \\
&= P(\omega : Y_t < y, M_t^Y < y) \\
&= F^Y(t, y, y),
\end{aligned}$$

and the result follows.   □

Barrier Options in the Black-Scholes Model

Consider again the situation with two assets, the riskless bond

$$S_t^0 = e^{rt}$$

and a risky asset S^1 with dynamics

$$dS_t^1 = S_t^1(\mu dt + \sigma dB_t).$$

(B_t) is a standard Brownian motion on a probability space $(\Omega, \mathcal{F}, P)$. Consider the risk-neutral probability P^θ and the P^θ-Brownian motion W^θ given by

$$dW_t^\theta = \theta dt + \sigma dB_t.$$

Here $\theta = \frac{r-\mu}{\sigma}$. Under P^θ,

$$dS_t^1 = S_t^1(rdt + \sigma dW_t^\theta),$$

so that

$$S_t^1 = S_0^1 \exp\left\{\left(r - \frac{\sigma^2}{2}\right)t + \sigma W_t^\theta\right\} = S_0^1 \exp\{Y_t\},$$

where

$$Y_t = \left(r - \frac{\sigma^2}{2}\right)t + \sigma W_t^\theta.$$

Write

$$\overline{S}_T^1 = \max\{S_t^1 : 0 \leq t \leq T\}, \qquad \underline{S}_T^1 = \min\{S_t^1 : 0 \leq t \leq T\}.$$

Clearly, with

$$M_T^Y = \max\{Y_t : 0 \leq t \leq T\}, \qquad m_T^Y = \min\{Y_t : 0 \leq t \leq T\},$$

we have

$$\overline{S}_T^1 = S_0^1 \exp\{M_T^Y\}, \qquad \underline{S}_T^1 = S_0^1 \exp\{m_T^Y\}.$$

Lemma 7.8.2. *Write, for given* $H > K > 0$,

$$d_1 = \frac{\log\left(\frac{K}{S_0^1}\right) - \left(r - \frac{\sigma^2}{2}\right)T}{\sigma\sqrt{T}}, \qquad d_2 = \frac{\log\left(\frac{KS_0^1}{H^2}\right) - \left(r - \frac{\sigma^2}{2}\right)T}{\sigma\sqrt{T}}.$$

Then

$$P^\theta\left(S_T^1 \le K, \overline{S}_T^1 \le H\right) = \Phi(d_1) - \left(\frac{H}{S}\right)^{\frac{2r}{\sigma^2}-1}\Phi(d_2).$$

Proof. We have, by continuity,

$$P^\theta(S_T^1 \le K, \overline{S}_T^1 \le H) = P^\theta(S_T^1 < K, \overline{S}_T^1 < H)$$
$$= P^\theta\left(Y_T \le \log\left(\frac{K}{S_0^1}\right), M_T^Y \le \log\left(\frac{H}{S_0^1}\right)\right).$$

The result follows from (7.53). $\qquad\square$

Remark 7.8.3. We assume that $H > K$ because if $H \le K$, then

$$P^\theta\left(S_T^1 \le K, \overline{S}_T^1 \le H\right) = P^\theta\left(S_T^1 \le H, \overline{S}_T^1 \le H\right),$$

which is a special case. Furthermore, if $S_0^1 > H$, this probability is zero.

Lemma 7.8.4. *Write*

$$d_3 = \frac{\log\left(\frac{S_0^1}{K}\right) + \left(r - \frac{\sigma^2}{2}\right)T}{\sigma\sqrt{T}}, \qquad d_4 = \frac{\log\left(\frac{H^2}{S_0^1 K}\right) + \left(r - \frac{\sigma^2}{2}\right)T}{\sigma\sqrt{T}}.$$

Then

$$P^\theta\left(S_T^1 \ge K, \underline{S}_T^1 \ge H\right) = \Phi(d_3) - \left(\frac{H}{S}\right)^{\frac{2r}{\sigma^2}-1}\Phi(d_4).$$

(Note that $d_3 = d_-$ as defined in (2.31).)

Proof. We have

$$P^\theta(S_T^1 \ge K, \underline{S}_T^1 \ge H) = P^\theta\left(Y_T \ge \log\left(\frac{K}{S_0^1}\right), m_T^Y \ge \log\left(\frac{H}{S_0^1}\right)\right)$$
$$= P^\theta\left(-Y_T \le \log\left(\frac{S_0^1}{K}\right), M_T^{-Y} \le \log\left(\frac{S_0^1}{H}\right)\right).$$

Now

$$-Y_t = \left(-r + \frac{\sigma^2}{2}\right)t + \sigma(-B_t),$$

and so has the same form as Y, because $-B$ is a standard Brownian motion. The result follows from (7.53). $\qquad\square$

Remark 7.8.5. Here $K > H$ and $S_0^1 > H$. If $K \leq H$ and $S_0^1 < H$, the same result is obtained with $K = H$ in (7.53). If $S_0^1 < H$, then the probability is zero.

Lemma 7.8.6. *Write*

$$d_5 = \frac{\log\left(\frac{K}{S_0^1}\right) - \left(r + \frac{\sigma^2}{2}\right)T}{\sigma\sqrt{T}}, \qquad d_6 = \frac{\log\left(\frac{KS_0^1}{H^2}\right) - \left(r + \frac{\sigma^2}{2}\right)T}{\sigma\sqrt{T}}.$$

Then

$$E^\theta\left(S_T^1 \mathbf{1}_{\{S_T^1 \leq K, \overline{S}_T^1 \leq H\}}\right) = S_0^1 \exp\{rT\}\left(\Phi(d_5) - \left(\frac{H}{S}\right)^{1 + \frac{2r}{\sigma^2}} \Phi(d_6)\right).$$

Proof. Write

$$\Gamma(t) = \exp\left\{\sigma W_t^\theta - \frac{1}{2}\sigma^2 t\right\},$$

and define a new probability P^θ by setting

$$\left.\frac{dP^\sigma}{dP^\theta}\right|_{\mathcal{F}_T} = \Gamma(T).$$

Girsanov's theorem states that, under P^σ, the process W^σ is a standard Brownian motion, where

$$dW^\sigma = dW^\theta - \sigma dt.$$

Consequently, under P^σ,

$$Y_t = \left(r + \frac{\sigma^2}{2}\right)t + \sigma W^\sigma(t).$$

Therefore, setting

$$A = \left\{S_t^1 \leq K, \overline{S}_T^1 \leq H\right\}, \quad B = \left\{Y_T \leq \log\left(\frac{K}{S_0^1}\right), M_T^Y \leq \log\left(\frac{H}{S_0^1}\right)\right\},$$

we obtain

$$E^\theta\left(S_T^1 \mathbf{1}_A\right) = S_0^1 e^{rT} E^\theta\left(\Gamma(T)\mathbf{1}_B\right) = S_0^1 e^{rT} E^\sigma\left(\mathbf{1}_B\right).$$

The result follows from Lemma 7.8.2.   □

Lemma 7.8.7. *Write*

$$d_7 = \frac{\log\left(\frac{S_0^1}{K}\right) + \left(r + \frac{\sigma^2}{2}\right)T}{\sigma\sqrt{T}}, \qquad d_8 = \frac{\log\left(\frac{H^2}{KS_0^1}\right) + \left(r + \frac{\sigma^2}{2}\right)T}{\sigma\sqrt{T}}.$$

Then

$$E^\theta\left(S_T^1 \mathbf{1}_{\{S_T^1 \geq K, \underline{S}_T^1 \geq H\}}\right) = S_0^1 e^{rT}\left(\Phi(d_7) - \left(\frac{H}{S_0^1}\right)^{1 + \frac{2r}{\sigma^2}} \Phi(d_8)\right).$$

(Note that $d_7 = d_+$, as defined in (2.31).)

Proof. The proof is similar to that of Lemma 7.8.6. □

In the following, we determine the expressions for prices $V(0)$ as functions $f(S, T)$ of the price $S = S_0^1$ at time 0 of the risky asset and the time T to expiration. The price at any time $t < T$ when the price is S_t^1 is then

$$V(t) = f(S_t^1, T - t).$$

Definition 7.8.8. A *down and out call option* with strike price K, expiration time T, and barrier H gives the holder the right (but not the obligation) to buy S^1 for price K at time T provided the price S^1 at no time falls below H (in which case the option ceases to exist).

Its price is sometimes denoted $C_{t,T}(K|H \downarrow O)$, and it corresponds to a payoff $\left(K - S_T^1\right)^+ 1_{\{\underline{S}(T) \geq H\}}$. The $\downarrow$ denotes 'down' and the O 'out'. From our pricing formula, we obtain, setting $U = S_T^1 \geq K, \underline{S}_T^1 \geq H$,

$$\begin{aligned}
C_{0,T}(K|H \downarrow O) \\
&= e^{-rT} E^\theta \left(\left(K - S_T^1\right)^+ 1_{\{\underline{S}_T^1 \geq H\}} \right) \\
&= e^{-rT} E^\theta \left(S_T^1 1_U \right) - e^{-rT} K E^\theta \left(1_U \right) \\
&= S_0^1 \left(\Phi(d_7) - \left(\frac{H}{S_0^1} \right)^{1 + \frac{2r}{\sigma^2}} \Phi(d_8) \right) \\
&\quad - e^{-rT} K \left(\Phi(d_3) - \left(\frac{H}{S_0^1} \right)^{\frac{2r}{\sigma^2} - 1} \Phi(d_4) \right)
\end{aligned} \tag{7.55}$$

by Lemmas 7.8.4 and 7.8.7.

Definition 7.8.9. An *up and out call option* gives the holder the right (but not the obligation) to buy S^1 for strike price K at time T provided that the price S_t^1 does not rise above H (in which case the option ceases to exist).

Its price is denoted by $C_{t,T}(K|H \uparrow O)$, and it corresponds to payoff $\left(K - S_T^1\right)^+ 1_{\{\overline{S}_T^1 \leq H\}}$. We have, setting $V = \left\{ S_T^1 \geq K, \overline{S}_T^1 \leq H \right\}$,

$$\begin{aligned}
C_{0,T}(K|H \uparrow O) &= e^{-rT} E^\theta \left(\left(S_T^1 - K \right) 1_V \right) \\
&= e^{-rT} E^\theta \left(S_T^1 1_V \right) - e^{-rT} K E^\theta \left(1_V \right).
\end{aligned}$$

Now, with $p = 0$ or $p = 1$, we have

$$E^\theta \left(\left(S_T^1 \right)^p 1_V \right) = E^\theta \left(\left(S_T^1 \right)^p 1_{\{\overline{S}_T^1 \leq H\}} \right) - E^\theta \left(\left(S_T^1{}^p \right) 1_{\{S_T^1 < K, \overline{S}_T^1 \geq H\}} \right),$$

and
$$E^\theta\left((S_T^1)^p \mathbf{1}_{\{\overline{S}_T^1 \leq H\}}\right) = E^\theta\left((S_T^1)^p \mathbf{1}_{S_T^1 \leq H, \overline{S}_T^1 \leq H}\right).$$

The price $C_{0,T}(K|H \uparrow O)$ is therefore again given by the formula of Lemmas 7.8.4 and 7.8.6.

Definition 7.8.10. An *up and in call option* gives the holder the right (but not the obligation) to buy S^1 at time T for strike price K provided that at some time before T the price S_t^1 becomes greater than H; otherwise, the option does not yet exist.

Its price is denoted by $C_{t,T}(K|H \uparrow I)$, and it corresponds to a payoff $\left(K - S_T^1\right)^+ \mathbf{1}_{\{\overline{S}_T^1 \geq H\}}$. Now

$$C_{t,T}(K|H \uparrow I) + C_{t,T}(K|H \uparrow O) = C_{t,T}(K),$$

where $C_{t,T}(K)$ is the usual European call option price given by

$$C_{t,T}(K) = S_t^1 N(d_+(t)) - Ke^{-r(T-t)} N(d_-(t)).$$

Here, as before,

$$d_+(t) = \frac{\log\left(\frac{S_t^1}{K}\right) + \left(r + \frac{\sigma^2}{2}\right)(T-t)}{\sigma\sqrt{T-t}}, \tag{7.56}$$

$$d_-(t) = \frac{\log\left(\frac{S_t^1}{K}\right) + \left(r - \frac{\sigma^2}{2}\right)(T-t)}{\sigma\sqrt{T-t}}. \tag{7.57}$$

Definition 7.8.11. A *down and in call option* gives the holder the right (but not the obligation) to buy S^1 for a strike price K at time T provided that at some time $t \leq T$ the price S_t^1 fell below H; otherwise, the option does not yet exist.

Its price is denoted by $C_{t,T}(K|H \downarrow I)$, and it corresponds to a payoff $\left(K - S_T^1\right)^+ \mathbf{1}_{\{\underline{S}_T^1 \leq H\}}$. Again,

$$C_{t,T}(K|H \downarrow I) + C_{t,T}(K|H \downarrow O) = C_{t,T}(K).$$

Remark 7.8.12. All the corresponding put options can be defined and priced similarly. To give one example, the down and out put has a price $P_{t,T}(K|H \downarrow O)$ and corresponds to a payoff $\left(K - S_T^1\right)^+ \mathbf{1}_{\{\underline{S}_T^1 \geq H\}}$. Then

$$\left(K - S_T^1\right)^+ \mathbf{1}_{\{\underline{S}_T^1 \geq H\}} = \left(S_T^1 - K\right)^+ \mathbf{1}_{\{\underline{S}_T^1 \geq H\}} - \left(K - S_T^1\right)\mathbf{1}_{\{\underline{S}_T^1 \geq H\}}$$

so that

$$P_{0,T}(K|H \downarrow O)$$

$$= C_{0,T}(K|H \downarrow O) - e^{-rT} E^{\theta} \left(S_T^1 1_{\{\underline{S}_T^1 \geq H\}} \right) + K e^{-rT} E^{\theta} \left(1_{\{\underline{S}_T^1 \geq H\}} \right).$$

Then we have

$$E^{\theta} \left((S_T^1)^p 1_{\{\underline{S}_T^1 \geq H\}} \right) = E^{\theta} \left((S_T^1)^p 1_{\{S_T^1 \geq H, \underline{S}_T^1 \geq H\}} \right)$$

for $p = 0$ or $p = 1$, and the option price follows from Lemmas 7.8.4 and 7.8.6.

Again we have the identity

$$P_{t,T}(K|H \downarrow O) + P_{t,T}(K|H \downarrow I) = P_{t,T}(K),$$

where $P_{t,T}(K)$ is the usual European put price given by the Black-Scholes formula. In fact, from call-put parity,

$$C_{t,T}(K) - P_{t,T}(K) = S_t^1 - e^{-r(T-t)} K$$

Definition 7.8.13. A *lookback call option* corresponds to a payoff function $S_T^1 - \underline{S}_T^1$, and a *lookback put option* corresponds to a payoff function $\overline{S}_T^1 - S_T^1$.

The price of a lookback put at time 0 is

$$V_p(0) = e^{-rT} E^{\theta} \left(\overline{S}_T^1 - S_T^1 \right) = e^{-rT} S_0^1 \left(E^{\theta} \left(\exp \{ M_T^Y \} \right) - e^{rT} \right),$$

where

$$Y_t = \left(r - \frac{\sigma^2}{2} \right) t + \sigma W_t^{\theta}.$$

From (7.55) the density of the random variable M_T^Y is, with $\mu = r - \frac{\sigma^2}{2}$,

$$f^M(c) = \int_{-\infty}^{\infty} f^Y(T, b, c) db$$

$$= \phi \left(\frac{c - \mu T}{\sigma \sqrt{T}} \right) - \frac{2\mu}{\sigma^2} e^{\frac{2\mu c}{\sigma^2}} \Phi \left(\frac{-c - \mu T}{\sigma \sqrt{T}} \right) + e^{\frac{2\mu c}{\sigma^2}} \phi \left(\frac{c + \mu T}{\sigma \sqrt{T}} \right).$$

Therefore, the lookback put price at time 0 is

$$V_p(0) = S_0^1 \left(e^{-rT} \int_{-\infty}^{\infty} f^M(c) dc - 1 \right).$$

Completing the square and integrating, we obtain, with $d = \frac{2r + \sigma^2}{2\sigma \sqrt{T}}$,

$$V_p(0) = S_0^1 \left(\Phi(-d) + e^{-rT} \Phi \left(-d + \sigma \sqrt{T} \right) \right.$$

$$\left. + \frac{\sigma^2}{2r} e^{-rT} \left(-\Phi \left(d - \frac{2r}{\sigma} \sqrt{T} \right) + e^{-rT} \Phi(d) \right) \right).$$

Similarly, it can be shown (see [49]) that the price of the lookback call option at time 0 is

$$V_C(0) = S_0^1 \left(\Phi(d) - e^{-rT} \Phi \left(-d + \sigma \sqrt{T} \right) \right.$$

$$\left. + \frac{\sigma^2}{2r} e^{-rT} \left(\Phi \left(-d + \frac{2r}{\sigma} \sqrt{T} \right) - e^{-rT} \Phi(-d) \right) \right).$$

7.9   The Black-Scholes Equation

In the Black-Scholes framework, the riskless bond has a price $S_t^0 = e^{rT}$ and the risky asset has dynamics

$$dS_t^1 = S_t^1 \left(r dt + \sigma dW_t^\theta \right)$$

under the risk-neutral measure P^θ. Consider a European claim with expiration time T of the form $h(S_T)$. Here h is C^2 and $|h(S)| \le K(1 + |S|^\beta)$ for some $\beta > 0$.

We have shown that the price of this option at time t is

$$V_{t,T} \left(S_t^1 \right) = E^\theta \left(e^{-r(T-t)} h \left(S_T^1 \right) | \mathcal{F}_t \right) = E^\theta \left(e^{-r(T-t)} h \left(S_T^1 \right) | S_t^1 \right).$$

Consequently,

$$e^{-rt} V_{t,T} \left(S_t^1 \right) = E^\theta \left(e^{-rT} h \left(S_T^1 \right) | \mathcal{F}_t \right),$$

and hence $\left(e^{-rt} V_{t,T} \left(S_t^1 \right) \right)$ is an $\left(\mathcal{F}_t, P^\theta \right)$-martingale. Now

$$S_T^1 = S_T^1 \exp \left\{ \left(r - \frac{\sigma^2}{2} \right) (T - t) + \sigma (W_T^\theta - W_t^\theta) \right\},$$

and h is C^2, so (by differentiating under the expectation) $V_{\cdot,T} (\cdot)$ is a $C^{1,2}$-function. Applying the Itô rule, we obtain

$$e^{-rt} V_{t,T} \left(S_t^1 \right)$$
$$= V_{0,T}(S_0^1) + \int_0^t \left(\frac{\partial V}{\partial u} + r S_u^1 \frac{\partial V}{\partial S} + \frac{\sigma^2}{2} \left(S_u^1 \right)^2 \frac{\partial^2 V}{\partial S^2} - rV \right) (u, S_u^1) e^{-ru} du$$
$$+ \int_0^t \sigma S_u^1 \frac{\partial V}{\partial S} (u, S_u^1) dW_u^\theta. \quad (7.58)$$

Note that $\left(e^{-rt} V_{t,T} \left(S_t^1 \right) \right)$ is a martingale; consequently the du-integral in (7.58) must be the identically zero process.

Consequently, the European option price $V_{t,T}(S)$ satisfies the partial differential equation

$$LV = \frac{\partial V}{\partial t} + r S \frac{\partial V}{\partial S} + \frac{\sigma^2}{2} S^2 \frac{\partial^2 V}{\partial S^2} - rV = 0 \text{ for } t \in [0, T] \quad (7.59)$$

with terminal condition $V_{T,T}(S) = h(S_T)$. This is often called the Black-Scholes equation.

The representation of the option price, together with

$$V_{t,T}(S) = E^\theta \left(e^{-r(T-t)} h \left(S_T^1 \right) | S_t^1 = S \right),$$

corresponds to the well-known Feynman-Kac formula (see [194]). As the solution (7.59), with the boundary condition $V_{T,T}(S) = H(S)$, is unique, the partial differential equation approach to option pricing investigates numerical solutions to this equation. However, for the vanilla European option, with $h(S) = (S - K)^+$ for a call or $(K - S)^+$ for a put, the exact solution is given by the Black-Scholes formula.

It is of interest to recall the original derivation of equation (7.59) by Black and Scholes [27] using a particular replicating portfolio in which the random component of the dynamics disappears. This approach has become widely known as *delta-hedging*. The terminology will become clear shortly.

Suppose, as above, that $V_{t,T}(S)$ represents the value at time t of a European call with expiration time T when the value of the underlying S^1 at time t is given by S. Consider the portfolio constructed at time t by buying one call for $V_{t,T}(S)$ and shorting Δ units of S. The value of this portfolio is

$$\pi(S,t) = V_{t,T}(S) - \Delta S.$$

Recall that the underlying stock has dynamics

$$dS^1 = \mu S_t^1 dt + \sigma S_t^1 dB_t$$

under the original measure P. Applying the Itô differentiation rule, we obtain

$$dV_{t,T}(S) = \frac{\partial V}{\partial t} dt + \frac{\partial V}{\partial S} dS + \frac{1}{2} \frac{\partial^2 V}{\partial S^2} \sigma^2 S^2 dt$$

$$= \left(\frac{\partial V}{\partial t} + \frac{1}{2} \frac{\partial^2 V}{\partial S^2} \sigma^2 S^2 \right) dt + \frac{\partial V}{\partial S} dS.$$

If the portfolio $\pi(S,t)$ is self-financing, we have

$$d\pi = \left(\frac{\partial V}{\partial t} + \frac{1}{2} \frac{\partial^2 V}{\partial S^2} \sigma^2 S^2 \right) dt + \frac{\partial V}{\partial S} dS - \Delta dS. \tag{7.60}$$

The right-hand side contains terms multiplying dt and the term $(\frac{\partial V}{\partial S} - \Delta) dS$, which represents the random part of the increment $d\pi$. If Δ is chosen to equal $\frac{\partial V}{\partial S}$, this random term vanishes and we have

$$d\pi = \left(\frac{\partial V}{\partial t} + \frac{1}{2} \frac{\partial^2 V}{\partial S^2} \sigma^2 S^2 \right) dt. \tag{7.61}$$

That is, the value of this increment is known. Consequently, to avoid arbitrage, it must be the same as what we would obtain by putting our money (of value π) in the bank. In other words,

$$d\pi = r\pi dt = r(V - \Delta S)dt = r\left(V - \frac{\partial V}{\partial S} S \right) dt = \left(\frac{\partial V}{\partial t} + \frac{1}{2} \frac{\partial^2 V}{\partial S^2} \sigma^2 S^2 \right) dt. \tag{7.62}$$

Equating the right-hand sides of (7.61) and (7.62) shows that $V = V_{t,T}(S)$ must satisfy the partial differential equation

$$\frac{1}{2} \frac{\partial^2 V}{\partial S^2} \sigma^2 S^2 + rS \frac{\partial V}{\partial S} + \frac{\partial V}{\partial t} - rV = 0, \tag{7.63}$$

with the terminal boundary condition $V_{T,T}(S) = (S - K)^+$.

This is the Black-Scholes equation (7.59). To solve it, and hence derive the Black-Scholes formula, one can apply a sequence of transformations to reduce this inhomogeneous linear parabolic equation to the well-known heat equation. We indicate the main steps in the solution. First write

$$V(S,t) = e^{-r(T-t)}U(S,t),$$

so that

$$\frac{\partial V}{\partial t} = e^{-r(T-t)}\left(\frac{\partial U}{\partial t} + rU\right).$$

Therefore (7.63) becomes

$$\frac{1}{2}\frac{\partial^2 U}{\partial S^2}\sigma^2 S^2 + rS\frac{\partial U}{\partial S} + \frac{\partial U}{\partial t} = 0. \tag{7.64}$$

Now let $\tau = T - t$, so that $\frac{\partial U}{\partial \tau} = -\frac{\partial U}{\partial t}$; hence

$$\frac{\partial U}{\partial \tau} = \frac{1}{2}\frac{\partial^2 U}{\partial S^2}\sigma^2 S^2 + rS\frac{\partial U}{\partial S}.$$

Setting $\xi = \log S$, so that $S = e^\xi$, and hence $\frac{1}{S} = e^{-\xi}$, leads to the equation

$$\frac{\partial U}{\partial \tau} = \frac{1}{2}\sigma^2 S^2\frac{\partial^2 U}{\partial \xi^2} + \left(r - \frac{1}{2}\sigma^2\right)\frac{\partial U}{\partial \xi}.$$

Note here that $S \geq 0$ corresponds to $\xi \geq \mathbb{R}$ and that the final equation has constant coefficients. The translation $x = \xi + \left(r - \frac{1}{2}\sigma^2\right)\tau$ and writing $U(S,t) = W(x,\tau)$ now suffices to reduce (7.64) to

$$\frac{\partial W}{\partial \tau} = \frac{1}{2}\sigma^2\frac{\partial^2 W}{\partial x^2}, \tag{7.65}$$

which is a variant of the heat equation. Looking for fundamental solutions of this equation in the form $W(x,\tau) = \tau^\alpha f(\eta)$, with $\eta = \frac{x-x'}{2\beta}$ and $\int_\mathbb{R} \tau^\alpha f(\eta)dx$ independent of τ, leads one to $\alpha = \beta = \frac{1}{2}$ and solutions of the form

$$W_f(x,\tau;x') = \frac{1}{\sqrt{2\pi\tau}\sigma}e^{-\frac{(x-x')^2}{2\sigma^2\tau}}. \tag{7.66}$$

Here the mean of this normal density has still to be chosen. The function W behaves as a Dirac δ-function when $x = x'$ and 'flattens' smoothly as x moves away from x. With a final condition of the form $V(S,T) = h(S_T)$ for a European claim, we can now write $W(x,0) = h(e^x)$ and show by differentiation that

$$W(x,\tau) = \int_{-\infty}^{\infty} W_f(x,\tau;x')h(e^{x'})dx'. \tag{7.67}$$

Retracing our steps, we then find that, with $x' = \log S'$,

$$V(S,t) =$$

$$\frac{e^{-r}(T-t)}{\sigma\sqrt{2\pi(T-t)}} \int_0^\infty \exp\left\{-\frac{\log\frac{S}{S'} + \left(r - \frac{1}{2}\sigma^2\right)(T-t)^2}{2\sigma^2(T-t)}\right\} h(S')\frac{dS'}{S'}.$$

For the European call and put, this value function reduces to the familiar ones derived earlier.

Remark 7.9.1. Our derivation of option pricing formulas via Itô calculus was made under the assumption that h is C^2. Approximating by C^2 functions establishes the result for payoff functions h that are not necessarily C^2 in S. In particular, the European call option $C_{t,T}(K)(S)$ is a solution of (7.59) with terminal condition

$$C_{T,T}(K)(S) = (S - K)^+.$$

Now, if $V(t,S)$ satisfies $LV = 0$, one may check that $L(S^{2-\frac{2r}{\sigma^2}}V(t,\frac{C}{S})) = 0$ for any constant $C > 0$.

The partial differential equation methods can also be applied to barrier options. From formula (7.55), we see that the price of the down and out option is, in fact,

$$C_{t,T}(K|H \downarrow O)(S) = C_{t,T}(K)(S) - \left(\frac{H}{S}\right)^{-1+\frac{2r}{\sigma^2}} C_{t,T}(K)\left(\frac{H^2}{S}\right).$$

Consequently, $C_{t,T}(K|H \downarrow O)(S)$ is a solution of (7.59) satisfying appropriate boundary conditions.

There are analogous representations for the other barrier options.

7.10 The Greeks

For the European call option, the value function has the form

$$V_{t,T}(S) = S\Phi(d_+(t)) - Ke^{-r(T-t)}\Phi(d_-(t)), \tag{7.68}$$

where $d_+(t)$ and $d_-(t)$ are defined by (7.56) and (7.57), respectively.

The European put price is, similarly,

$$P_{t,T}(S) = Ke^{-r(T-t)}\Phi(-d_-(t)) - S\Phi(-d_+(t)).$$

We investigate the sensitivity of these prices with respect to the parameters appearing in these equations. The purpose of hedging is to reduce (if not eliminate) the sensitivity of the value of a portfolio to changes in the underlying by means of diversifying the position. Thus analysis of the rate of change in V with respect to the underlying is fundamental. More generally, we shall add to this by considering the nature of the various partial derivatives of the value function V with respect to the parameters occurring in the Black-Scholes formulas. We begin by studying the sensitivity to changes in the underlying, S.

Delta

Differentiation with respect to S in the above expressions (which must be done with some care since $d_+(t)$ also depends on S!) yields

$$\Delta_C = \frac{\partial V}{\partial S}(S, t) = \Phi(d_+(t))$$

for the delta of the call and

$$\Delta_P = \frac{\partial P}{\partial S}(S, t) = -\Phi(-d_+(t))$$

for the delta of the put with the same strike and expiry.

Exercise 7.10.1. Carry out the differentiation to verify these results. *Hint:* Much effort can be avoided by observing that

$$\frac{1}{2}(d_+(t)^2 - d_-(t)^2) = \log\left(\frac{S_t}{K}\right) + r(T - t).$$

Remark 7.10.2. Note that since $\Phi(-x) = 1 - \Phi(x)$, it follows at once that $\Delta_P = \Delta_C - 1$. This relation is also immediate from call-put parity.

Clearly, each Δ measures the sensitivity of the value function V with respect to the price of the underlying, S. *Delta-hedging* is (at least in theory) the simplest way to eliminate risk: rebalancing the portfolio by adjusting the stock holdings in line with changes in the partial derivative $\frac{\partial V}{\partial S}$ will provide a risk-neutral position at each time point. Perfect hedging is only possible in idealised markets, but the technique is used in practice to indicate the direction in which investment decisions should be taken. Note that we can also use it for a portfolio of options since the linearity of differentiation will ensure that if Δ_i corresponds to the value function V_i of the ith option, then the whole portfolio $V = \sum_{i=1}^n V_i$ has Δ equal to $\frac{\partial V}{\partial S} = \sum_{i=1}^n \Delta_i$.

Gamma

The gamma, Γ, of the option value enters into more sophisticated hedging strategies. It is the second derivative of the option value with respect to the underlying, i.e., writing $\phi(x) = \frac{1}{\sqrt{2\pi}}e^{-\frac{x^2}{2}}$ for the standard normal density, we have

$$\Gamma_C = \frac{\partial^2 V}{\partial S^2} = \frac{1}{S\sigma\sqrt{T - t}}\phi(d_+(t)).$$

For the European put, we must have $\Gamma_P = \Gamma_C$ since $\Delta_P = \Delta_C - 1$.

Heuristically, the gamma measures 'how often' we need to adjust Δ to ensure that it will compensate for changes in the underlying, S, and also by how much. Keeping Γ near 0 helps one to keep the amounts and frequency of hedging under control, thus reducing the transaction costs associated with the hedging strategy. Because Γ_C is strictly positive, it also follows that the price of the call option is a strictly convex function of the price of the underlying.

Theta

The theta, Θ, measures the sensitivity of the option price to the expiration time T. For the European call, we have

$$\Theta_C = \frac{\partial V}{\partial T} = \frac{-S\phi(y_1(T-t))\sigma}{2\sqrt{T-t}} - rKe^{-r(T-t)}\Phi(d_-(t)).$$

Likewise, for the European put,

$$\Theta_P = \frac{\partial P}{\partial T} = \frac{-S\phi(y_1(T-t))\sigma}{2\sqrt{T-t}} + rKe^{-r(T-t)}\Phi(-d_-(t)).$$

Note that Θ is always negative for European calls. For this reason, theta is often called the *time decay* of the option (even though the value of a put does not necessarily decay over time). This parameter is frequently expressed in terms of the time to maturity, that is, V is then differentiated with respect to $\tau = T - t$ instead.

Since $\Phi(-x) = 1 - \Phi(x)$, it follows from the above expressions that

$$\Theta_P = \Theta_C + rKe^{-r(T-t)}. \tag{7.69}$$

We can, again, deduce this relation in much simpler fashion without calculating these values.

Exercise 7.10.3. Deduce the identity (7.69) from call-put parity.

The decay of the option value as the time to expiry goes to 0, even when the price of the underlying remains constant, complements the sensitivity of the option value to changes in S. Changes in option value are thus determined (if volatility and the riskless rate remain constant) by the phenomenon of time decay and by Δ.

Rho

The sensitivity of the option price to changes in the riskless short rate r is denoted by ρ. For the European call, it is

$$\rho_C = \frac{\partial V}{\partial r} = (T-t)Ke^{-r(T-t)}\Phi(d_-(t)),$$

and for the European put,

$$\rho_P = \frac{\partial P}{\partial r} = -(T-t)Ke^{-r(T-t)}\Phi(-d_-(t)).$$

Thus ρ is always positive for calls and negative for puts. This fits with the intuition that stock prices will normally rise with rising external interest rates.

Vega

The sensitivity of the option price to changes in the volatility σ is called *vega*, although there is no such Greek letter. At one time, this derivative was denoted by kappa, κ. For the European call, we have

$$\frac{\partial V}{\partial \sigma} = S\sqrt{T-t}\phi(d_+(t)),$$

while, for the put,

$$\frac{\partial P}{\partial \sigma} = S\sqrt{T-t}\phi(-d_+(t)).$$

Note that since $\phi(-x) = \phi(x)$ these two values are actually equal (i.e., the vega of a European put equals that of the European call with the same strike and expiry). Moreover, vega declines as t approaches T and is proportional to S. It turns out that vega peaks when the option is at the money (i.e., $S = K$) but since the payoff is small for values of S_T near K, it is necessary to normalise vega by the option value (i.e., consider relative price changes) in order to draw significant conclusions about the sensitivity of an investment to changes in volatility.

Nevertheless, it is instructive in the Black-Scholes setting to compare risk and return for options against holdings in the stock, as was done for the single-period binomial model in Chapter 1. We show that, just as in the discrete setting, the volatility of the call is proportional to that of the stock and that the constant of proportionality is again bounded below by 1.

To determine the volatility of the call, we first apply the Itô formula to the value process $V = V_{t,T}(S)$ as in (7.68). We obtain

$$
\begin{aligned}
dV &= \frac{\partial V}{\partial t}dt + \frac{\partial V}{\partial S}dS + \frac{1}{2}\sigma^2 S^2 \frac{\partial^2 V}{\partial S^2}dt \\
&= \left(\frac{\partial V}{\partial t} + \frac{\partial V}{\partial S}\mu S + \frac{1}{2}\sigma^2 S^2 \frac{\partial^2 V}{\partial S^2}\right)dt + \frac{\partial V}{\partial S}\sigma S dB,
\end{aligned}
\tag{7.70}
$$

which shows that the volatility of the call (i.e., the coefficient of the random term in $\frac{dV}{V}$) should be taken to be

$$\sigma_C = \frac{1}{V}\sigma S\frac{\partial V}{\partial S} = \frac{S}{V}\Delta\sigma_S, \tag{7.71}$$

where we have written σ_S for σ, the volatility of the stock price S. This shows that the two volatilities are proportional, with $E_C = \frac{S}{V}\Delta$ as the constant of proportionality. Clearly, $E_C > 1$, since

$$\frac{S_t\phi(d_+(t))}{S_t\phi(d_+(t)) - Ke^{-r(T-t)}\phi(d_-(t))} > 1.$$

Moreover, we can rewrite (7.70) in the form $dV_t = V_t(\mu_C dt + \sigma_C dB_t)$, where

$$\mu_C = \frac{1}{V_t}\left(\frac{\partial V_t}{\partial t} + \frac{\partial V_t}{\partial S_t}\mu S_t + \frac{1}{2}\sigma^2 S_t^2 \frac{\partial^2 V_t}{\partial S_t^2}\right).$$

By the Black-Scholes equation, this can be written as

$$\mu_C = \frac{1}{V_t}\left(rV_t - rS_t\frac{\partial V_t}{\partial S_t} + \mu S_t\frac{\partial V_t}{\partial S_t}\right) = \frac{S_t}{V_t}\frac{\partial V_t}{\partial S_t}\mu + r\left(1 - \frac{S_t}{V_t}\frac{\partial V_t}{\partial S_t}\right).$$

Hence, writing μ_S for μ and observing that $E_C = \frac{S}{V}\Delta = \frac{S_t}{V_t}\frac{\partial V_t}{\partial S_t}$, we have

$$\mu_C - r = E_C(\mu_S - r). \tag{7.72}$$

This is the exact analogue of equation (1.22) obtained for the single-period binomial model.

Chapter 8

The American Put Option

8.1 Extended Trading Strategies

As in Chapter 7, we suppose there is an underlying probability space $(\Omega, \mathcal{F}, Q)$. The time parameter t will take values in $[0, T]$. There is a filtration $\mathbb{F} = (\mathcal{F}_t)$ that satisfies the 'usual conditions' as described in Definition 6.1.1. We assume as before that the market is frictionless; that is, there are no transaction costs or taxes, no restrictions on short sales and trading can take place at any time t in $[0, T]$.

We suppose there is a savings account S_t^0 with constant interest rate r, such that

$$dS_t^0 = rS_t^0 dt. \tag{8.1}$$

As usual, we take $S^0(0) = 1$.

In addition, we suppose there is a risky asset S_t^1 whose dynamics are given by the usual log-normal equation:

$$dS_t^1 = S_t^1(\mu dt + \sigma dW_t). \tag{8.2}$$

Here, W is a standard Brownian motion on $(\Omega, \mathcal{F}, Q)$, μ is the appreciation rate (drift), and σ is the volatility of S_t^1.

A trading strategy is an adapted process $\pi = (\pi^0, \pi^1)$ satisfying

$$\int_0^T (\pi_u^i)^2 (S_u^i)^2 du < \infty \text{ a.s.}$$

The amount (π^i) is the amount held, or shorted, in units of the savings account $(i = 0)$ or stock $(i = 1)$. A short position in the savings account is a loan.

A consumption process is a progressive, continuous non-decreasing process C.

223

What investment and consumption processes are admissible? Such a triple of processes (π^0, π^1, C) is admissible if the corresponding wealth process is self-financing. The wealth process is

$$V_t(\pi) = \pi_t^0 S_t^0 + \pi_t^1 S_t^1.$$

We saw in Section 7.4 that this is self-financing if

$$\pi_t^0 S_t^0 + \pi_t^1 S_t^1 = \pi_0^0 + \pi_0^1 S_0^1 + \int_0^t \pi_u^0 dS_u^0 + \int_0^t \pi_u^1 dS_u^1 - C_t \text{ for } t \in [0, T] \quad (8.3)$$

with $C_0 = 0$ a.s. Note that equation (8.3) states that all changes in total wealth come from changes in the stock price plus interest on the savings account less the amount consumed, C_t.

For the pricing models we consider throughout this chapter, we shall assume the existence of an EMM $\widetilde{Q}$ without further comment. For the dynamics (8.1), (8.2), we have seen that the martingale measure $\widetilde{Q}$ is defined by setting

$$\left. \frac{d\widetilde{Q}}{dQ} \right|_{\mathcal{F}_t} = \Lambda_t = \exp\left\{ \frac{r - \mu}{\sigma} W_t - \frac{1}{2} \left(\frac{r - \mu}{\sigma} \right)^2 t \right\}.$$

Under $\widetilde{Q}$, $\widetilde{W}_t$ is a standard Brownian motion, where

$$\widetilde{W}_t = W_t - \left(\frac{r - \mu}{\sigma} \right) t, \qquad dS_t^1 = S_t^1(rdt + \sigma d\widetilde{W}_t). \quad (8.4)$$

In the remainder of this chapter we shall work under probability $\widetilde{Q}$, so the stock price has dynamics (8.4) and the wealth process $V_t(\pi)$ satisfies

$$V_t(\pi) = V_0(\pi) + \int_0^t rV_u(\pi)du + \int_0^t \sigma \pi_u^1 S_u^1 d\widetilde{W}_u - C_t \text{ a.s.} \quad (8.5)$$

Definition 8.1.1. A *reward function* ψ is a continuous, non-negative function on $\mathbb{R}^+ \times [0, T]$. We suppose ψ is in $C^{1,0}$ and piecewise in $C^{2,1}$. The latter condition means there is a partition of $\mathbb{R}^+$ into intervals in the interior of which ψ is $C^{2,1}$ in x. We require that, where defined, all the functions $\psi, \frac{\partial \psi}{\partial x}, \frac{\partial^2 \psi}{\partial x^2}, \frac{\partial \psi}{\partial t}$ have polynomial growth as $x \to \infty$.

Definition 8.1.2. An *American option* with reward ψ is a security that pays the amount $\psi(S_t, t)$ when exercised at time t.

Example 8.1.3. Recall, as in Chapter 1, that examples of American options are the American call option, with $\psi(S_t, t) = (S_t - K)^+$, the American put option, with $\psi(S_t, t) = (K - S_t)^+$, and the American straddle (bottom version) with $\psi(S_t, t) = |S_t - K|$.

If one sells such a claim, one accepts the obligation to pay $\psi(S_t, t)$ to the buyer at any time $t \in [0, T]$. The final time T is the expiration time.

Having introduced this new financial instrument, the American option, into the market, it is expedient to extend the notion of a trading strategy. As we shall concentrate on put options, $P(x, t) = P(x) = P_t = P$ will denote the value process of the American option.

Definition 8.1.4. For any stopping time $\tau \in \mathcal{T}_{0,T}$, a *buy-and-hold strategy* in the option P is a pair (π^2, τ), where π^2 is the process

$$\pi^2(t) = k \mathbf{1}_{[0,\tau]}(t), t \in [0, T].$$

The associated position in P is then $\pi^2(t)P(x, t)$. This means that k units of the American option security are purchased (or shorted if $k < 0$) at time 0 and held until time τ. Denote by Π^+ (resp. Π^-) the set of buy-and-hold strategies in P for which $k \geq 0$ (resp. $k < 0$).

Write $\hat{\pi}$ for a triple (π^0, π^1, π^2).

An extended admissible trading strategy in (S^0, S^1, P) is then a collection $(\pi^0, \pi^1, \pi^2, \tau)$ such that (π^0, π^1) is an admissible trading strategy in (S^0, S^1), (π^2, τ) is a buy-and-hold strategy in P, and, on the interval $(\tau, T]$, we have

$$\pi_t^0 = \pi_\tau^0 + \frac{\pi_\tau^1 S_\tau^1}{S_\tau^0} + \frac{\pi_\tau^2 \psi(S_\tau, \tau)}{S_\tau^0}, \qquad \pi_t^1 = \pi_t^2 = 0.$$

This means that, using the extended strategy $\hat{\pi} = (\pi^0, \pi^1, \pi^2)$, at time τ we liquidate the stock and option accounts and invest everything in the riskless bond (savings account). The buy-and-hold strategy $(\hat{\pi}, \tau)$ is now self-financing if, with a consumption process C, we have

$$\pi_t^0 S_t^0 + \pi_t^1 S_t^1 = \pi_0^0 + \pi_0^1 S_0^1 + \int_0^t \pi_u^0 dS_u^0 + \int_0^t \pi_u^1 dS_u^1 - C_t \text{ a.s. for } t \in [0, \tau]$$

and

$$\int_\tau^t dC_u = 0 \text{ a.s. for } t \in (\tau, T].$$

That is, C is constant on $(\tau, T]$.

Notation 8.1.5. Denote the set of extended admissible trading strategies in (S^0, S^1, P) by $\mathcal{A}$.

Definition 8.1.6. There is said to be *arbitrage* in the market if either there exists $(\pi^2, \tau) \in \Pi^+$ with (π^0, π^1, C) such that

$$(\pi, \tau) \in \mathcal{A}, \qquad \pi_0^0 + \pi_0^1 S_0^1 + \pi_0^2 V_0 < 0, \qquad \pi_T^0 S_t^0 \geq 0 \text{ a.s.} \qquad (8.6)$$

or, there exists $(\pi^2, \tau) \in \Pi^-$, with (π^0, π^1, C), such that

$$(\pi, \tau) \in \mathcal{A}, \qquad \pi_0^0 + \pi_0^1 S_0^1 + \pi_0^2 V_0 < 0, \qquad \pi_T^0 S_t^0 \geq 0 \text{ a.s.} \qquad (8.7)$$

Statement (8.6) means it is possible to hold an American option and find an exercise policy that gives riskless profits.

Conversely, statement (8.7) means it is possible to sell the American option and be able to make riskless profits for every exercise policy option of the buyer.

Statements (8.6) and (8.7) define arbitrage opportunities for the buyer or seller, respectively, of an American option. Our assumption is that arbitrage is not possible; the fundamental question is: what price should be paid today (time t) for such an option?

Our discussion concentrates on the American put option. (We showed in Chapter 1, using simple arbitrage arguments, that the price of an American call on a stock that does not pay dividends is equal to the price of the European call (see [224]).

8.2 Analysis of American Put Options

Notation 8.2.1. Let $\mathcal{T}_{t_1,t_2}$ denote the set of all stopping times that take values in $[t_1, t_2]$.

Lemma 8.2.2. *Consider the process*

$$X_t = ess \sup_{\tau \in \mathcal{T}_{t,T}} \widetilde{E} \left(e^{-r(\tau-t)} (K - S_\tau)^+ | \mathcal{F}_t \right) \text{ for } t \in [0, T]; \qquad (8.8)$$

(X_t is the supremum of the random variables $\widetilde{E} \left(e^{-r(\tau-t)} (K - S_\tau)^t | \mathcal{F}_t \right)$ for $\tau \in \mathcal{T}_{t,T}$ in the complete lattice $L^1(\Omega, \mathcal{F}_t, \widetilde{Q})$.) Then there are admissible strategies $\left(\pi_t^0, \pi_t^1 \right)$ and a consumption process C such that, with $V_t(\pi)$ given by (8.5), we have

$$X_t = V_t(\pi).$$

Proof. Karatzas ([182])
    Define
$$J_t = ess \sup_{\tau \in \mathcal{T}_{t,T}} \widetilde{E} \left(e^{-r\tau} (K - S_\tau)^+ | \mathcal{F}_t \right) \text{ a.s.}$$

Then J is a supermartingale, and, in fact, J is the smallest supermartingale dominating the discounted reward $(e^{-r\tau} (K - S_\tau)^+)$. The process J is called the *Snell envelope* (see Chapter 5 for the discrete case).

Recall (see Remark 5.3.12, and refer for more details to [109, Chapter 8]) that a right-continuous supermartingale X is said to be of class D if the set of random variables X_τ is uniformly integrable, where τ is any stopping time.

Furthermore, J is right-continuous, has left limits, is regular and is of class D (in fact J is bounded). Consequently, J has a Doob-Meyer decomposition as the difference of a (right-continuous) martingale M and a predictable increasing process A;

$$J_t = M_t - A_t. \qquad (8.9)$$

Here M is a $\left(\widetilde{Q}, \mathcal{F}_t\right)$-martingale and A is a unique, predictable continuous non-decreasing process with $A_0 = 0$. From the martingale representation theorem, we can write

$$M_t = J_0 + \int_0^t \eta_u d\widetilde{W}_u$$

for some progressively measurable process η with

$$\int_0^T \eta_u^2 du < \infty \text{ a.s.}$$

Consequently,

$$X_t = e^{rt} J_t, \qquad dX_t = re^{rt} J_t dt + e^{rt} \eta_t d\widetilde{W}_t - e^{rt} dA_t.$$

Therefore $X_t = V_t(\pi)$ if we take

$$\pi_t^0 = e^{rt} J_t - e^{rt} \sigma^{-1} \eta_t, \quad \pi_t^1 = e^{rt} \eta_t \sigma^{-1} (S_t^1)^{-1}, \quad dC_t = e^{rt} dA_t. \quad (8.10)$$

$\square$

Optimal Stopping Times

Remark 8.2.3. Note that

$$X_t \geq (K - S_t)^+ \text{ a.s. for } t \in [0, T], \qquad X_T = (K - S_T)^+ \text{ a.s.} \quad (8.11)$$

Also, a stopping time τ^* is said to be *optimal* if

$$J_t = \widetilde{E}\left(e^{-r\tau^*}(K - S_{\tau^*})^+ | \mathcal{F}_t\right).$$

We can now verify that the price X_0 of the put option obtained in this way is the unique price that will preclude arbitrage. First we quote results, entirely analogous to those established in Chapter 5 for the discrete case, that characterise optimal stopping times in this model.

Notation 8.2.4. Write

$$\rho_t = \inf\left\{u \in [t, T] : J_u = e^{-ru}(K - S_u)^+\right\}.$$

That is, ρ_t is the first time in $[t, T]$ that J falls to the level of the discounted reward.

From the work of El Karoui [99], we know the following.

a) ρ_t is the optimal stopping time on $[t, T]$.

b) A, in the decomposition (8.9), is constant on the interval $[t, \rho_t]$.

c) The stopped process $J_{t \wedge \rho_t}$ is a martingale on $[t, T]$.

Theorem 8.2.5. *Taking the price of the American put option at time $t = 0$ to be X_0 is necessary and sufficient for there to be no arbitrage.*

Proof. Suppose the market price of the American put option were $Y_0 > X_0$. Consider the trading strategies π^0, π^1, and C given by (8.10). For any stopping time $\tau \in \mathcal{T}_{0,T}$, and with $k = -1$, consider the buy-and-hold strategy

$$\pi_t^2 = -\mathbf{1}_{[0,\tau]}(t).$$

Construct the extended trading strategy $\widehat{\pi} = (\widehat{\pi}_t^0, \widehat{\pi}_t^1, \widehat{\pi}_t^2)$ by setting

$$\widehat{\pi}_t^0 = \begin{cases} \pi_t^0 & \text{if } t \in [0, \tau], \\ \pi_\tau^0 + \pi_\tau^1 e^{-r\tau} S_\tau^1 - (K - S_\tau)^+ e^{-r\tau} & \text{if } t \in (\tau, T], \end{cases}$$

and

$$\widehat{\pi}_t^1 = \pi_t^1 \mathbf{1}_{[0,\tau]}(t), \qquad\qquad \widehat{\pi}_t^2 = \pi_t^2 = -\mathbf{1}_{[0,\tau]}(t),$$

with a consumption process $\widehat{C}_t = C_{t \wedge \tau}$. From the hedging property,

$$X_\tau = \pi_\tau^0 e^{r\tau} + \pi_\tau^1 S_\tau^1 \geq (K - S_\tau)^+ \text{ a.s.}$$

we see that

$$e^{rT} \widehat{\pi}_T^0 \geq 0 \text{ a.s.}$$

However, by definition,

$$\widehat{\pi}_0^0 + \widehat{\pi}_0^1 S_0 + \widehat{\pi}_0^2 Y_0 = X_0 - Y_0 < 0.$$

We would therefore have an arbitrage opportunity.

Now suppose $Y_0 < X_0$. Take π^0, π^1, and C as in (8.10), and use the optimal stopping time ρ_0 of Notation 8.2.4. As before, construct an extended trading strategy by setting

$$\widehat{\pi}_t^0 = \begin{cases} -\pi_t^0, & \text{if } t \in [0, \rho_0], \\ -\pi_{\rho_0}^0 - \pi_{\rho_0}^1 e^{-r\rho_0} S_{\rho_0}^1 + (K - S_{\rho_0})^+ e^{-r\rho_0}, & \text{if } t \in (\rho_0, T]. \end{cases}$$

and

$$\widehat{\pi}_t^1 = -\pi_t^1 \mathbf{1}_{[0,\rho_0]}(t), \qquad\qquad \widehat{\pi}_t^2 = \mathbf{1}_{[0,\rho_0]}(t),$$

with the consumption process $\widehat{C}_t = -C_{t \wedge \rho_0}$. However, we know $C = \widehat{C} \equiv 0$ on $[0, \rho_0]$ (see the remarks after Notation 8.2.4) and, from the definition of ρ_0,

$$\pi_{\rho_0}^0 e^{r\rho_0} + \pi_{\rho_0}^1 S_{\rho_0}^1 = (K - S_{\rho_0})^+.$$

Therefore $\widehat{\pi}_T^0 S_t^0 = 0$, but

$$\widehat{\pi}_0^0 + \widehat{\pi}_0^1 S^0(0) + \widehat{\pi}_0^2 Y_0 = Y_0 - X_0 < 0.$$

Again there is arbitrage.

Finally (see Lemma 8.2.2), we know that $X_t = V_t(\pi)$ is a martingale under $\widetilde{Q}$ up to time ρ_0 so X_0 is the fair price at time 0 for the American put option. $\square$

Continuation and Stopping Regions: The Critical Price

Definition 8.2.6. For $t \in [0, T]$ and $x \in \mathbb{R}^+$, define

$$P(x, t) = \sup_{\tau \in \mathcal{T}_{t,T}} \widetilde{E}\left(e^{-r(\tau - t)}(K - S_\tau)^+ \mid S_t = x\right). \tag{8.12}$$

Then $P(x, t)$ is the value function and represents the fair, or arbitrage-free, price of the American put at time t.

From [203, Theorem 3.1.10] we can state the following.

Theorem 8.2.7. *The first optimal stopping time after time t is*

$$\rho_t = \inf\left\{u \in [t, T] : P(S_u, u) = (K - S_u)^+\right\}.$$

It is important to determine the principal analytical properties of the process defined in (8.12).

Lemma 8.2.8. *For every* $t \in [0, T]$, *the American put value* $P(x, t)$ *is convex and non-decreasing in* $x > 0$. *The function* $P(x, t)$ *is non-increasing in* t *for every* $x \in \mathbb{R}^+$. *The function* $P(x, t)$ *is continuous on* $\mathbb{R}^+ \times [0, T]$.

Proof. The convexity of $P(\cdot, t)$ follows from the supremum operation, and the non-increasing properties of $P(\cdot, t)$ and $P(x, \cdot)$ are immediate from the definition.

For $(t_i, x_i) \in \mathbb{R}^+ \times [0, T], i = 1, 2$ we have

$$\begin{aligned}
P(x_2, t_2) - P(x_1, t_1) &= \sup_{\tau \in \mathcal{T}_{t_2,T}} \widetilde{E}\left(e^{-r(\tau - t_2)}(K - S_\tau)^+ \mid S_{t_2} = x_2\right) \\
&\quad - \sup_{\tau \in \mathcal{T}_{t_2,T}} \widetilde{E}\left(e^{-r(\tau - t_1)}(K - S_\tau)^+ \mid S_{t_1} = x_2\right) \\
&\quad + \sup_{\tau \in \mathcal{T}_{t_2,T}} \widetilde{E}\left(e^{-r(\tau - t_1)}(K - S_\tau)^+ \mid S_{t_1} = x_2\right) \\
&\quad - \sup_{\tau \in \mathcal{T}_{t_1,T}} \widetilde{E}\left(e^{-r(\tau - t_1)}(K - S_\tau)^+ \mid S_{t_1} = x_1\right).
\end{aligned}$$

Therefore, with $t_1 \leq t_2$,

$$\begin{aligned}
&|P(x_2, t_2) - P(x_1, t_1)| \\
&\leq \widetilde{E}\left(\sup_{t_2 \leq s \leq T} \left|e^{-r(s - t_2)}\left(K - S_s^{t_2, x_2}\right) - e^{-r(s - t_1)}\left(K - S_s^{t_1, x_2}\right)^+\right|\right) \\
&\quad + \widetilde{E}\left(\sup_{t_1 \leq s \leq t_2} \left|e^{-r(s - t_1)}\left(K - S_s^{t_1, x_1}\right)^+ - e^{-r(t_2 - t_1)}\left(K - S_s^{t_1, x_2}\right)^+\right|\right),
\end{aligned}$$

and the result follows from the continuity properties of the flow. $\qquad\square$

Definition 8.2.9. Consider the two sets

$$\mathcal{C} = \left\{ (x,t) \in \mathbb{R}^+ \times [0,T) : P(x,t) > (K-x)^+ \right\},$$
$$\mathcal{S} = \left\{ (x,t) \in \mathbb{R}^+ \times [0,T) : P(x,t) = (K-x)^+ \right\}.$$

$\mathcal{C}$ is called the *continuation region* and $\mathcal{S}$ is the *stopping region*.

Then $\rho_t = \inf \{ u \in [t,T] : S_u \notin \mathcal{C} \}$. We now establish some properties of P and $\mathcal{C}$.

Lemma 8.2.10. *We have $P(x,t) > 0$ for all $x \geq 0, t \in [0,T]$.*

Proof. Note that $(K-x)^+ > 0$ for $x < K$. Now fix t and consider the solution of (8.2) such that $S_t^1 = x$. Write $\tau_{\frac{K}{2}} = \inf \{ u \geq t : S_u^1 \leq \frac{K}{2} \} \wedge T$. Then, if $x \geq K$, from (8.12)

$$P(x,t) \geq \frac{K}{2} E \left(e^{-\tau_{K/2}} 1_{\{\tau_{K/2} < T\}} \right) > 0.$$

$\square$

The following two results are adapted from Jacka [164].

Lemma 8.2.11. *For each $t > 0$, the t-section of $\mathcal{C}$ is given by*

$$\mathcal{C}_t = \{ x : (x,t) \in \mathcal{C} \}$$
$$= \left\{ x : (x,t) \in \mathbb{R}^+ \times [0,T), P(x,t) > (K-x)^+ \right\}$$
$$= (S_t^*, \infty)$$

for some S_t^* such that $0 < S_t^* < K$.

Proof. Clearly $0 \notin \mathcal{C}_t$.

We shall show that if $x < y$ and $x \in \mathcal{C}_t$, then $y \in \mathcal{C}_t$. Write

$$\tau = \inf \{ s \geq 0 : (S_s(x), s) \notin \mathcal{C} \},$$

so τ is the optimal stopping time for $S_s(x)$. Now τ is also a stopping time for $S(y)$, so

$$P(y,t) - P(x,t) = P(y,t) - E \left(e^{-r\tau} \left(K - S_\tau(x) \right)^+ \right)$$
$$\geq E \left(e^{-r\tau} \left((K - S_\tau(y))^+ - (K - S_\tau(x))^+ \right) \right)$$
$$= E \left(e^{-r\tau} \{ (K - S_\tau(y)) - (K - S_\tau(x)) \} \right)$$
$$+ E \left(e^{-r\tau} \left\{ (K - S_\tau(y))^- - (K - S_\tau(x))^- \right\} \right). \quad (8.13)$$

Now

$$S_\tau(y) = y \exp \left\{ \left(r - \frac{\sigma^2}{2} \right) \tau + \sigma \widetilde{W}_\tau \right\}$$

and similarly for $S_\tau(x)$; therefore the second expectation in (8.13) is non-negative and

$$P(y,t) - P(x,t) \geq E\left(e^{-r\tau}\left(S_\tau(x) - S_\tau(y)\right)\right)$$

$$= (x - y)E\left(\exp\left\{-\frac{\sigma^2}{2}\tau + \sigma\widetilde{W}_\tau\right\}\right)$$

$$= (x - y). \tag{8.14}$$

Therefore,

$$P(y,t) \geq (x - y) + P(x,t) > (x - y) + (K - x)^+ \geq K - y$$

because $x \in C_t$ (implying that $P(x,t) > (K - x)^+$). Now $P(y,t) > 0$, so $P(y,t) > (K - y)^+$ and $y \in C_t$.

Clearly $S_t^* \leq K$ for all $t > 0$ because if $x > K$, then $(K - x)^+ = 0$, although $P(x,t) > 0$. □

Corollary 8.2.12. *From* (8.14) *we see that* $\frac{\partial P(x,t)}{\partial x} \geq -1$ *for* $x,y \in C_t$.

Proposition 8.2.13. *The boundary* S^* *is increasing in* t *and is bounded above by* K.*(*S^* *is also known as the* critical price.)

Proof. Clearly, for $0 \leq s \leq t \leq T, P(x,s) \geq P(x,t)$. Therefore

$$(K - S_{t+s}^* - \varepsilon)^+ < P(S_{t+s}^* + \varepsilon, t + s) \leq P(S_{t+s}^* + \varepsilon, t) \text{ for } t > 0, s \geq 0, \varepsilon > 0,$$

so $S_{t+s}^* + \varepsilon \in C_t$ and $S_t^* \leq S_{t+s}^*$ for $\varepsilon > 0$ and $s > 0$.

Now $(K - x)^+$ is zero for $x \geq K$. But $P(x,t) > 0$ from Lemma 8.2.10, so $S_t^* < K$. □

Exercise 8.2.14. Sketch (in three dimensions) the value function and the continuation and stopping regions for an American put option with strike K and expiry T in the Black-Scholes model. Indicate the critical boundary in terms of t-sections and sketch the critical price as a function of t.

8.3 The Perpetual Put Option

We now discuss the limiting behaviour of S_t^* by introducing the 'perpetual' American put option; this is the situation when $T = \infty$. The mathematics involves deeper results from analysis and optimal stopping, particularly when we discuss free boundaries and smooth pasting. Perpetual put options are a mathematical idealisation: no such options are traded in real markets.

Theorem 8.3.1. *Consider the function*

$$P(x) = \sup_{\tau \in \mathcal{T}_{0,\infty}} \widetilde{E}_x\left(e^{-r\tau}(K - S_\tau)^+ \mathbf{1}_{\{\tau < \infty\}}\right).$$

Then

$$P(x) = \begin{cases} (K - S^*) \left(\frac{x}{S^*}\right)^{-\gamma} & \text{if } x > S^*, \\ K - x & \text{if } x \leq S^*, \end{cases}$$

where $S^* = \frac{K\gamma}{1+\gamma}$ *and* $\gamma = \frac{2r}{\sigma^2}$.

Proof. From the definition, it is immediate that $P(x)$ is convex, decreasing on $[0, \infty)$, and satisfies $P(x) > (K - x)^+$. Furthermore,

$$P(x) \geq E\left(e^{-rT}(K - S_T)^+\right) \text{ for all } T > 0.$$

This implies that $P(x) > 0$ for all $x \geq 0$.
Write $S^* = \sup\{x \geq 0 : P(x) = K - x\}$. Then clearly

$$P(x) = K - x \text{ for } x \leq S^*, \qquad P(x) > (K - x)^+ \text{ for } x > S^*. \tag{8.15}$$

However, from the results for the Snell envelope, (see [99]), we know that

$$P(x) = E\left(\left(Ke^{-r\rho_x} - S_{\rho_x}\right)^+ \mathbf{1}_{\{\rho_x < \infty\}}\right).$$

Here

$$\rho_x = \rho_0(x) = \inf\{t \geq 0 : P(S_t) = (K - S_t)^+\},$$

with $\inf \emptyset = \infty$.
Recall that

$$S_t = x \exp\left\{\left(r - \frac{\sigma^2}{2}\right)t + \sigma B_t\right\}.$$

We have seen that ρ_x is an optimal stopping time. Now from the inequalities (8.15), ρ_x is also given by

$$\rho_x = \inf\{t \geq 0 : S_t \leq S^*\} = \inf\left\{t \geq 0 : \left(r - \frac{\sigma^2}{2}\right)t + \sigma B_t \leq \log\left(\frac{S^*}{x}\right)\right\}.$$

For any $z \in \mathbb{R}^+$, define the stopping time

$$\tau_{x,z} = \inf\{t \geq 0 : S_t \leq z\}.$$

Then $\rho_x = \tau_{x,S^*}$. For any fixed $x \in \mathbb{R}^+$, consider the function

$$u(z) = E\left(e^{-r\tau_{x,z}} \mathbf{1}_{\{\tau_{x,z} < \infty\}} (K - S_{\tau_{x,z}})^+\right).$$

As τ_{x,S^*} is an optimal stopping time, the function u is maximized when $z = S^*$.

Now, if $z > x$, clearly $\tau_{x,z} = 0$ and $u(z) = (K - x)^+$. If $z \leq x$, then $\tau_{x,z} = \inf\{t \geq 0 : S_t = z\}$, as the trajectories of S are continuous. Therefore

$$u(z) = (K - z)^+ E\left(e^{-r\tau_{x,z}} \mathbf{1}_{\{\tau_{x,z} < \infty\}}\right) = (K - z)^+ E\left(e^{-r\tau_{x,z}}\right)$$

(as $e^{-r\infty} = 0$). Now

$$\tau_{x,z} = \inf\left\{t \geq 0 : \left(r - \frac{\sigma^2}{2}\right)t + \sigma B_t = \log\left(\frac{z}{x}\right)\right\}$$

$$= \inf\left\{t \geq 0 : \gamma t + B_t = \frac{1}{\sigma}\log\left(\frac{z}{x}\right)\right\},$$

where $\gamma = \sigma^{-1}\left(r - \frac{\sigma^2}{2}\right)$.

For any $b \in \mathbb{R}$, write, as in Corollary 7.2.6,

$$T(b) = \inf\left\{t \geq 0 : \gamma t + B_t = b\right\}.$$

Then

$$u(z) = \begin{cases} (K - x)^+ & \text{if } z > x, \\ (K - z)E\left(\exp\left\{-\frac{rT}{\sigma}\log\left(\frac{z}{x}\right)\right\}\right) & \text{if } z \in [0, x] \cap [0, K], \\ 0 & \text{if } z \in [0, x] \cap [K, \infty). \end{cases}$$

The maximum value of u is therefore attained in the interval $[0, x] \cap [0, K]$.
Now, from Corollary 7.2.6,

$$E\left(e^{-\alpha T(b)}\right) = \exp\left\{\gamma b - |b|\sqrt{\gamma^2 + 2\alpha}\right\}.$$

Therefore

$$u(z) = (K - z)\left(\frac{z}{x}\right)^\lambda \quad \text{for all } z \in [0, x] \cap [0, K],$$

where $\lambda = \frac{2r}{\sigma^2}$.

This function has derivative

$$u'(z) = \frac{z^{\lambda-1}}{x^\lambda}(\lambda K - (\lambda + 1)z).$$

Therefore, it follows that

$$\max_z u(z) = \begin{cases} u(x) = K - x & \text{if } x \leq \frac{\lambda K}{\lambda+1}, \\ u\left(\frac{\lambda K}{\lambda+1}\right) & \text{if } x > \frac{\lambda K}{\lambda+1}. \end{cases}$$

The stated results are then established. $\qquad\qquad\qquad\qquad\qquad\square$

Remark 8.3.2. Consider the free boundary problem

$$-ru + Sr\frac{du}{dS} + \frac{1}{2}\sigma^2 S^2\frac{d^2u}{dS^2} = 0, \qquad u(\infty) = 0, \qquad (8.16)$$

with free "boundary" S^* given by

$$u(S^*) = (K - S^*)^+, \qquad \left.\frac{\partial u}{\partial S}\right|_{S=S^*} = -1. \qquad (8.17)$$

It is known (see [22]) that the American put price $P(S)$ and the critical price S^* of Theorem 8.3.1 give the solution of this boundary value problem. In fact, any solution of the homogeneous equation (8.16) is of the form

$$a_1 S^{\gamma_1} + a_2 S^{\gamma_2},$$

where γ_1, γ_2 are the roots of the quadratic equation

$$\frac{1}{2}\sigma^2\gamma(\gamma - 1) + r\gamma - r = 0,$$

i.e.,

$$\gamma = \frac{-r + \frac{\sigma^2}{2} \pm \sqrt{r^2 + \frac{\sigma^4}{4} + r\frac{\sigma^2}{2}}}{\sigma^2}.$$

Discarding the positive root, because of the condition at $S = \infty$, we see the solution is of the form

$$u(S) = a_1 S^{-\frac{2r}{\sigma^2}}.$$

The conditions (8.17) give

$$S^* = \frac{2rK}{2r + \sigma^2}, \qquad\qquad a_1 = (K - S^*)(S^*)^{\frac{2r}{\sigma^2}},$$

in agreement with Theorem 8.3.1.

8.4   Early Exercise Premium

We return to consideration of the general American put option.

Theorem 8.4.1. *For $t \in [0, T]$, the Snell envelope J has the decomposition*

$$J_t = \widetilde{E}\left(e^{-rT}(K - S_T)^+ \,|\mathcal{F}_t\right) + \widetilde{E}\left(\int_t^T e^{-ru} rK \mathbf{1}_{\{S_u < S_u^*\}} du \,|\mathcal{F}_t\right) \quad a.s.$$

Proof. Suppose $\rho_t = \inf\{u \in [t, T) : S_u \leq S_u^*\} \wedge T$. Then ρ_t is the optimal stopping time in $[0, T]$, and

$$J_t = \widetilde{E}\left(e^{-r\rho_t}(K - S_{\rho_t})^+ \,|\mathcal{F}_t\right).$$

Write J_t in the form

$$\widetilde{E}\left(e^{-rT}(K - S_T)^+ \,|\mathcal{F}_t\right) + \widetilde{E}\left(e^{-r\rho_t}(K - S_{\rho_t})^+ - e^{-rT}(K - S_T)^+ \,|\mathcal{F}_t\right).$$

The first term is the value of the associated European option with exercise time T. The second term is the early exercise premium representing the advantage the American option has over the European. Using the generalized Itô rule for convex functions (see [194]), it can be represented as

$$\widetilde{E}\left(\int_{\rho_t}^T e^{-ru} rK \mathbf{1}_{\{S_u < K\}} du - \int_{\rho_t}^T e^{-ru} dL_u^K(S) \,|\mathcal{F}_t\right).$$

Here $L_u^K(S)$ is the local time of S at level K in the interval $[0, u]$.

Consider the anticipating right-continuous process of finite variation

$$D_t = \int_{\rho_0}^{\rho_t} e^{-ru} r K \mathbf{1}_{\{S_u < K\}} du - \int_{\rho_0}^{\rho_t} e^{-ru} dL_u^K(S).$$

From [109], we know there is a unique predictable process D^p, the dual predictable projection of D, such that

$$\widetilde{E}(D_T - D_t | \mathcal{F}_t) = \widetilde{E}(D_T^p - D_t^p | \mathcal{F}_t).$$

Consequently,

$$\begin{aligned}
J_t &= \widetilde{E}\left(e^{-rT}(K - S_T)^+ | \mathcal{F}_t\right) + \widetilde{E}(D_T - D_t | \mathcal{F}_t) \\
&= \widetilde{E}\left(e^{-rT}(K - S_T)^+ | \mathcal{F}_t\right) + \widetilde{E}(D_T^p - D_t^p | \mathcal{F}_t) \\
&= \widetilde{E}\left(e^{-rT}(K - S_T)^+ + D_T^p | \mathcal{F}_t\right) - D_t^p.
\end{aligned}$$

This expresses the supermartingale J as the difference of a martingale and a predictable process. From the uniqueness of the decomposition of the special semimartingale J, we see that $D^p = A$, so D^p is non-decreasing.

Write $D_t = A_t + B_t$, where

$$A_t = \int_{\rho_0}^{\rho_t} e^{-ru} r K \mathbf{1}_{\{S_u < K\}} \mathbf{1}_{\{S_u \le S_u^*\}} du - \int_{\rho_0}^{\rho_t} e^{-ru} \mathbf{1}_{\{S_u \le S_u^*\}} dL_u^K(S),$$

$$B_t = \int_{\rho_0}^{\rho_t} e^{-ru} r K \mathbf{1}_{\{S_u < K\}} \mathbf{1}_{\{S_u > S_u^*\}} du - \int_{\rho_0}^{\rho_t} e^{-ru} \mathbf{1}_{\{S_u > S_u^*\}} dL_u^K(S).$$

Now $S_t^* < K$ for $t \in [0, T)$ and dL^K does not charge $\{S < K\}$, that is, the dL^K-measure of this set is zero. Therefore

$$A_t = \int_{\rho_0}^{\rho_t} e^{-ru} r K \mathbf{1}_{\{S_u \le S_u^*\}} du = \int_0^t e^{-ru} r K \mathbf{1}_{\{S_u < S_u^*\}} du \text{ a.s.,}$$

so A is predictable and non-decreasing. Consequently, $A^p = A$.

The dual predictable projection of B is more difficult to determine. Although not necessary (see [295]), we shall assume that the critical price boundary S_t^* is continuous. Write

$$\chi(\omega) = \{t \in [\rho_0(\omega), T) : S_t(\omega) > S_t^*\}$$

for the excursion intervals of the stock process into the continuation region. From the continuity of S_t^* and the a.s.-continuity of S_t, the random set χ is a countable union of open sets.

Choose $\varepsilon > 0$ and note that, for every choice, the number of excursions (N^ε) in χ with duration greater than ε is finite. Label these intervals (a_n, b_n) with $a_n < b_n < a_{n+1} < b_{n+1}$ and put

$$N_t^\varepsilon = \sup\{1 \le n \le N^\varepsilon : a_n \le t\}.$$

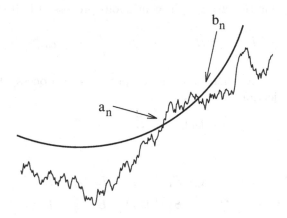

Figure 8.1: Excursion intervals

Consider the approximate process

$$B_t^\varepsilon = \sum_{n=1}^{N_t^\varepsilon} \left(\int_{a_n+\varepsilon}^{b_n} e^{-ru} r K \mathbf{1}_{\{S_u < K\}} du - \int_{a_n+\varepsilon}^{b_n} e^{-ru} dL_u^K(S) \right).$$

Using dominated convergence, B_t^ε converges to B_t as $\varepsilon \to 0$ for almost every ω and also in L^1.

However, B^ε is constant off $\{t \in [0, T) : S_t \geq S_t^*\}$, so its dual predictable projection $(B^\varepsilon)^p$ is also constant off this set. Now, in [104] it is shown that because $(J_{u \wedge \rho_t})_{t \leq u \leq T}$ is a martingale, the process $(B^\varepsilon)^p$ is *non-increasing*. The limit process B^p inherits both these properties. Now $D^p = A^p + B^p$ is *non-decreasing*, so we must have $B^p \equiv 0$. Consequently,

$$\widetilde{E}\left(D_T^p - D_t^p \,|\, \mathcal{F}_t\right) = \widetilde{E}\left(\int_t^T e^{-ru} r K \mathbf{1}_{\{S_u < S_u^*\}} du \,|\, \mathcal{F}_t \right) \quad \text{a.s.}$$

and the result follows.   □

Remark 8.4.2. The supermartingale property of the Snell envelope requires B^p to be a process with non-decreasing sample paths. On the other hand, the minimal property of the Snell envelope implies B^p should have non-increasing sample paths. Consequently, we must have $B^p \equiv 0$.

D^p can be thought of as the (predictable) hedging process that covers the non-adapted process D. Also $D_t^p = \int_0^t r K \mathbf{1}_{\{S_u < S_u^*\}} du$ a.s., so D^p is absolutely continuous, non-decreasing, and constant off $\{t \in [0, T) : S_t < S_t^*\}$.

Recall $P(x, t) = X_t$, as defined in (8.8). The following result is immediate.

Corollary 8.4.3. *The value* $P(x, t)$ *of the American put has the decomposition*

$$P(x, t) = p(x, t) + e(x, t) \text{ for } (x, t) \in \mathbb{R}^+ \times [0, T],$$

where

$$p(x, t) = \widetilde{E}_x \left(e^{-r(T-t)} (K - S_T)^+ \right),$$

$$e(x, t) = \widetilde{E}_x \left(\int_t^T e^{-r(u-t)} rK \mathbf{1}_{\{S_u < S_u^*\}} du \right)$$

with $S_t = x$.

Here $p(x, t)$ is the value of the European put with exercise date T, and $e(x, t)$ is the early exercise premium; it measures the advantage of being able to stop at any time between t and T.

Indeed, $e^{-r\Delta} rK$ represents the discounted gain of exercising compared with continuing when the stock price belongs to the stopping region S over the time $[u, u + \Delta]$.

From the above representation, we can deduce the following result.

Lemma 8.4.4. *We have*

$$P(\cdot, t) \in C^1 \left(\mathbb{R}^+ \right) \text{ for } t \in [0, T).$$

Remark 8.4.5. We can also consider the effect of delaying exercise of the option as follows. Write

$$J_t = e^{-rt} (K - S_t)^+ + \widetilde{E} \left(e^{-r\rho_t} (K - S_{\rho_t})^+ - e^{-rt} (K - S_t)^+ | \mathcal{F}_t \right)$$

$$= e^{-rt} (K - S_t)^+$$

$$+ \widetilde{E} \left(-\int_t^{\rho_t} e^{-ru} Kr \mathbf{1}_{\{S_u < K\}} du + \int_t^{\rho_t} e^{-ru} dL_u^K(S) | \mathcal{F}_t \right).$$

Paralleling the computations of Theorem 8.4.1, we obtain the representation of $P(x, t)$ in terms of the delayed exercise value:

$$P(x, t) = (K - x)^+$$

$$+ \widetilde{E}_x \left(\int_t^T e^{-r(u-t)} dL_u^K(S) - \int_t^T e^{-r(u-t)} rK \mathbf{1}_{\{S_u^* < S_u < K\}} du \right).$$

The delayed exercise value describes the gain relative to stopping now; the early exercise premium describes the gain relative to stopping at the final expiration time T.

8.5   Relation to Free Boundary Problems

McKean [218] and van Moerbeke [295] established the following representation for P. It relates the value function P of the American option to the solution of a free boundary problem. Such a problem consists of a partial differential equation, its Dirichlet conditions, and a Neumann condition that determines an unknown stopping boundary, or 'free boundary', S_t^*. Write

$$L = \frac{\sigma^2}{2} x^2 \frac{\partial^2}{\partial x^2} + rx \frac{\partial}{\partial x} + \frac{\partial}{\partial t}.$$

From the martingale property of $(J_{u \wedge \rho_t})$ and the smoothness of P on the continuation region $\mathcal{C}$, it can be shown that

$$L(e^{-rt} P(x,t)) = 0.$$

For a proof, see [295, Lemma 5]. The Dirichlet and optimality conditions for P are given in the next result.

Theorem 8.5.1. *The American put $P(x,t)$ and the boundary S_t^* satisfy*

$$\lim_{x \downarrow S_t^*} P(x,t) = K - S_t^* \text{ for } t \in [0,T),$$

$$\lim_{t \to T} P(x,t) = (K - x)^+ \text{ for } x \geq 0,$$

$$\lim_{x \to \infty} P(x,t) = 0 \text{ for } t \in [0,T),$$

and

$$P(x,t) \geq (K - x)^+ \text{ for } (x,t) \in [0,\infty) \times [0,T).$$

Proof. The first result follows from the optimality of S^*. The second is a consequence of the continuity of P.

Write $S_t(x)$ for the solution of (8.2) with $S_0(x) = x$ and define the stopping time τ_K by $\tau_K = \inf \{t : S_t(x) \leq K\}$. Now $\tau_K \to \infty$ a.s. as $x \to \infty$ and, for $x > K$,

$$0 < P(x,t) \leq K P(\tau_K \leq t).$$

Therefore, $\lim_{x \to \infty} P(x,t) = 0$.

The final condition restates the hedging property. $\square$

These conditions do not determine the free boundary, or critical price, S^*. An additional 'smooth pasting' condition is required.

Proposition 8.5.2. *The derivative $\frac{\partial P(x,t)}{\partial x}$ is continuous across the free boundary S^*; i.e.,*

$$\lim_{x \downarrow S_t^*} \frac{\partial P(x,t)}{\partial x} = -1 = \left. \frac{\partial (K - S)^+}{\partial S} \right|_{S = S_t^*}.$$

Proof. We adapt McKean's argument. Lemma 8.5.6 below shows that, in the sense of distributions,

$$L(e^{-rt}P(x,t)) \leq 0 \text{ for } (x,t) \in \mathbb{R}^+ \times [0,T). \tag{8.18}$$

Introduce the change of variable $\xi = \log x$, and write

$$\widehat{P}(\xi,t) = P(e^{\xi},t).$$

Then (8.18) implies that

$$\frac{\sigma^2}{2}\widehat{P}_{\xi\xi} \leq \left(\frac{\sigma^2}{2} - r\right)\widehat{P}_{\xi} - \widehat{P}_t + r\widehat{P}. \tag{8.19}$$

In the new variable ξ, the free boundary S^* becomes $\xi_t^* = \log(S_t^*)$.

Integrate (8.19) over a region $\mathcal{R}$ in (ξ,t) space, where $\mathcal{R}$ has width ε on either side of ξ_t and is over the interval $[t_1,t_2]$. Consequently,

$$\frac{\sigma^2}{2}\int_{t_1}^{t_2}\left(\widehat{P}_{\xi}(\xi_t^* + \varepsilon, t) - \widehat{P}_{\xi}(\xi_t^* - \varepsilon, t)\right)dt$$

$$\leq \left(\frac{\sigma^2}{2} - r\right)\int_{t_1}^{t_2}\left(\widehat{P}(\xi_t^* + \varepsilon, t) - \widehat{P}(\xi_{t_1}^* - \varepsilon, t)\right)dt + \int_{\mathcal{R}}\left(r\widehat{P} - \widehat{P}_t\right)d\xi dt.$$

For a fixed ξ, consider the horizontal line in $\mathcal{R}$ that goes from time $t^-(\xi)$ to time $t^+(\xi)$. There is an interval $\mathcal{R}_\xi$ of ξ-space such that the final integral can be written

$$\int_{\mathcal{R}}r\widehat{P}d\xi dt - \int_{\mathcal{R}_\xi}\left(\widehat{P}(\xi,t^*(\xi)) - \widehat{P}(\xi,t^-(\xi))\right)d\xi.$$

On the (transformed) stopping region $\mathcal{S}$, we have $\widehat{P}_{\xi} = \varepsilon^{-\xi}$. Therefore, from the dominated convergence theorem, using the continuity of $\widehat{P}$, as $\varepsilon \downarrow 0$ we have

$$\int_{t_1}^{t_2}\left(\lim_{\xi\downarrow\xi^*}\widehat{P}_{\xi} + e^{\xi^*}\right)dt \leq 0. \tag{8.20}$$

From Corollary 8.2.12, we know that $\frac{\partial P(x,t)}{\partial x} \geq -1$ in $\mathcal{C}$. Since

$$\frac{\partial P}{\partial x} = \frac{\partial \widehat{P}}{\partial \xi} \cdot \frac{\partial \xi}{\partial x},$$

we have $\widehat{P}_{\xi} \geq -e^{-\xi}$. Consequently, from (8.20) we must have $\lim_{\xi\downarrow\xi^*}\widehat{P}_{\xi} + e^{\xi^*} = 0$, and the slope exhibits the smooth pasting condition across ξ^*. $\square$

The results of Theorem 8.5.1 and Proposition 8.5.2 show that the American put value P can be expressed as the solution of a free boundary problem. McKean was the first to discuss the problem and provide this formulation.

Using the regularity we have now established for P, the following result can be proved.

Theorem 8.5.3. *We have $P = p + e$, where p is the European put function and e is the early exercise premium as in Corollary 8.4.3. The critical price S^* is determined by the equation*

$$P(S_u^*, u) = K - S_u^* \text{ for } u \in [t, T)$$

together with $S_T^* = K$.

Proof. The function $P(x, t)$ is in $C^{1,0}$ and piecewise in $C^{2,1}$ on $\mathbb{R}^+ \times [0, T)$. Regularity of the boundary S_t^* implies the derivative P_t is continuous across S^* and so, in fact, in all $\mathbb{R}^+ \times [0, T)$. An extension of the Itô differentiation rule (due to Krylov [203, Theorem 2.10.1]) implies that

$$e^{-r(T-t)} P(S_T, T) = P(S_t, t) + \int_t^t e^{-r(u-t)} \sigma S_u P_x(S_u, u) d\widetilde{W}_u$$

$$+ \int_t^T L(e^{-r(u-t)} P)(S_u, u) du \text{ for } t \in [0, T]. \quad (8.21)$$

We have already noted that $L(e^{-r(u-t)} P)(x, u) = 0$ when $(x, u) \in \mathcal{C}$.
     When $(x, u) \in \mathcal{S}$, we have

$$P(x, u) = K - x, \qquad L(e^{-r(u-t)}(K - x)) = -e^{-r(u-t)} rK.$$

Substituting in (8.21), for $t \in [0, T]$,

$$e^{-r(T-t)} P(S_T, T) = P(S_t, t) + \int_t^T e^{-r(u-t)} \sigma S_u P_x(S_u, u) d\widetilde{W}_u$$

$$- \int_t^T e^{-r(u-t)} rK \mathbf{1}_{\{S_u < S_u^*\}} du.$$

The derivative P_x is bounded, so the stochastic integral is a martingale. With $S_t = x$ and $P(x, T) = (K - s)^+$, we have

$$P(x, t) = \widetilde{E}_x \left(e^{-r(T-t)} (K - S_T)^+ \right) + \widetilde{E}_x \left(\int_t^T e^{-r(T-t)} rK \mathbf{1}_{\{S_u < S_u^*\}} du \right).$$

The equation for S^* follows from the first statement of Theorem 8.5.1. $\square$

Definition 8.5.4. A function $g(x, t) \in C^{3,1}(\mathbb{R} \times [0, T))$ has *Tychonov growth* if $g, g_t, g_x, g_{xx}, g_{xx}$, and g_{xxx} have growth at most $\exp(o(x^2))$ uniformly on compact sets as $|x|$ goes to infinity.

If we assume the equation for S^* has a C^1 solution, the following uniqueness result is a consequence of Theorem 8.5.1; its proof can be found in [295].

Theorem 8.5.5. *Suppose* $\mathcal{D} \subset \mathbb{R}^+ \times [0, T)$ *is an open domain with a continuously differentiable boundary* c.

Furthermore, suppose $f \in C^{3,1}$, *that* $g(x, t) = f(e^x, t)$ *has Tychonov growth and* $L(e^{-rt} f(x, t)) = 0$ *on* $\mathcal{D}$, *and*

$$f(x, T) = (K - x)^+ \text{ for } x \in \mathbb{R}^+,$$
$$f(x, t) > (K - x)^+ \text{ for } (x, t) \in \mathcal{D},$$
$$f(x, t) = (K - x)^+ \text{ for } (x, t) \in \mathbb{R}^+ \times [0, T) \cap \mathcal{D}^c,$$
$$\lim_{x \downarrow c(t)} f_x(x, t) = -1 \text{ for } t \in [0, T).$$

Then $f(x, t) = P(x, t)$, *the American put function,* $\mathcal{D} = \mathcal{C}$, *the continuation region, and* $c(t) = S_t^*$, *the optimal stopping boundary.*

We require the following extension of the harmonic property of P on $\mathcal{C}$.

Lemma 8.5.6. *On* $\mathbb{R}^+ \times [0, T]$, *we have* $L(e^{-rt} P(x, t)) \leq 0$ *in the sense of Schwartz distributions. (This shows that the American put value function* P *is 'r-excessive'.)*

Proof. Choose $\varepsilon > 0$, and consider the set of stopping times

$$V_\varepsilon = \left\{ \tau : t \leq \tau \leq T, \widetilde{E}\left(e^{-r(\tau - t)}(K - S_\tau)^+ | S_t \right) \geq P(S_t, t) - \varepsilon \right\}.$$

For all $t \in [0, T)$, this set is non-empty. Choose $\tau_\varepsilon \in V_\varepsilon$ and write $\widetilde{E}_x(\cdot)$ for the $\widetilde{Q}$-expectation given $S_0 = x$. Then

$$\widetilde{E}_x\left(e^{-r\tau_\varepsilon}(K - S_{\tau_\varepsilon})^+ \right) = \widetilde{E}_x\left(e^{-rt} \widetilde{E}\left(e^{-r(\tau_\varepsilon - t)}(K - S_{\tau_\varepsilon})^+ | S_t \right) \right)$$
$$\geq \widetilde{E}_x\left(e^{-rt} P(S_t, t) \right) - \varepsilon e^{-rt}.$$

However, by definition,

$$P(S_0, 0) = P(x, 0) \geq \widetilde{E}_x\left(e^{-r\tau_\varepsilon}(K - S_{\tau_\varepsilon})^+ \right) \geq \widetilde{E}_x\left(e^{-rt} P(S_t, t) \right) - \varepsilon e^{-rt}.$$

Letting $\varepsilon \downarrow 0$ gives
$$P(x, 0) \geq \widetilde{E}_x\left(e^{-rt} P(S_t, t) \right).$$

This inequality implies the result, as any excessive function is the limit of an increasing sequence of infinitely differentiable excessive functions (see Port and Stone [243]). $\square$

Lemma 8.5.7. *The American put function* $P(x, t)$ *satisfies*

$$\left(L\left(e^{-rt} P(x, t) \right) \right) \left((K - x)^+ - P(x, t) \right) = 0 \text{ for } (x, t) \in \mathbb{R}^+ \times [0, T].$$

Proof. In the continuation region, we know that $L(e^{-rt} P(x, t)) = 0$. In the stopping region, $P(x, t) = (K - x)^+$. $\square$

Definition 8.5.8. For any $m \in \mathbb{Z}^+$ and $\lambda > 0$, write $H^{m,\lambda}$ for the space of measurable, real-valued functions f on $\mathbb{R}$ whose derivatives, in the sense of distributions, up to and including the mth order, belong to $L^2(\mathbb{R}, e^{-\lambda|x|}dx)$. Write

$$\|f\| = \left(\sum_{i=0}^{m} \int_R \left| \partial^i f(x) \right|^2 e^{-\lambda|x|} dx \right)^{\frac{1}{2}}.$$

The space $L^2([0,T], H^{m,\lambda})$ is the set of measurable functions $g : [0,T] \to H^{m,\lambda}$ such that

$$\int_{[0,T]} \|g(t)\|^2 \, dt < \infty.$$

In [169], Jaillet, Lamberton, and Lapeyre extend the work of Bensoussan and Lions ([24]) to show that the American put value function is characterized by a variational inequality. Their result is as follows.

Theorem 8.5.9. *Consider a continuous function f defined on $\mathbb{R}^+ \times [0,T]$ that satisfies*

$$f(e^{\cdot}, \cdot) \in L^2([0,T], H^{2,\lambda}),$$
$$f_t(e^{\cdot}, \cdot) \in L^2([0,T], H^{0,\lambda})$$
and
$$f(x,t) \geq (K - x)^+ \text{ for } (x,t) \in \mathbb{R}^+ \times [0,T),$$
$$f(x,T) = (K - x)^+ \text{ for } x \in \mathbb{R}^+,$$
$$L\left(e^{-rt} f(x,t) \right) \leq 0$$
$$\left(L\left(e^{-rt} f(x,t) \right) \right) \left(f(x,t) - (K - x)^+ \right) = 0$$

where $(x,t) \in \mathbb{R}^+ \times [0,T]$.

Then f is unique and equals the American put value function P.

Remark 8.5.10. The application of variational inequalities gives rise to a numerical algorithm. In fact, the early numerical work of Brennan and Schwartz [33] was justified for the American put, using variational inequalities, by Jaillet, Lamberton, and Lapeyre [169].

The most widely used numerical technique for calculating the American option value is dynamic programming. The risky asset price S is modelled as evolving on a binomial tree in discrete time. The Bellman equation is then solved recursively by evaluating

$$P_i = \max \left\{ (K - S_i)^+, e^{-r\Delta} \widetilde{E} \left(P_{i+1} \,|\, \mathcal{F}_i \right) \right\}$$

with final condition $P_T = (K - S_T)^+$.

8.6 An Approximate Solution

We have seen that the American put function $P(x,t)$ can be written

$$P(x,t) = p(x,t) + e(x,t),$$

where

$$p(x,t) = \tilde{E}_x \left(e^{-r(T-t)} (K - S_T)^+ \right)$$

is the European put value, and

$$e(x,t) = \tilde{E}_x \left(\int_t^T e^{-r(u-t)} r K \mathbf{1}_{\{S_u < S_u^*\}} du \right)$$

is the 'early exercise' premium.

The early exercise premium involves the critical price, or free boundary, S^*, and is consequently difficult to evaluate.

Allegretto, Barone-Adesi, and Elliott [5] proposed an approximation for $e(x,t)$ of the form

$$\varepsilon(x,t) = A(t) \left(\frac{x}{S_t^*} \right)^{q(t)},$$

where A and q are functions of t that are to be determined.

Now we know that, in the continuation region $\mathcal{C}$, we have

$$L[e^{rt} P(x,t)] = 0, \qquad L \left(e^{-rt} P(x,t) \right) = 0. \qquad (8.22)$$

Also, at the critical price,

$$P(S_t^*,t) = (K - S_t^*)^+, \qquad \left. \frac{\partial P}{\partial x} \right|_{x=S_t^*} = -1. \qquad (8.23)$$

Now $LP(x,t) = 0$ in $\mathcal{C}$ and $LP(x,t) = 0$ in $\mathcal{C}$, so

$$L[e^{-rt} e(x,t)] = 0, \qquad (8.24)$$

in $\mathcal{C}$. Substituting $P = p + A(t) \left(\frac{S}{S_t^*} \right)^{q(t)}$ in (8.23), we have

$$p(S_t^*,t) + A(t) = K - S_t^*$$

and

$$\frac{A(t)q(t)}{S_t^*} - e^{-(\mu-r)(T-t)} \Phi(-d_1(S_t^*,t)) = -1, \qquad (8.25)$$

where Φ is the standard normal distribution and

$$d_1(x,t) = \frac{\log \left(\frac{x}{S_t^*} \right) + \left(\mu + \frac{\sigma^2}{2} \right)(T - t)}{\sigma \sqrt{T - t}}.$$

However, we also would like $L[e^{-rt}\varepsilon(x,t)] = 0$. This is the case if

$$\frac{1}{2}\sigma^2 q(t)(q(t)-1)A(t)\left(\frac{x}{S_t^*}\right)^{q(t)} - rA(t)\left(\frac{x}{S_t^*}\right)^{q(t)} + A(t)\mu q(t)\left(\frac{x}{S_t^*}\right)^{q(t)}$$

$$+ \frac{\partial}{\partial t}\left(\frac{x}{S_t^*}\right)^{q(t)} = 0. \quad (8.26)$$

Now

$$\frac{\partial}{\partial t}\left(A(t)\left(\frac{x}{S_t^*}\right)^{q(t)}\right) = \frac{dA(t)}{dt}\left(\frac{x}{S_t^*}\right)^{q(t)}$$

$$- \frac{dS_t^*}{dt}\left(\frac{A(t)q(t)}{x}\right)\left(\frac{x}{S_t^*}\right)^{q(t)+1} + \frac{dq(t)}{dt}A(t)\left(\frac{x}{S_t^*}\right)^{q(t)}\log\left(\frac{x}{S_t^*}\right).$$

Substituting this into (8.26) and dividing by $A(t)\left(\frac{x}{S_t^*}\right)^{q(t)}$ implies that

$$\frac{1}{2}\sigma^2 q(t)(q(t)-1) - r + \mu q(t) + \left(\frac{1}{A(t)}\frac{dA(t)}{dt} - \frac{q(t)}{S_t^*}\frac{dS_t^*}{dt}\right)$$

$$+ \log\left(\frac{x}{S_t^*}\right)\frac{dq(t)}{dt} = 0. \quad (8.27)$$

However, this equation indicates that q is not independent of x, and so $e(x,t)$ is not of the form given by $\varepsilon(x,t)$. Nonetheless, a useful approximation is obtained by neglecting the last term of (8.27). That is, we suppose $q(t)$ is a solution of

$$\frac{1}{2}\sigma^2 q(t)(q(t)-1) - r + \mu q(t) + \left(\frac{1}{A(t)}\frac{dA(t)}{dt} - \frac{q(t)}{S_t^*}\frac{dS_t^*}{dt}\right) = 0. \quad (8.28)$$

This approximation is reasonable when $\log\left(\frac{x}{S_t^*}\right)\cdot\frac{dq(t)}{dt}$ is small. This is the case when x is in a neighbourhood of S_t^* or when $\frac{dq(t)}{dt}$ is small (at long maturities).

From equation (8.25), we have

$$\frac{dA(t)}{dt} = \left(e^{(\mu-r)(T-t)}\Phi\left(-d_1(S_t^*,t)\right)-1\right)\frac{dS_t^*}{dt} - \frac{\partial p(x,t)}{\partial t}.$$

From the second equation of (8.25),

$$\frac{1}{A(t)}\frac{dA(t)}{dt} - \frac{q(t)}{S_t^*}\frac{dS_t^*}{dt} = -\frac{1}{A(t)}\cdot\frac{\partial p(S_t^*,t)}{\partial t}.$$

Writing

$$g(t) = \frac{1}{A(t)}\frac{\partial p(S_t^*,t)}{\partial t}, \qquad M = \frac{2r}{\sigma^2}, \qquad N = \frac{2b}{\sigma^2}, \qquad G(t) = \frac{2q(t)}{\sigma^2},$$

equation (8.28) becomes

$$q(t)^2 + (N-1)q(t) - (M - G(t)) = 0.$$

To satisfy the boundary condition of zero at $x = \infty$, we consider only the root

$$q(t) = \frac{1}{2}\left(1 - N - \sqrt{(1-N)^2 + 4(M + G(t))}\right).$$

With this value of $q(t)$, an approximation for the early exercise premium is

$$\varepsilon(x, t) = A(t)\left(\frac{x}{S_t^*}\right)^{q(t)}.$$

To summarise, we have the following system of three equations in three unknowns $A(t), q(t)$, and S_t^*:

$$S_t^* = \frac{(K - p(S_t^*, t))q(t)}{-1 + q(t) + e^{(\mu - r)(T-t)}\Phi(-d_1(S_t^*, t))}, \qquad (8.29)$$

$$A(t) = -p(S_t^*, t) - S_t^* + K, \qquad (8.30)$$

$$0 = q(t)^2 + (N-1)q(t) - M + G(t). \qquad (8.31)$$

For a fixed value of t, these equations can be solved using the following iterative procedure.

(i) Give a trial value of S_t^*.

(ii) Calculate $A(t)$ from (8.30).

(iii) Calculate $q(t)$ from (8.31).

(iv) Calculate a new value of S_t^* from (8.29).

Using the new value for S_t^*, steps (ii)-(iv) are repeated.

This algorithm was investigated numerically in [5] and shown to give satisfactory results.

Chapter 9

Bonds and Term Structure

9.1 Market Dynamics

Suppose $(\Omega, \mathcal{F}, P)$ is a probability space, $(B_t)_{0 \le t \le T}$ is a Brownian motion, and $(\mathcal{F}_t)$ is the (complete, right-continuous) filtration generated by B. We first review the martingale pricing results of Chapter 7.

Consider again the case of a bond S^0 and a single risky asset S^1. We suppose

$$S_t^0 = \exp\left\{ \int_0^t r_u du \right\}, \quad S_t^1 = S_0^1 + \int_0^t \mu(u) S_u^1 du + \int_0^t \sigma(u) S_u^1 dB_u.$$

Here r, μ and σ are adapted (random) processes. In particular, r is a stochastic interest rate in general. Consider a self-financing trading strategy (H^0, H^1). The corresponding wealth process is

$$X_t = H_t^0 S_t^0 + H_t^1 S_t^1,$$

and

$$dX_t = r H_t^0 S_t^0 dt + H_t^1 dS_t^1 = r \left(X_t - H_t^1 S_t^1 \right) dt + H_t^1 dS_t^1.$$

With $\theta_t = \frac{\mu(t) - r_t}{\sigma(t)}$ (which requires $\sigma \ne 0$), the process W^θ, where

$$dW_t^\theta = \theta_t dt + dB_t,$$

is a Brownian motion under the measure P^θ. Consequently, under P^θ the discounted wealth process is $\frac{X_t}{S_t^0}$, and

$$d\left(\frac{X_t}{S_t^0} \right) = H_t^1 \sigma(t) \frac{S_t^1}{S_t^0} dW_t^\theta.$$

That is, for any self-financing strategy, the discounted wealth process $\left(\frac{X_t}{S_t^0}\right)$ is a martingale under the martingale measure P^θ.

Consider a contingent claim $h \in L^2(\Omega, \mathcal{F}_T)$. Then

$$M_t = E^\theta\left(\frac{h}{S_t^0} \,|\, \mathcal{F}_t\right)$$

is a martingale. Using the martingale representation result Theorem 7.3.9,

$$M_t = M_0 + \int_0^t \phi_u dW_u^\theta.$$

If we take

$$H_t^1 = \frac{S_t^0 \phi_t}{\sigma(t) S_t^1}, \qquad\qquad X_0 = M_0 = E^\theta\left(\frac{h}{S_t^0}\right),$$

and write

$$M_t = \frac{X_t}{S_t^0} = X_0 + \int_0^t H_u^1 \sigma(t) \frac{S_u^1}{S_u^0} dW_u^\theta,$$

then, with

$$H_t^0 = \frac{X_t}{S_t^0} - H_t^1 \frac{S_t^1}{S_t^0},$$

the pair (H^0, H^1) is a self-financing strategy that hedges the claim h. That is,

$$X_T = H_T^0 S_t^0 + H_T^1 S_T^1 = h.$$

The fair price for the claim at time 0 is $E^\theta\left(\frac{h}{S_0^0}\right)$; the price at time $t \in [0, T]$ is $X_h(t) = X_t$, and this equals

$$S_t^0 E^\theta\left(\frac{h}{S_0^0} \,|\, \mathcal{F}_t\right) = S_t^0 E^\theta\left(\frac{X_T}{S_T^0} \,|\, \mathcal{F}_t\right)$$

because $\left(\frac{X_t}{S_t^0}\right)$ is a martingale under P^θ.

Suppose we have a market with several risky assets $S_t^0, S_t^1, \ldots, S^N(t)$ with dynamics

$$dS_t^0 = r_t S_t^0 dt, \qquad\qquad S_0^0 = 1,$$

$$dS_t^i = S_t^i \left(\mu_i(t)dt + \sum_{j=1}^m \sigma_{ij}(t)dW_j(t)\right), \quad S_0^i = s_i \text{ for } i = 1, 2, \ldots, n.$$

Here $(W_t) = (W_1(t), \ldots, W_m(t))$ is an m-dimensional Brownian motion on $(\Omega, \mathcal{F}, P)$. The risk-neutral pricing formula holds as long as there is a unique risk-neutral measure P^θ, as introduced in Chapter 7. In such an example the price at time $t \leq T$ of a claim $h \in L^2(\mathcal{F}_T)$ is

$$X_t = S_t^0 E^\theta\left(h \cdot (S_T^0)^{-1} \,|\, \mathcal{F}_t\right).$$

Notation 9.1.1. From now on in this chapter, we assume we are working in a market where there is a unique risk-neutral measure P^θ. The superscript θ will be dropped. For simplicity, we suppose there is a single risky asset that has dynamics (under $P^\theta = P$)

$$dS_t^1 = r_t S_t^1 dt + \sigma(t) S_t^1 dW_t.$$

Further, we suppose the martingale representation result holds, so that every $(\mathcal{F}_t, P)$-martingale has a representation as a stochastic integral with respect to W.

Definition 9.1.2. A *zero coupon bond* maturing at time T is a claim that pays 1 at time T.

From the pricing formula, its value at time $t \in [0, T]$ is

$$B(t, T) = S_t^0 E\left(\frac{1}{S_T^0} \,|\mathcal{F}_t\right).$$

As $S_t^0 = \exp\left\{\int_0^t r_u du\right\}$, this is

$$B(t, T) = E\left(\exp\left\{-\int_t^T r_u du\right\} |\mathcal{F}_t\right).$$

Consequently, given $B(t, T)$ dollars at time t, one can construct a self-financing hedging strategy (H_t^0, H_t^1) such that the corresponding wealth process

$$X_t = H_t^0 S_t^0 + H_t^1 S_t^1$$

has value 1 at time T.

If the instantaneous rate r_t is deterministic, then

$$B(t, T) = \exp\left\{-\int_t^T r_u du\right\}$$

and

$$dB(t, T) = r_t B(t, T) dt,$$

so H_t^1 is identically 0.

Definition 9.1.3. The *T-forward price* $F(t, T)$ for the risky asset S^1 is a price agreed at time $t \leq T$ (and thus $\mathcal{F}_t$-measurable) that will be paid for S^1 at time T.

Such a price $F(t, T)$ is characterised by requiring that the claim $S_T^1 - F(t, T)$ has (discounted) value 0 under the risk-neutral (martingale) measure P. Therefore,

$$0 = E\left(\frac{S_T^1 - F(t, T)}{S_T^0}\,|\mathcal{F}_t\right) = E\left(\frac{S_T^1}{S_T^0}\,|\mathcal{F}_t\right) - \frac{F(t, T)}{S_t^0} E\left(\frac{S_t^0}{S_T^0}\,|\mathcal{F}_t\right)$$

$$= \frac{S_t^1}{S_t^0} - \frac{F(t,T)}{S_t^0} B(t,T)$$

because the discounted price $\frac{S^1}{S^0}$ is a martingale under the measure P. Therefore $F(t,T) = \frac{S_t^1}{B(t,T)}$.

Remark 9.1.4. The forward price can be defined for other claims. Indeed, suppose $h \in L^2(\Omega, \mathcal{F}_T)$ is a contingent claim with exercise date T. The *T-forward price* for h, denoted by $F(h,t,T)$, is the $\mathcal{F}_t$-measurable random variable that has the property that

$$E\left(\frac{h - F(h,t,T)}{S_T^0} | \mathcal{F}_t\right) = 0.$$

Consequently,

$$F(h,t,T) = \frac{S_t^0 E\left(\left(S_T^0\right)^{-1} h | \mathcal{F}_t\right)}{B(t,T)} = \frac{X_h(t)}{B(t,T)},$$

where $X_h(t)$ is, from the pricing discussion, the fair price for h at time t.

In particular, h could be the zero coupon bond of maturity $T^* \geq T$. Then

$$F\left(B(T,T^*),t,T\right) = \frac{B(t,T^*)}{B(t,T)}.$$

Definition 9.1.5. Define a new probability measure Q_T, equivalent to P, on $(\Omega, \mathcal{F}_T)$ by setting

$$\frac{dQ_T}{dP}\bigg|_{\mathcal{F}_T} = \frac{\left(S_T^0\right)^{-1}}{E\left(\left(S_T^0\right)^{-1}\right)} = \frac{1}{S_T^0 B(0,T)}.$$

The measure Q_T is called the *forward measure* for the settlement date T. It was introduced in [140] and [170]. Define

$$\Gamma_t = E\left(\frac{dQ_T}{dP} | \mathcal{F}_t\right) = E\left(\frac{1}{S_T^0 B(0,T)} | \mathcal{F}_t\right) = \frac{B(t,T)}{S_t^0 B(0,T)}.$$

The process Γ is a $(P, \mathcal{F}_t)$-martingale, so there is an integrand $\gamma(s,T)$ such that

$$\Gamma_t = 1 + \int_0^t \gamma(s,T) dW_s.$$

Now $\Gamma_s > 0$ a.s. for all s; define

$$\beta(s,T) = \Gamma_s^{-1} \gamma(s,T).$$

Then

$$\Gamma_t = 1 + \int_0^t \Gamma_s \beta(s,T) dW_s,$$

and so

$$\Gamma_t = \exp\left\{\int_0^T \beta(s,T)dW_s - \frac{1}{2}\int_0^T \beta(s,T)^2 ds\right\}.$$

The next lemma shows how the forward price can be expressed in terms of the forward measure.

Lemma 9.1.6. *Suppose that $h \in L^2(\Omega, \mathcal{F}_T)$ is a contingent claim with exercise time T. Then*

$$F(h,t,T) = E_{Q_T}(h\,|\mathcal{F}_t).$$

Consequently, the forward price of h is a Q_T-martingale.

Proof. Using Bayes' rule, we have

$$E_{Q_T}(h\,|\mathcal{F}_t) = \frac{E\left(\Gamma_T h\,|\mathcal{F}_t\right)}{E\left(\Gamma_T\,|\mathcal{F}_t\right)} = E\left(\Gamma_t^{-1}\Gamma_T h\,|\mathcal{F}_t\right).$$

Substituting the expressions for Γ, the result follows. $\qquad\square$

Remark 9.1.7. Consider the T-forward price for the contingent claim $h \in L^2(\Omega, \mathcal{F}_T)$ at time 0,

$$F(h,0,T) = \frac{X_h(0)}{B(0,T)}.$$

By definition, $F(h,0,T)$ is the price, agreed at time 0, that one will pay at time T for the claim h. The related claim $V = h - F(h,0,T)$ has price 0 at time 0. However, at later times $t \in [0,T]$, this claim V does not have value 0. Indeed, using the pricing formula, at time t it has value

$$V_t = S_t^0 E\left(\frac{h - F(h,0,T)}{S_T^0}\,\Big|\mathcal{F}_t\right) = X_h(t) - \frac{F(h,0,T)}{B(t,T)}.$$

One can hedge this claim as follows. At time 0, one shorts $F(h,0,T)$ zero coupon bonds with maturity T. This provides an amount

$$\frac{F(h,0,T)}{B(0,T)} = \frac{X_h(0)}{B(0,T)} \cdot B(0,T) = X_h(0),$$

where $X_h(0)$ is the price of the claim h at time 0. Consequently, this amount $X_h(0)$ can be used at time 0 to buy the claim h. This strategy requires no initial investment. If this position is held until time T, it is then worth

$$X_h(T) - \frac{F(h,0,T)}{B(T,T)} = h - F(h,0,T).$$

9.2   Future Price and Futures Contracts

Suppose a contingent claim h has a price $\$h$ at time T. (By abuse of notation, we write h for the claim and its price at time T.)

Clearly, at time T, one need not pay anything for the right to buy the claim for $\$h$. Therefore, at time T, the price of the claim is $G(h, T, T) = h$. (Note that this assumes there are no transaction costs, and we are not discussing problems of delivering the claim itself-we are thinking of a cash settlement.)

Suppose initially there are only a finite number of trading times $t_0, \ldots, t_n$ with $0 = t_0 < t_1 < \cdots < t_n = T$. Furthermore, suppose that r_u is constant on each interval $[t_i, t_{i+1})$. Then

$$S^0_{t_{j+1}} = \exp \int_0^{t_{j+1}} r_u du = \exp \left\{ \sum_{i=0}^{n} r_{t_i}(t_{i+1} - t_i) \right\},$$

and $S^0_{t_{j+1}}$ is $\mathcal{F}_{t_j}$-measurable.

Consider the time t_{n-1} and suppose the price agreed at time t_{n-1} for the claim, (to be delivered at time $t_n = T$) is

$$G(h, t_{n-1}, T).$$

Then the difference in the price agreed at time t_{n-1} and the price at $t_n = T$ is

$$G(h, t_n, T) - G(h, t_{n-1}, T).$$

At time t_{n-1}, one estimates $G(h, t_{n-1}, T)$, given the information $\mathcal{F}_{t_{n-1}}$, so that this difference, discounted and conditioned on $\mathcal{F}_{t_{n-1}}$, is zero; that is, so that the claim $G(h, t_n, T) - G(h, t_{n-1}, T)$ has value zero at time t_{n-1}, namely

$$S^0_{t_{n-1}} E\left(\frac{G(h, t_n, T) - G(h, t_{n-1}, T)}{S^0_{t_n}} \Big| \mathcal{F}_{t_{n-1}} \right) = 0.$$

Similarly, at time t_{n-2}, one estimates $G(h, t_{n-2}, T)$ so that

$$S^0_{t_{n-2}} E\left(\frac{G(h, t_{n-1}, T) - G(h, t_{n-2}, T)}{S^0_{t_{n-1}}} \Big| \mathcal{F}_{t_{n-2}} \right) = 0.$$

Here $G(h, t_{n-2}, T)$ is the estimate at time t_{n-2} of the price of the claim h at time T.

Consequently, the value at time $t = t_k$ of the sum of future adjustments is

$$S^0_{t_k} E\left(\sum_{j=k}^{n-1} \frac{G(h, t_{j+1}, T) - G(h, t_j, T)}{S^0_{t_{j+1}}} \Big| \mathcal{F}_{t_k} \right) = 0.$$

The continuous-time version of this condition gives

$$S^0_t E\left(\int_t^T (S^0_u)^{-1} dG(h, u, T) | \mathcal{F}_t \right) = 0 \text{ for } t \in [0, T]. \tag{9.1}$$

Write
$$M_t = \int_0^t \left(S_u^0\right)^{-1} dG(h, u, T).$$

Then (9.1) implies that
$$E\left(M_t \mid \mathcal{F}_s\right) = M_s \text{ for } 0 \le s \le t \le T.$$

That is, M is an $(\mathcal{F}_t, P)$-martingale. Consequently,
$$\int_0^t S_u^0 dM_u = G(h, t, T) - G(h, 0, T)$$

is an $(\mathcal{F}_t, P)$-martingale. Therefore, as $G(h, T, T) = h$,
$$G(h, t, T) = E\left(h \mid \mathcal{F}_t\right)$$

is the 'future price' at time t for the claim h. This motivates the following definition.

Definition 9.2.1. The T-*future price* G at time t of the $\mathcal{F}_T$-measurable contingent claim h is
$$G(h, t, T) = E\left(h \mid \mathcal{F}_t\right).$$

By definition, $G(h, t, T)$ is a martingale under P.

Lemma 9.2.2. *a)* $\left(S_T^0\right)^{-1}$ *and h are (conditionally) uncorrelated if and only if $F(h, t, T) = G(h, t, T)$.*

b) If $\left(S_T^0\right)^{-1}$ and h are positively correlated conditional on $\mathcal{F}_t$, then
$$G(h, t, T) \le F(h, t, T).$$

Proof. The T-future price is
$$G(h, t, T) = E\left(h \mid \mathcal{F}_t\right).$$

The T-forward price is
$$F(h, t, T) = \frac{X_h(t)}{B(t, T)} = E_Q\left(h \mid \mathcal{F}_t\right) = \frac{E\left(\frac{h}{S_T^0} \mid \mathcal{F}_t\right)}{E\left(\left(S_T^0\right)^{-1} \mid \mathcal{F}_t\right)}.$$

Part (a) follows immediately. Part (b) states that
$$E\left[\left((S_t^0)^{-1} - E(S_t^{0-1} \mid \mathcal{F}_t)\right)(h - E\left(h \mid \mathcal{F}_t\right)) \mid \mathcal{F}_t\right] \ge 0,$$
and the result follows. $\qquad\qquad\qquad\qquad\qquad\qquad\qquad\qquad\square$

Remark 9.2.3. The hypothesis of the second part of the lemma arises when the stock price tends to rise with a fall in the interest rate and conversely. Holding a futures contract is not advantageous if there is a positive correlation between $\left(S_t^0\right)^{-1}$ and h. Therefore, a buyer of a futures contract is compensated by the lower future price compared with the forward price.

Futures Contracts

We have noticed that forward contracts possibly have non-zero value. In contrast, a futures contract is constructed so that the risk of default inherent in a forward contract is eliminated.

The value at time 0 of a forward contract, entered into at time 0, is 0. However, at later times $t \in [0, T]$ it has value

$$V_t = X_h(t) - \frac{F(h, 0, T)}{B(t, T)}.$$

In contrast to a forward contract, the value of a futures contract is maintained at zero at all times. Consequently, either party to the contract can close his or her position at any time. This is done by *marking to market*.

To describe this process, suppose again that trading takes place only at the finite number of times $t_0, \ldots, t_n$ with $0 = t_0 < t_1 < \cdots < t_n = T$ and that r_u is constant on each interval $[t_i, t_{i+1})$.

At time t_k, the future price of the claim h is $G(h, t_k, T) = E\left(h \mid \mathcal{F}_{t_k}\right)$. Suppose we buy a futures contract at this price. At the time t_{k+1}, the future price of h is $G(h, t_{k+1}, T)$.

If $G(h, t_{k+1}, T) > G(h, t_k, T)$, the buyer of the futures contract receives a payment of

$$G(h, t_{k+1}, T) - G(h, t_k, T).$$

If $G(h, t_{k+1}, T) < G(h, t_k, T)$, then the buyer of the futures contract makes a payment of

$$G(h, t_k, T) - G(h, t_{k+1}, T).$$

To make or receive these payments, a *margin account* is held by the broker.

At the final time $T = t_n$, the buyer of the futures contract will have received payments

$$G(h, t_{k+1}, T) - G(h, t_k, T), G(h, t_{k+2}, T) - G(h, t_{k+1}, T),$$
$$\ldots, G(h, t_n, T) - G(h, t_{n-1}, T)$$

at times $t_{k+1}, \ldots, t_n = T$. The value at time $t = t_k$ of this sequence of payments is

$$S_t^0 E \left(\sum_{i=k}^{n-1} \frac{G(h, t_{i+1}, T) - G(h, t_i, T)}{S_{t_{i+1}}^0} \mid \mathcal{F}_t \right).$$

The future price $G(h, t, T)$ is such that the cost of entering a futures contract at any time is zero. Consequently, the value of this sequence of payments at time t must be 0.

With a continuum of trading times, the sum above becomes a stochastic integral and the condition is

$$S_t^0 E \left(\int_t^T \left(S_u^0 \right)^{-1} dG(h, u, T) \mid \mathcal{F}_t \right) = 0.$$

Now by definition $G(h, t, T) = E(h \,|\, \mathcal{F}_t)$ is a martingale. This integral is therefore a stochastic integral with respect to a martingale and so, under standard conditions, has conditional expectation zero.

With a T-forward contract, the only payment is at time T: the buyer agrees at time 0 to pay $F(h, 0, T)$ for the claim h at time T.

With a T-futures contract, the buyer receives a (positive or negative) cash flow from time 0 to time T. If he still holds the contract at time T, he pays an amount h at time T for the claim, which has value h. Between time 0 and time T, the buyer has received an amount

$$\int_0^T dG(h, u, T) = G(h, T, T) - G(h, 0, T) = h - G(h, 0, T).$$

Therefore, at time T he has *paid* an amount $-(h - G(h, 0, T)) + h = G(h, 0, T)$ for the claim that has value h at time T.

9.3 Changing Numéraire

Consider again the situation described in Notation 9.1.1, where, under a risk-neutral measure P, there is a risky asset S^1 with dynamics

$$dS_t^1 = r_t S_t^1 dt + \sigma(t) S_t^1 dW_t.$$

Here W is a Brownian motion on a probability space $(\Omega, \mathcal{F}, P)$ with a filtration $(\mathcal{F}_t)_{0 \le t \le T^*}$. In general, $(\mathcal{F}_t)$ may be larger than the filtration generated by W. The short-term rate r and volatility σ are adapted (random) processes. The value of a dollar in the money market is, as before, $S_t^0 = \exp\left\{ \int_0^t r_u du \right\}$. We note that

$$d\left(\frac{S_t^1}{S_t^0} \right) = \frac{S_t^1}{S_t^0} \sigma(t) dW_t,$$

so the discounted asset price is a martingale.

When we consider the discounted price $\frac{S_t^1}{S_t^0}$, we are saying that, at time t, one unit of stock is worth $\frac{S_t^1}{S_t^0}$ units of the money market account. Similarly, from the expression after Definition 9.1.2 at time t, with $T \le T^*$, the T-maturity bond is worth $\frac{B(t,T)}{S_t^0}$ units of the money market account; again this discounted price is $E\left((S_t^0)^{-1} \,|\, \mathcal{F}_t \right)$ and so is a martingale.

Now any strictly positive price process could play the role of S^0, and other assets can be expressed in terms of this process. As in the discrete-time setting, we have the following

Definition 9.3.1. A strictly positive process Z is the *numéraire* of the model if all asset prices are expressed in terms of Z.

For example, the T-maturity bond price $B(t,T)$ could be taken as the numéraire for $t \leq T$. In terms of $B(t,T)$, at time t, the risky asset is worth $\frac{S_t^1}{B(t,T)} = F(t,T)$ units of $B(t,T)$, where $B(t,T)$ is the forward price of Definition 9.1.3. Of course, the price of the bond itself in terms of the numéraire $B(t,T)$ is just $\frac{B(t,T)}{B(t,T)} = 1$ unit.

We could also, for example, take S^1 to be the numéraire. Then the price at time t of a T-maturity bond in units of S_t^1 is $\frac{B(t,T)}{S_t^1} = \frac{1}{F(t,T)}$.

Definition 9.3.2. Suppose Z is a strictly positive process so Z can be taken as a numéraire. A probability measure P_Z on $(\Omega, \mathcal{F}, P)$ is said to be *risk-neutral* for Z if the price of any asset divided by Z (i.e., expressed in units of Z) is a martingale under P_Z.

We assumed in Notation 9.1 that the original measure P was risk-neutral for the numéraire S^0.

Theorem 9.3.3. *Suppose Z is a numéraire, so it is the strictly positive price process of some asset. Define a new probability measure P_Z on $(\Omega, \mathcal{F}, P)$ by putting for any $A \in \mathcal{F}_{T^*}$*

$$P_Z(A) = \frac{1}{Z_0} \int_A \frac{Z_T}{S_T^0} dP.$$

Then P_Z is equivalent to P and is a risk-neutral measure for the numéraire Z.

Proof. Note that, for $A \in \mathcal{F}_{T^*}$,

$$P(A) = Z_0 \int_A S_T^0 Z_T^{-1} dP_Z,$$

so P and P_Z have the same null sets.

From the definition of P, $\frac{Z}{S^0}$ is a martingale under P. Consequently,

$$P_Z(\Omega) = \frac{1}{Z_0} \int_\Omega \frac{Z_T}{S_T^0} dP = \frac{1}{Z_0} E\left(\frac{Z_T}{S_T^0}\right) = \frac{Z_0}{Z_0 S_0^0} = 1$$

because $\frac{Z}{S^0}$ is a P-martingale. Consequently, P_Z is a probability measure.

Now suppose X is an asset price process, so $\frac{X}{S^0}$ is a P-martingale. We shall show that $\frac{X}{Z}$ is then a P_Z-martingale. We have

$$M_t = \frac{Z_t}{Z_0 S_t^0} = \frac{1}{Z_0} E\left(\frac{Z_{T^*}}{S^0(T^*)} \Big| \mathcal{F}_t\right)$$

because $\frac{Z}{S^0}$ is a P-martingale. From Lemma 7.2.2, $\frac{X}{Z}$ is a P_Z-martingale if and only if $\frac{X}{Z} M = \frac{1}{Z_0} \frac{X}{Z} \frac{Z}{S^0}$ is a P-martingale. The result follows. $\square$

Remark 9.3.4. Note that, if we take the numéraire Z to be the bond price $B(t,T)$ for $0 < T \leq T^*$, then the risk-neutral measure $P_{B(t,T)}$ for this bond has a density

$$\frac{B(T,T)}{B(0,T)S_T^0} = \frac{1}{B(0,T)S_T^0}.$$

Consequently, the risk-neutral measure for the bond B is just the forward measure given in Definition 9.1.5. Note that, as the bond is not defined after time T, the measure change is defined only on $\mathcal{F}_T$, that is, only up to time T.

With the T-maturity bond as numéraire, we have seen that the price of the risky asset S^1 is given by its forward price

$$F(t,T) = \frac{S_t^1}{B(t,T)} \text{ for } 0 \leq t \leq T.$$

Now $F(t,T)$ must be a martingale under the risk-neutral measure $P_{B(t,T)}$ for B and consequently the differential $dF(t,T)$ must be of the form

$$dF(t,T) = \sigma_F(t,T)F(t,T)dW_B(t) \text{ for } 0 \leq t \leq T. \tag{9.2}$$

We note that this is a differential without any bounded variation dt terms and $W_B(t), 0 \leq t \leq T$, is a process that is a standard Brownian motion under the measure $P_B(t,T)$. As usual, $\sigma_F(t,T)$ can be taken to be non-negative.

Suppose now the price S^1 of the risky asset is taken as the numéraire. Of course, in terms of S^1, the price of the risky asset S^1 is always 1 unit. The risk-neutral measure for the numéraire S^1 is defined by

$$P_S(A) = \frac{1}{S_0^1} \int_A \frac{S^1(T^*)}{S^0(T^*)} dP \text{ for } A \in \mathcal{F}_{T^*}.$$

In terms of units of S^1, the value of a T-maturity bond is just

$$\frac{B(t,T)}{S_t^1} = \frac{1}{F(t,T)} \text{ for } 0 \leq t \leq T \leq T^*.$$

However, this is to be a martingale under P_S, so it has a differential

$$d\left(\frac{1}{F(t,T)}\right) = \sigma_{F^{-1}}(t,T)\left(\frac{1}{F(t,T)}\right) dW_S(t) \text{ for } 0 \leq t \leq T \leq T^*. \tag{9.3}$$

Again there will be no dt terms in the differential, and $(W_S(t))_{0 \leq t \leq T}$ is a standard Brownian motion under P_S. Again, $\sigma_{F^{-1}}(t,T)$ can be taken as non-negative.

Theorem 9.3.5. $\sigma_F(t,T) = \sigma_{F^{-1}}(t,T)$.

Proof. Applying the Itô rule to (9.3), we have

$$d\left(\frac{1}{F(t,T)}\right) = -\frac{1}{F(t,T)^2}\sigma_F(t,T)F(t,T)dW_B(t)$$

$$+ \frac{1}{F(t,T)^3}\sigma_F(t,T)^2 F(t,T)^2 dt$$

$$= \sigma_F(t,T)\left(\frac{1}{F(t,T)}\right)(-dW_B(t) + \sigma_B(t,T)dt). \qquad (9.4)$$

We know that $(W_B(t))$ is a standard Brownian motion under $P_{B(t,T)}$, as is $(-W_B(t))$. Therefore, under $P_{B(t,T)}$ the process $\frac{1}{F(t,T)}$ has volatility $\sigma_F(t,T)$ and mean rate of return $\sigma_F(t,T)^2$. Changing the measure from $P_{B(t,T)}$ to P_S transforms $\frac{1}{F(t,T)}$ into a P_S-martingale. Consequently, under P_S the mean rate of return of $\frac{1}{F(t,T)}$ is zero, but the volatility is not changed. In fact, from (9.4)

$$d\left(\frac{1}{F(t,T)}\right) = \sigma_{F^{-1}}(t,T)\frac{1}{F(t,T)}dW_S(t) \text{ for } 0 \le t \le T \le T^*. \qquad (9.5)$$

Comparing (9.4) and (9.5), we see that

$$\sigma_F(t,T) = \sigma_{F^{-1}}(t,T), \qquad W_S(t) = -W_B(t) + \int_0^t \sigma_F(s,T)ds.$$

$\square$

9.4   A General Option Pricing Formula

Following El Karoui, Geman, and Rochet [102] the risk-neutral measures for the numéraires S^1 and B can be used to express the price of a European call option:

$$V_0 = E\left((S_T^0)^{-1}(S_T^1 - K)^+\right)$$

$$= E\left((S_T^0)^{-1}S_T^1 \mathbf{1}_{\{S_T^1 > K\}}\right) - KE\left((S_T^0)^{-1}\mathbf{1}_{\{S_T^1 > K\}}\right)$$

$$= S_0^1 \int_{\{S_T^1 > K\}} \frac{S_T^1}{S_0^1 S_T^0}dP - KB(0,T)\int_{\{S_T^1 > K\}} \frac{1}{B(0,T)S_T^0}dP$$

$$= S_0^1 P_S\left(S_T^1 > K\right) - KB(0,T)P_B\left(S_T^1 > K\right)$$

$$= S_0^1 P_S\left(F(T,T) > K\right) - KB(0,T)P_B\left(F(T,T) > K\right)$$

$$= S_0^1 P_S\left(\frac{1}{F(T,T)} < \frac{1}{K}\right) - KB(0,T)P_B\left(F(T,T) > K\right).$$

Let us suppose that $\sigma_F(t,T)$ is a constant σ_F. Then, from (9.3), recalling that $\sigma_F = \sigma_{F^{-1}}$, we have

$$\frac{1}{F(T,T)} = \frac{B(0,T)}{S_0^1}\exp\left\{\sigma_F W_S(T) - \frac{1}{2}\sigma_F^2 T\right\}$$

where W_S is a standard Brownian motion under P_S. Consequently,

$$P_S\left(\frac{1}{F(T,T)} < \frac{1}{K}\right) = P_S\left(\sigma_F W_S(T) - \frac{1}{2}\sigma_F^2 T < \log\left(\frac{S_0^1}{KB(0,T)}\right)\right)$$
$$= P_S\left(\frac{W_S(T)}{\sqrt{T}} < \frac{1}{\sigma_F\sqrt{T}}\log\left(\frac{S_0^1}{KB(0,T)}\right) + \frac{1}{2}\sigma_F\sqrt{T}\right).$$

Now $\frac{W_S(T)}{\sqrt{T}}$ is a standard normal random variable. Writing, as usual, Φ for the standard normal distribution, we have

$$P_S\left(\frac{1}{F(T,T)} < \frac{1}{K}\right) = \Phi(h_1),$$

where

$$h_1 = \frac{1}{\sigma_F\sqrt{T}}\left(\log\left(\frac{S_0^1}{KB(0,T)}\right) + \frac{1}{2}\sigma_F^2 T\right).$$

From (9.2), we have that, with $P_B = P_B(t,T)$,

$$F(T,T) = \frac{S_0^1}{B(0,T)}\exp\left\{\sigma_F W_B(T) - \frac{1}{2}\sigma_F^2 T\right\},$$

where W_B is a standard Brownian motion under P_B. Therefore

$$P_B\left(F(T,T) > K\right) = P_B\left(\sigma_F W_B(T) - \frac{1}{2}\sigma_F^2 T > \log\left(\frac{KB(0,T)}{S_0^1}\right)\right)$$
$$= P_B\left(\frac{W_B(T)}{\sqrt{T}} < \frac{1}{\sigma_F\sqrt{T}}\left(\log\left(\frac{KB(0,T)}{S_0^1}\right) + \frac{1}{2}\sigma_F^2 T\right)\right)$$
$$= P_B\left(-\frac{W_B(T)}{\sqrt{T}} < \frac{1}{\sigma_F\sqrt{T}}\left(\log\left(\frac{S_0^1}{KB(0,T)}\right) - \frac{1}{2}\sigma_F^2 T\right)\right).$$

Again, $\frac{W_B(T)}{\sqrt{T}}$ is a standard normal random variable, so that

$$P_B\left(F(T,T) > K\right) = \Phi(h_2),$$

where

$$h_2 = \frac{1}{\sigma_F\sqrt{T}}\left(\log\left(\frac{S_0^1}{KB(0,T)}\right) - \frac{1}{2}\sigma_F^2 T\right).$$

Consequently, the price of the European call is

$$V_0 = S_0^1\Phi(h_1) - KB(0,T)\Phi(h_2).$$

If r is constant, then $B(0,T) = e^{-rT}$ and this formula reduces to the Black-Scholes formula of Theorem 7.6.2.

A modification of this argument shows that for any intermediate time $0 \le t \le T$, the value of the European call, with strike price K and expiration time T, is

$$V_t = S_t^1\Phi(h_1(t)) - KB(0,T)\Phi(h_2(t)), \tag{9.6}$$

where now, recalling that $F(t,T) = \frac{S_t^1}{B(t,T)}$,

$$h_1(t) = \frac{1}{\sigma_F \sqrt{T-t}} \left(\log\left(\frac{F(t,T)}{K}\right) + \frac{1}{2}\sigma_F^2(T-t) \right),$$

$$h_2(t) = \frac{1}{\sigma_F \sqrt{T-t}} \left(\log\left(\frac{F(t,T)}{K}\right) - \frac{1}{2}\sigma_F^2(T-t) \right).$$

Formula (9.6) suggests the European call can be hedged, at each time t, by holding $\Phi(h_1(t))$ units of S^1 and shorting $K\Phi(h_2(t))$ bonds.

We shall establish that this is a self-financing strategy. However, first we show that a change of numéraire does not change a trading strategy.

Lemma 9.4.1. *Suppose* $S^1, S^2, \ldots, S^d$ *are the price processes of k assets. Consider a self-financing strategy* $(\theta^1, \theta^2, \ldots, \theta^d)$, *where* θ_t^i *represents the number of units of asset i held at time t. Suppose Z is a numéraire and* $\widehat{S}^i = \frac{S^i}{Z}$, $1 \le i \le d$, *is the price of asset i in units of Z. Then* θ^i *represents the number of units of* $\widehat{S}^i$ *in the portfolio, evaluated in terms of the new numéraire (there are no other riskless assets).*

Proof. The wealth process is

$$X_t = \sum_{i=1}^{d} \theta_t^i S_t^i.$$

As the strategy is self-financing, we have

$$dX_t = \sum_{i=1}^{d} \theta_t^i dS_t^i.$$

Write $\widehat{X}_t = \frac{X_t}{Z_t}$ for the wealth process expressed in terms of the numéraire Z. Then

$$\begin{aligned}
d\widehat{X} &= Z^{-1}dX + Xd\left(\frac{1}{Z}\right) + d\left\langle X, \frac{1}{Z}\right\rangle \\
&= \frac{1}{Z}\sum_{i=1}^{d}\theta^i dS^i + \left(\sum_{i=1}^{d}\theta^i S^i\right)d\left(\frac{1}{Z}\right) + \sum_{i=1}^{d}\theta^i d\left\langle S^i, \frac{1}{Z}\right\rangle \\
&= \sum_{i=1}^{d}\theta^i d\widehat{S}^i.
\end{aligned}$$

$\square$

Corollary 9.4.2. *In Lemma 9.4.1, the strategy* $(\theta^1, \theta^2, \ldots, \theta^d)$ *determined the wealth process X. Suppose now that components* $\theta^1, \theta^2, \ldots, \theta^{d-1}$ *are given, together with the wealth process X. Then*

$$\theta_t^d = \frac{1}{S_t^d}\left(X_t - \sum_{i=1}^{d-1}\theta_t^i S_t^i\right)$$

and

$$dX_t = \sum_{i=1}^{d} \theta_t^i dS_t^i = \sum_{i=1}^{d-1} \theta_t^i dS_t^i + \frac{1}{S_t^d}\left(X_t - \sum_{i=1}^{d-1} \theta_t^i S_t^i\right) dS_t^d.$$

In terms of the numéraire Z, we still have

$$\theta_t^d = \frac{1}{S_t^d}\left(X_t - \sum_{i=1}^{d-1} \theta_t^i S_t^i\right) = \frac{1}{\widehat{S}_t^d}\left(\widehat{X}_t - \sum_{i=1}^{d-1} \theta_t^i \widehat{S}_t^i\right)$$

and

$$d\widehat{X}_t = \sum_{i=1}^{d-1} \theta_t^i d\widehat{S}_t^i + \frac{1}{\widehat{S}_t^d}\left(\widehat{X}_t - \sum_{i=1}^{d-1} \theta_t^i \widehat{S}_t^i\right) d\widehat{S}_t^d.$$

Let us return to the price (9.6) at time t for a European call option.

Theorem 9.4.3. *Holding $\Phi(h_1(t))$ units of S^1 and shorting $K\Phi(h_2(t))$ bonds at each time $t \in [0, T]$ is a self-financing strategy for the European call option with strike price K and expiration time T.*

Proof. This result could be established using Theorem 9.4.1. Alternatively, suppose we start with an initial investment of \$$V_0$ and hold $\Phi(h_1(t))$ units of S^1 at each time t. To maintain this position, we short as many bonds as necessary.

If we can show that the number of bonds we must short at time t is $K\Phi(h_2(t))$, then the value of our portfolio is indeed

$$\Phi(h_1(t))S_t^1 - KB(t,T)\Phi(h_2(t)),$$

which equals V_t, the price of the call option at time $t \in [0, T]$, and we have a hedge.

Let us write $\theta_t^1 = \Phi(h_1(t))$ so that at time t we hold θ_t^1 units of S^1.

Suppose X_t is the value of our portfolio at time t. Then we invest $X_t - \theta_t^1 S_t^1$ in the bond and the number of bonds in the portfolio is

$$\theta_t^2 = \frac{X_t - \theta_t^1 S_t^1}{B(t,T)}.$$

Then

$$dX_t = \theta_t^1 dS_t^1 + \theta_t^2 dB(t,T) = \theta_t^1 dS_t^1 + \frac{X_t - \theta_t^1 S_t^1}{B(t,T)} dB(t,T).$$

We must show that, if $X_0 = V_0$, then

$$X_t = V_t \text{ for } 0 \le t \le T.$$

To establish this, it is easier to work with $B(t,T)$ as numéraire. In terms of this zero coupon bond, the asset values S^1, B, and X become

$$\widehat{S}_t^1 = \frac{S_t^1}{B(t,T)} = F(t,T),$$

$$\widehat{B}(t,T) = 1,$$
$$\widehat{X}_t = \Phi(h_1(t))F(t,T) + \widehat{X}_t - \theta_t^1 S_t^1,$$

and $d\widehat{X}_t = \Phi(h_1(t))dF(t,T)$.

The option value is

$$V_t = \Phi(h_1(t))S_t^1 - KB(t,T)\Phi(h_2(t)),$$

and in terms of the numéraire $B(t,T)$ this becomes

$$\widehat{V}_t = \Phi(h_1(t))F(t,T) - K\Phi(h_2(t)).$$

Consequently,

$$dV_t = \Phi(h_1(t))dF(t,T) + F(t,T)d\Phi(h_1(t)) - Kd\Phi(h_2(t))$$
$$+ d\langle\Phi(h_1(t)), F(t,T)\rangle.$$

Recall the dynamics (9.2) given by

$$dF(t,T) = \sigma_F F(t,T)dW_B(t).$$

Now consider $(\Phi(h_1(t)))$, where

$$h_1(t) = \frac{1}{\sigma_F\sqrt{T-t}}\left(\log\left(\frac{F(t,T)}{K}\right) + \frac{1}{2}\sigma_F^2(T-t)\right).$$

The Itô rule gives, after some cancellation, with ϕ as the standard normal density,

$$d\Phi(h_1(t)) = \phi(h_1)\cdot\frac{1}{\sigma_F\sqrt{T-t}}\cdot\frac{1}{F}dF - \phi(h_1)\frac{\sigma_F}{2\sqrt{T-t}}dt,$$

where ϕ is the standard normal density function.

Also, $F\phi(h_1) = K\phi(h_2)$, and some elementary but tedious calculations confirm that

$$Fd\Phi(h_1) - Kd\Phi(h_2) + d\langle\Phi(h_1), F\rangle = 0.$$

The result follows. $\square$

9.5   Term Structure Models

Again let W be a standard Brownian motion on $(\Omega, \mathcal{F}, P)$ and $(\mathcal{F}_t)_{0\leq t\leq T}$ the (completed) filtration generated by W.

The instantaneous interest rate r_t is an adapted, measurable process and the numéraire asset S_t^0 has value

$$S_t^0 = \exp\left\{\int_0^t r_u du\right\} \text{ for } 0\leq t\leq T.$$

We have seen that the price at time $t \in [0, T]$ of a zero coupon bond maturing at time T is

$$B(t, T) = S_t^0 E\left((S_t^0)^{-1} | \mathcal{F}_t\right).$$

If r is non-random, then

$$B(t, T) = \exp\left\{-\int_t^T r_u du\right\}.$$

Zero coupon bonds are traded in the market, and their prices can be used to calibrate the model. They are known as 'zeros'.

Definition 9.5.1. A *term structure model* is a mathematical model for the prices $B(t, T)$ for all t and T with $0 \le t \le T \le T_2$.

The *yield* $R(t, T) = \frac{\log B(t, T)}{T - t}$ provides a *yield curve* for each fixed time t as the graph of $R(t, T)$ against T, which displays the average return of bonds after elimination of the distorting effects of maturity. We expect different yields at different maturities, reflecting market beliefs about future changes in interest rates. While greater uncertainty about interest rates in the distant future will tend to lead to increases in yield with maturity, high current rates (which may be expected to fall) can produce 'inverted' yield curves, in which long bonds will have a lower yield than short ones. A satisfactory term structure model should be able to handle both situations.

Remark 9.5.2. Recall that we are working under a martingale, or risk-neutral, measure P and that

$$B(t, T) = S_t^0 E\left((S_t^0)^{-1} | \mathcal{F}_t\right).$$

That is,

$$\frac{B(t, T)}{S_t^0} = E\left((S_t^0)^{-1} | \mathcal{F}_t\right),$$

and so $\left(\frac{B(t,T)}{S_t^0}\right)$ is a martingale under P.

If the market measure P does not have the property that all processes $\left(\frac{B(t,T)}{S_t^0}\right)$ are martingales, then the term structure model is free of arbitrage only if there is an equivalent measure $\widetilde{P}$ such that, under $\widetilde{P}$, all processes $\left(\frac{B(t,T)}{S_t^0}\right)$ are martingales for all maturity times T.

$B(t, T)$ is a positive process for all T, so that, using the martingale representation theorem, dynamics for $B(t, T)$ can be expressed in a log-normal form

$$dB(t, T) = \mu(t, T)B(t, T)dt + \sigma(t, T)B(t, T)dW_t \text{ for } t \in [0, T].$$

Consequently,

$$d\left(\frac{B(t,T)}{S_t^0}\right) = (\mu(t,T) - r_t)\frac{B(t,T)}{S_t^0}dt + \sigma(t,T)\frac{B(t,T)}{S_t^0}dW_t$$

and $\left(\frac{B(t,T)}{S_t^0}\right)$ is a martingale under P if and only if $\mu(t,T) = r_t$.

The statement that

$$B(t,T) = E\left(\exp\left\{-\int_t^T r_u du\right\}|\mathcal{F}_t\right)$$

is sometimes called the *Local Expectations Hypothesis*.

The assumption that holding a discount bond to maturity gives the same return as rolling over a series of single-period bonds is called the *Return to Maturity Expectations Hypothesis*. In continuous time, it would state that, under some probability P',

$$\frac{1}{B(t,T)} = E_{P'}\left(\exp\left\{\int_t^T r_u du\right\}|\mathcal{F}_t\right).$$

The *Yield to Maturity Expectations Hypothesis* states that the yield from holding a bond equals the yield from rolling over a series of single-period bonds. In continuous time, this would imply

$$B(t,T) = \exp\left\{-E_{P'}\left(\int_t^T r_u du|\mathcal{F}_t\right)\right\}$$

for some probability P'. A fuller discussion of these concepts can be found in the papers of Frachot and Lesne [137], [138].

9.6   Short-rate Diffusion Models

Vasicek's Model

Vasicek [296] proposed a mean-reverting version of the Ornstein-Uhlenbeck process for the short term rate r. Specifically, under the risk-neutral measure P, r is given by

$$dr_t = a(b - r_t)dt + \sigma dW_t$$

for $r_0 > 0$, $a > 0$, $b > 0$, and $\sigma > 0$. Then

$$r_t = e^{-at}\left(r_0 + b\left(e^{at} - 1\right) + \sigma\int_0^t e^{au}dW_u\right).$$

Consequently, r_t is a normal random variable with mean

$$E\left(r_t\right) = e^{-at}\left(r_0 + b\left(e^{at} - 1\right)\right)$$

and variance

$$\text{Var}(r_t) = \sigma^2 \left(\frac{1 - e^{-2at}}{2a} \right).$$

However, a normal random variable can be negative with positive probability so this model for r is not too realistic (unless the probability of being negative is small). Nonetheless, its simplicity validates its discussion.

As $t \to \infty$, we see that r_t converges in law to a Gaussian random variable with mean b and variance $\frac{\sigma^2}{2a}$.

The price of a zero coupon bond in the Vasicek model is therefore

$$B(t,T) = E \left(\exp \left\{ - \int_t^T r_u du \right\} | \mathcal{F}_t \right)$$

$$= e^{-b(T-t)} E \left(\exp \left\{ - \int_t^T X_u du \right\} | \mathcal{F}_t \right), \quad (9.7)$$

where $X_u = r_u - b$. Now X is the solution of the classical Ornstein-Uhlenbeck equation

$$dX_t = -aX_t dt + \sigma dW_t, \quad (9.8)$$

with $X_0 = r_0 - b$. Write

$$Z(t,x) = E \left(\exp \left\{ - \int_0^t X(u,x) du \right\} \right), \quad (9.9)$$

where $X(u,x)$ is the solution of (9.2) with $X(0,x) = x$. Now

$$X(u,x) = e^{-au} \left(x + \int_0^u \sigma e^{as} dW_s \right),$$

so $X(u,x)$ is a Gaussian process with continuous sample paths. Consequently, $\left(\int_0^t X(u,x) du \right)$ is a Gaussian process; this can be established by considering moment-generating functions $\exp \{ u_1 X(t_1) + \cdots + u_n X(t_n) \}$.

If Y is a Gaussian random variable with $E(Y) = m$ and $\text{Var}(Y) = \gamma^2$, we know that

$$E(e^Y) = e^{-m + \frac{1}{2}\gamma^2}.$$

Now

$$E(X(u,x)) = xe^{-au}, \qquad E \left(\int_0^t X(u,x) du \right) = \frac{x}{a} (1 - e^{-at}),$$

and

$$\text{Cov}[X(t,x), X(u,x)] = \sigma^2 e^{-a(u+t)} E \left(\int_0^t e^{as} dW_s \int_0^u e^{as} dW_s \right)$$

$$= \sigma^2 e^{-a(u+t)} \int_0^{u \wedge t} e^{2as} ds$$

$$= \frac{\sigma^2}{2a} e^{-a(u+t)} \left(e^{2a(u \wedge t)} - 1 \right). \tag{9.10}$$

Therefore,

$$\mathrm{Var}\left(\int_0^t X(u,x)du \right) = \mathrm{Cov}\left(\int_0^t X(u,x)du, \int_0^t X(s,x)ds \right)$$

$$= \int_0^t \int_0^t \mathrm{Cov}[X(u,x), X(s,x)]duds$$

$$= \int_0^t \int_0^t \frac{\sigma^2}{2a} e^{-a(u+s)} \left(e^{2a(u \wedge s)} - 1 \right) duds$$

$$= \frac{\sigma^2}{2a^3} \left(2at - 3 + 4e^{-at} - e^{-2at} \right).$$

Consequently,

$$Z(t,x) = E\left(\exp\left\{ -\int_0^t X(u,x)du \right\} \right)$$

$$= \exp\left\{ -\frac{x}{a} \left(1 - e^{-at} \right) + \frac{1}{4} \frac{\sigma^2}{a^3} \left(2at - 3 + 4e^{-at} - e^{-2at} \right) \right\}.$$

Using the time homogeneity of the X process,

$$B(t,T) = e^{-b(T-t)} Z(T-t, r_t - b).$$

This can be written as

$$B(t,T) = \exp\left\{ -(T-t)R(T-t, r_t) \right\},$$

where $R(T-t, r_t)$ can be thought of as the interest rate between times t and T. With $R_\infty = b - \frac{\sigma^2}{2a^2}$, we can write

$$R(t,r) = R_\infty - \frac{1}{at} \left((R_\infty - r)\left(1 - e^{-at}\right) - \frac{\sigma^2}{4a^2}\left(1 - e^{-at}\right)^2 \right).$$

Note that $R_\infty = \lim_{t \to \infty} R(t,r)$, so R_∞ can be thought of as the long-term interest rate. However, R_∞ does not depend on the instantaneous rate r_t. Practitioners consider this to be a weakness of the Vasicek model.

Exercise 9.6.1. Let $0 \le t \le T \le T^*$ and consider a call option with expiry T and strike K on the zero coupon bond $B(t, T^*)$. Show that this option will be exercised if and only if $r(T) < r^*$, where, with R_∞ as above,

$$r^* = R_\infty \left(1 - \frac{\alpha(T^* - T)}{1 - e^{-\alpha(T^* - T)}} \right) - \frac{\sigma^2[1 - e^{-\alpha(T^* - T)}]}{4\alpha^2}$$

$$- \log(K) \left(\frac{\alpha}{1 - e^{-\alpha(T^* - T)}} \right). \tag{9.11}$$

The Hull-White Model

In its simplest form this model is a generalisation of the Vasicek model using deterministic, time-varying coefficients. It is popular with practitioners. Its more general form includes a term r_t^β in the volatility, in which case it generalises the Cox-Ingersoll-Ross model discussed in the next section.

In this model, the short rate process is supposed given by the stochastic differential equation

$$dr_t = (\alpha(t) - \beta(t)r_t)\,dt + \sigma(t)dW_t \tag{9.12}$$

for $r_0 > 0$. Here, α, β, and σ are deterministic functions of t.

Write $b(t) = \int_0^t \beta(u)du$, so b is also non-random. We solve (9.12) by variation of constants to obtain

$$r_t = e^{-b(t)}\left(r_0 + \int_0^t e^{b(u)}\alpha(u)du + \int_0^t e^{b(u)}\sigma(u)dW_u\right).$$

Again, r_t is a deterministic quantity plus the stochastic integral of a deterministic function.

Consequently, r_t is a Gaussian Markov process with mean

$$E(r_t) = m(t) = e^{-b(t)}\left(r_0 + \int_0^t e^{b(u)}\alpha(u)du\right)$$

and covariance

$$\text{Cov}(r_t, r_s) = e^{-b(s)-b(t)}\int_0^{s \wedge t} e^{2b(u)}\sigma^2(u)du.$$

Again we can argue that $\int_0^T r_t dt$ is normal. Its mean is

$$E\left(\int_0^T r_t dt\right) = \int_0^T e^{-b(t)}\left(r_0 + \int_0^t e^{b(u)}\alpha(u)du\right)dt$$

and its variance is

$$\text{Var}\left(\int_0^T r_t dt\right) = \int_0^T e^{2b(u)}\sigma^2(u)\left(\int_u^T e^{-b(s)}ds\right)^2 du.$$

The price of a zero coupon bond for this model is

$$B(0,T) = E\left(\exp\left\{-\int_0^T r_t dt\right\}\right).$$

The quantity in the exponential is Gaussian, so we have

$$B(0,T) = \exp\left\{-E\left(\int_0^T r_t dt\right) + \frac{1}{2}\text{Var}\left(\int_0^T r_t dt\right)\right\}$$

$$= \exp\left\{-r_0 \int_0^T e^{-b(t)}dt - \int_0^T \int_0^t e^{-b(t)+b(u)}\alpha(u)dudt\right.$$

$$\left. + \frac{1}{2}\int_0^T e^{2b(u)}\sigma^2(u)\left(\int_u^T e^{-b(s)}ds\right)^2 du\right\}$$

$$= \exp[-r_0 C(0,T) - A(0,T)],$$

where

$$C(0,T) = \int_0^T e^{-b(t)}dt$$

and

$$A(0,T) = \int_0^T \int_0^t e^{-b(t)+b(u)}\alpha(u)dudt$$

$$- \frac{1}{2}\int_0^T e^{2b(u)}\sigma^2(u)\left(\int_u^T e^{-b(s)}ds\right)^2 du.$$

Note that the first term in A can be written, using Fubini's theorem, as

$$\int_0^T \int_u^T e^{-b(t)+b(u)}\alpha(u)dudt = \int_0^T e^{b(u)}\alpha(u)\left(\int_u^T e^{-b(s)}ds\right)du.$$

Therefore

$$A(0,T) = \int_0^T \left(e^{b(u)}\alpha(u)\gamma(u) - \frac{1}{2}e^{2b(u)}\sigma^2(u)\gamma^2(u)\right)du,$$

where

$$\gamma(u) = \int_u^T e^{-b(s)}ds.$$

The price at time t of a zero coupon bond is

$$B(t,T) = E\left(\exp\left\{-\int_t^T r_u du\right\}|\mathcal{F}_t\right) = E\left(\exp\left\{-\int_t^T r_u du\right\}|r_t\right),$$

where the final step follows because r is Markov. Write

$$C(t,T) = e^{b(t)}\int_t^T e^{-b(u)}du = e^{b(t)}\gamma(t)$$

and

$$A(t,T) = \int_t^T \left(e^{b(u)}\alpha(u)\gamma(u) - \frac{1}{2}e^{2b(u)}\sigma^2(u)\gamma^2(u)\right)du.$$

Then it can be shown that

$$B(t,T) = \exp\left\{-r_t C(t,T) - A(t,T)\right\}. \tag{9.13}$$

Now α, β, and γ are deterministic functions of time, t; consequently $C(t,T)$ and $A(t,T)$ are also functions only of t. Write $C_t(t,T)$ and $A_t(t,T)$ for their derivatives in t. From (9.13), we have

$$dB(t,T) = B(t,T)\Big[-C(t,T)\left(\alpha(t) - \beta(t)r_t\right)dt$$
$$- C(t,T)\sigma(t)dW_t - \frac{1}{2}C^2(t,T)\sigma^2(t)dt - r_tC_t(t,T)dt - A_t(t,T)dt\Big].$$
$$(9.14)$$

We are working under the risk-neutral measure, so

$$dB(t,T) = r_tB(t,T)dt + \Delta(t)dW_t, \qquad (9.15)$$

where Δ is some coefficient function. Comparing (9.14) and (9.15), we see that we must have

$$r_t = -C(t,t)\left(\alpha(t) - \beta(t)r_t\right) - \frac{1}{2}C^2(t,T)\sigma^2(t) - r_tc_t(t,T) - A(t,T). \quad (9.16)$$

Consequently,

$$dB(t,T) = r_tB(t,T)dt - B(t,T)\sigma(t)C(t,T)dW_t.$$

The volatility of the zero coupon bond is $\sigma(t)C(t,T)$.

Some Normal Densities

Consider times $0 \leq t \leq T_1 < T_2$. In the Hull-White framework, we have seen that $r(T_1)$ is Gaussian with

$$E\left(r(T_1)\right) = m_1 = e^{-b(T_1)}\left(r_0 + \int_0^{T_1} e^{b(u)}\alpha(u)du\right),$$

$$\mathrm{Var}[r(T_1)] = \sigma_1^2 = e^{-2b(T_1)}\left(\int_0^{T_1} e^{2b(u)}\sigma^2(u)du\right).$$

Also, $\int_0^{T_1} r_u du$ is Gaussian with

$$E\left(\int_0^{T_1} r_u du\right) = m_2 = \int_0^{T_1} e^{-b(v)}\left(r_0 + \int_0^v e^{b(u)}\alpha(u)du\right)dv,$$

$$\mathrm{Var}\int_0^{T_1} r_u du = \sigma_2^2 = \int_0^{T_1} e^{2b(u)}\sigma^2(u)\left(\int_u^{T_1} e^{-b(s)}ds\right)^2 du.$$

The covariance of $r(T_1)$ and $\int_0^{T_1} r_u du$ is

$$E\left(\left(\int_0^{T_1}\left(r_u - E\left(r_u\right)\right)du\right)\left(r(T_1) - E\left(r(T_1)\right)\right)\right)$$

$$= \int_0^{T_1} E\left((r_u - E(r_u))(r(T_1) - E(r(T_1)))\right) du$$

$$= \int_0^{T_1} \text{Cov}(r_u, r(T_1)) du$$

$$= \int_0^T \left(e^{-b(u)-b(T_1)} \int_0^T e^{2b(s)} \sigma^2(s) ds\right) du$$

$$= \rho \sigma_1 \sigma_2,$$

say.

Bond Options

Consider a European call option on the zero coupon bond that has strike price K and expiration time T_1. The bond matures at time $T_2 > T_1$.

The calculations above imply that $(r(T_1), \int_0^{T_1} r_u du)$ is Gaussian with density

$$f(x, y) = \frac{1}{2\pi \sigma_1 \sigma_2 \sqrt{1 - \rho^2}}$$

$$\times \exp\left\{-\frac{1}{2(1-\rho^2)}\left(\frac{(x-m_1)^2}{\sigma_1^2} - \frac{2\rho(x-m_1)(y-m_2)}{\sigma_1 \sigma_2} + \frac{(y-m_2)^2}{\sigma_2^2}\right)\right\}.$$

The price of the European option on B with expiration time T_1 and strike K at time 0 is

$$V_0 = E\left(e^{-\int_0^{T_1} r_u du} (B(T_1, T_2) - K)^+\right)$$

$$= E\left(e^{-\int_0^{T_1} r_u du} (\exp\{-r(T_1)C(T_1, T_2) - A(T_1, T_2)\} - K)^+\right)$$

$$= \int_{-\infty}^{\infty} \int_{-\infty}^{\infty} e^{-y} (\exp\{-xC(T_1, T_2) - A(T_1, T_2)\} - K)^+ f(x, y) dx dy.$$

To determine the price of the bond option at time $t \leq T_1 < T_2$, we note that the random variable $\left(r(T_1), \int_t^{T_1} r_u du\right)$ is Gaussian with a density similar to $f(x, y)$, except that $m_1, m_2, \sigma_2, \sigma_2,$ and ρ are replaced by

$$m_1(t) = E(r(t_1)|r_t) = e^{-b(T_1)}\left(e^{b(t)} r_t + \int_t^{T_1} e^{b(u)} \alpha(u) du\right),$$

$$\sigma_1^2(t) = E\left((r(T_1) - m_1(t))^2 |r_t\right) = e^{-2b(T_1)} \int_t^{T_1} e^{2b(u)} \sigma^2(u) du,$$

$$m_2(t) = E\left(\int_t^{T_1} r_u du |r_t\right)$$

$$= \int_t^{T_1} \left(r_t e^{-b(v)+b(t)} + e^{-b(v)} \int_t^v e^{b(u)} \alpha(u) du\right) dv,$$

$$\sigma_2^2(t) = E\left(\left(\int_t^{T_1} r_u du - m_2(r)\right)^2 | r_t\right)$$

$$= \int_t^{T_1} e^{2b(v)} \sigma^2(v) \left(\int_v^{T_1} e^{-b(s)} ds\right)^2 dv,$$

and

$$\rho(t)\sigma_1(t)\sigma_2(t) = E\left(\left(\int_t^{T_1} r_u du - m_2(t)\right)(r(T_1) - m_1(t)) | r_t\right)$$

$$= \int_t^{T_1} e^{-b(u)-b(T_1)} \int_t^u e^{2b(s)} \sigma^2(s) ds du.$$

These quantities now depend on r_t and so are stochastic as is, therefore, the corresponding option price

$$E\left(e^{-\int_t^{T_1} r_u du} (B(T_1, T_2) - K)^+ | \mathcal{F}_t\right)$$

$$= E\left(e^{-\int_t^{T_1} r_u du} (\exp\{-r(T_1)C(T_1, T_2) - A(T_1, T_2)\} - K)^+ | r_t\right).$$

This price can be expressed in terms of an integration with respect to a density analogous to $f_t(x, y)$ in which $m_1, \sigma_1, m_2, \sigma_2, \rho$ are replaced by $m_1(t), \sigma_1(t), m_2(t), \sigma_2(t), \rho(t)$, respectively.

The Hull-White model leads to a closed form expression for the option on the bond. Also, the parameters of the model can be estimated so the initial yield curve is matched exactly. However, it is a 'one-factor' model and

$$B(t, T) = \exp\{-r_t C(t, T) - A(t, T)\},$$

so all bond prices for all T are perfectly correlated. Further, the short rate r_t is normally distributed. This means it can take negative values with positive probability, and the bond price can exceed 1.

The Cox-Ingersoll-Ross Model

We have noted in the Vasicek and Hull-White models for r_t that, because r_t is Gaussian, there is a positive probability that $r_t < 0$.

The Cox-Ingersoll-Ross model for r_t provides a stochastic differential equation for r_t, the solution of which is always non-negative. To describe this process, recall the Ornstein-Uhlenbeck equation (9.8)

$$dX_t = -aX_t dt + \sigma dW_t \tag{9.17}$$

with solution

$$X(t, x) = e^{-at}\left(x + \int_0^t \sigma e^{as} dW_s\right).$$

Here W is a standard Brownian motion on a probability space $(\Omega, \mathcal{F}, P)$.

In fact, suppose we have n independent Brownian motions $W_1, \ldots, W_n$ on $(\Omega, \mathcal{F}, P)$ and n Ornstein-Uhlenbeck processes $X_1, \ldots, X_n$ given by equations

$$dX_i(t) = -\frac{1}{2}\alpha X_i(t)dt + \frac{1}{2}\sigma dW_i(t),$$

so that

$$X_i(t) = e^{-\frac{1}{2}\alpha t}\left(X_i(0) + \frac{1}{2}\sigma\int_0^t e^{\frac{1}{2}\beta s}dW_i(s)\right).$$

Consider the process

$$r_t = X_1^2(t) + X_2^2(t) + \cdots + X_n^2(t).$$

From Itô's differential rule,

$$dr_t = \sum_{i=1}^n 2X_i(t)\left(-\frac{1}{2}\alpha X_i(t)dt + \frac{1}{2}\sigma dW_i(t)\right) + \sum_{i=1}^n \frac{1}{4}\sigma^2 dt$$

$$= -\alpha r_t dt + \sigma\left(\sum_{i=1}^n X_i(t)dW_i(t)\right) + \frac{n\sigma^2}{4}dt$$

$$= \left(\frac{n\sigma^2}{4} - \alpha r_t\right)dt + \sigma\sqrt{r_t}\sum_{i=1}^n \frac{X_i(t)}{\sqrt{r_t}}dW_i(t).$$

Consider the process

$$W_t = \sum_{i=1}^n \int_0^t \frac{X_i(u)}{\sqrt{r_u}}dW_i(u).$$

Then W is a continuous martingale and

$$W_t^2 = 2\int_0^t W_u dW_u + \sum_{i=1}^n \int_0^t \frac{X_i^2(u)}{r_u}du = 2\int_0^t W_u dW_u + t,$$

so $W_t^2 - t$ is a martingale. From Lévy's characterisation, therefore, W is a standard Brownian motion and we can write

$$dr_t = \left(\frac{n\sigma^2}{4} - \alpha r_t\right)dt + \sigma\sqrt{r_t}dW_t.$$

It is known (see [240], for example) that if $n = 1$, then $P(r_t > 0) = 1$ but

$$P\left(\text{there are infinitely many times } t > 0 \text{ for which } r_t = 0\right) = 1.$$

However, if $n \geq 2$, then

$$P\left(\text{there is at least one time } t > 0 \text{ for which } r_t = 0\right) = 0.$$

Definition 9.6.2. A *Cox-Ingersoll-Ross (CIR) process* is the process defined by an equation of the form

$$dr_t = (a - br_t)dt + \sigma\sqrt{r_t}dW_t, \qquad (9.18)$$

where $a > 0, b > 0$, and $\sigma > 0$ are constant. With $n = \frac{4a}{\sigma^2}$, we can interpret r_t as $\sum_{i=1}^{n} X_i^2(t)$ for Ornstein-Uhlenbeck processes X_i as above. However, equation (9.18) makes sense whether or not n is an integer.

Remark 9.6.3. Geman and Yor [142] explore the relationship between the Vasicek and CIR models and show in particular that the CIR process is a Bessel process.

Similarly to the results for integer n, we quote the following ([240]). If $a < \frac{\sigma^2}{2}$, so $n < 2$, then

$$P \text{ (there are infinitely many times } t > 0 \text{ for which } r_t = 0) = 1.$$

Consequently, this range for a is not too useful. If $a \geq \frac{\sigma^2}{2}$, so $n \geq 2$, then

$$P \text{ (there is at least one time } t > 0 \text{ for which } r_t = 0) = 0.$$

Write $r_{0,t}(x)$ for the solution of (9.18) for which $r_0 = x$. The next result describes the law of the pair of random variables $\left(r_{0,t}(x), \int_0^t r_{0,u}(x)du\right)$. Note that ϕ and ψ are functions of t only, reminiscent of the A and C functions in the Hull-White model.

Theorem 9.6.4. *For any* $\lambda > 0$ *and* $\mu > 0$, *we have*

$$E\left(e^{-\lambda r_{0,t}(x)}e^{-\mu \int_0^t r_{0,u}(x)du}\right) = e^{-a\phi_{\lambda,\mu}(t)}e^{-x\psi_{\lambda,\mu}(t)},$$

where

$$\phi_{\lambda,u}(t) = -\frac{2}{\sigma^2}\log\left(\frac{2\gamma e^{t(b+\gamma)/2}}{\sigma^2\lambda(e^{\gamma t} - 1) + \gamma - b + e^{\gamma t}(\gamma + b)}\right),$$

$$\psi_{\lambda,u}(t) = \frac{\lambda(\gamma + b) + e^{\gamma t}(\gamma - b) + 2\mu(e^{\gamma t} - 1)}{\sigma^2\lambda(e^{\gamma t} - 1) + \gamma - b + e^{\gamma t}(\gamma + b)}$$

and $\gamma = \sqrt{b^2 + 2\sigma^2\mu}$.

Proof. Suppose $0 \leq t \leq T$. From the uniqueness of solutions of (9.18), we have the following 'flow' property:

$$r_{0,T}(x) = r_{t,T}(r_{0,t}(x)).$$

Consider the expectation

$$E\left(e^{-\lambda r_{t,T}(r_{0,t}(x))}e^{-\mu \int_t^T r_{0,u}(x)d\mu} \,|\, \mathcal{F}_t\right).$$

From the Markov property, this is the same as conditioning on $r_{0,t}(x)$, so write

$$V\left(t, r_{0,t}(x)\right) = E\left(e^{-\lambda r_{0,T}(x)} e^{-\mu \int_t^T r_{0,u}(x)du} \,|r_{0,t}(x)\right).$$

Now

$$e^{-\mu \int_0^t r_{0,u}(x)du} V\left(t, r_{0,t}(x)\right) = E\left(e^{-\lambda r_{0,T}(x)} e^{-\mu \int_0^T r_{0,u}(x)du} \,|\mathcal{F}_t\right)$$

and so is a martingale. However, applying the Itô differentiation rule, we obtain

$$e^{-\mu \int_0^t r_{0,u}du} V\left(t, r_{0,t}(x)\right)$$

$$= V(0, x) + \int_0^t \left(\frac{\partial V}{\partial u}\left(u, r_{0,u}(x)\right) - \mu r_{0,u}(x) V\left(u, r_{0,u}(x)\right)\right.$$

$$+ \frac{\partial V}{\partial \xi}\left(u, r_{0,u}(x)\right)\left(a - b r_{0,u}(x)\right)$$

$$\left. + \frac{1}{2}\frac{\partial^2 V}{\partial \xi^2}\left(u, r_{0,u}(x)\right) \sigma^2 r_{0,u}(x)\right) e^{-\mu \int_0^u r_{0,s}(x)ds} du$$

$$+ \int_0^t e^{-\mu \int_0^u r_{0,s}(x)ds} \frac{\partial V}{\partial \xi}\left(u, r_{0,u}(x)\right) \sigma \sqrt{r_{0,u}(x)} dW_u.$$

As the left-hand side is a martingale and the right-hand side is an Itô process, the du integral must be the zero process. Consequently,

$$\frac{\partial V}{\partial t}(t, y) - \mu y V(t, y) + \frac{\partial V}{\partial y}(t, y)(a - by) + \frac{1}{2}\frac{\partial^2 V}{\partial y^2}(t, y)\sigma^2 y = 0$$

with

$$V(t, y) = E\left(e^{-\lambda r_{t,T}(y)} e^{-\mu \int_t^T r_{t,u}(y)du}\right).$$

Because the coefficients of (9.18) are independent of t, the solution of (9.18) is stationary and we can write

$$V(t, y) = E\left(e^{-\lambda r_{0,T-t}(y)} e^{-\mu \int_0^{T-t} r_{0,u}(y)du}\right).$$

Define

$$F(t, y) = E\left(e^{-\lambda r_{0,t}(y)} e^{-\mu \int_0^t r_{0,u}(y)du}\right),$$

so that $V(t, y) = F(T - t, y)$ and F satisfies

$$\frac{\partial F}{\partial t} = \frac{\partial F}{\partial y}(a - by) - \mu y F + \frac{1}{2}\sigma^2 y \frac{\partial^2 F}{\partial y^2} \tag{9.19}$$

with $F(0, y) = e^{-\lambda y}$.

Motivated by the formula of the Hull-White model, we look for a solution of (9.19) in the form

$$F(t, y) = e^{-a\phi(t) - x\psi(t)}.$$

This is the case if $\phi(0) = 0$ and $\psi(0) = \lambda$ with

$$\phi'(t) = \psi(t), \qquad -\psi'(t) = \frac{\sigma^2}{2}\psi^2(t) + b\psi(t) - \mu.$$

Solving these equations gives the expressions for ϕ and ψ. $\qquad\square$

Remark 9.6.5. Taking $\mu = 0$, we obtain the Laplace transform of $r_t(x)$:

$$E\left(e^{\lambda r_t(x)}\right) = (2\lambda K + 1)^{-2a/\sigma^2} \exp\left\{\frac{-\lambda K z}{2\lambda K + 1}\right\},$$

where

$$K = \frac{\sigma^2}{4b}\left(1 - e^{-bt}\right), \qquad z = \frac{4bx}{\sigma^2(e^{bt} - 1)}.$$

Consequently, the Laplace transform of $\frac{r_t(x)}{K}$ is given by $g_{\frac{4a}{\sigma^2}, z}$, where

$$g_{\delta, z} = \frac{1}{(2\lambda + 1)^{\delta/2}} \exp\left\{-\frac{\lambda z}{2\lambda + 1}\right\}.$$

However, consider the chi-square density $f_{\delta, z}$, having δ degrees of freedom and decentral parameter z, given by

$$f_{\delta, z}(x) = \frac{e^{-z/2}}{2z^{\frac{\delta}{4} - \frac{1}{2}}} e^{-x/2} x^{\frac{\delta}{4} - \frac{1}{2}} I_{\frac{\delta}{2} - 1}(\sqrt{xz}) \text{ for } x > 0.$$

Here I_ν is the modified Bessel function of order ν, given by

$$I_\nu(x) = \left(\frac{x}{2}\right)^\nu \sum_{n=0}^{\infty} \frac{\left(\frac{x}{2}\right)^{2n}}{n!\Gamma(\nu + n + 1)}.$$

Then it can be shown that $g_{\delta, z}$ is the Laplace transform of the law of a random variable having density $f_{\delta, z}(x)$. Consequently, $\frac{r_t(x)}{K}$ is a random variable having a chi-square density with δ degrees of freedom.

Recall that we are working under the risk-neutral probability P. The price of a zero coupon bond at time 0 is

$$B(0, T) = E\left(\exp\left\{-\int_0^T r_u(x)du\right\}\right) = e^{-a\phi_{0,1}(0,T) - r_0(x)\psi_{0,1}(0,T)}.$$

Here

$$\phi_{0,1}(T) = -\frac{2}{\sigma^2}\log\left(\frac{2\gamma e^{T(\gamma + b)/2}}{\gamma - b + e^{\gamma T}(\gamma + b)}\right), \quad \psi_{0,1}(T) = \frac{2(e^{\gamma T} - 1)}{\gamma - b + e^{\gamma T}(\gamma + b)}$$

with $\gamma = \sqrt{b^2 + 2\sigma^2}$. The price of a zero coupon bond at time t is similarly, because of stationarity,

$$B(t,T) = e^{-a\phi_{0,1}(T-t)-r_t(x)\psi_{0,1}(T-t)}.$$

Suppose $0 \le T \le T^*$. Consider a European call option with expiration time T and strike price K on the zero coupon bond $B(t,T^*)$. At time 0, this has a price

$$
\begin{aligned}
V_0 &= E\left(e^{-\int_0^T r_u(x)du}(B(T,T^*)-K)^+\right) \\
&= E\left(E\left(e^{-\int_0^T r_u(x)du}(B(T,T^*)-K)^+|\mathcal{F}_t\right)\right) \\
&= E\left(e^{-\int_0^T r_u(x)du}\left(e^{-a\phi_{0,1}(T^*-T)-r_T(x)\psi_{0,1}(T^*-T)}-K\right)^+\right).
\end{aligned}
$$

Write

$$r^* = \frac{-a\phi_{0,1}(T^*-T)+\log K}{\psi_{0,1}(T^*-T)}.$$

Then

$$
\begin{aligned}
V_0 = E\left(e^{-\int_0^T r_u(x)du}B(T,T^*)\mathbf{1}_{\{r_T(x)<r^*\}}\right) \\
- KE\left(e^{-\int_0^T r_u(x)du}\mathbf{1}_{\{r_T(x)<r^*\}}\right).
\end{aligned}
$$

Now

$$E\left(e^{-\int_0^T r_u(x)du}B(T,T^*)\right) = B(0,T^*), \quad E\left(e^{-\int_0^T r_u(x)du}\right) = B(0,T).$$

Define two new probability measures P_1 and P_2 by setting

$$\frac{dP_1}{dP}\bigg|_{\mathcal{F}_T} = \frac{e^{-\int_0^T r_u(x)du}B(T,T^*)}{B(0,T^*)}, \quad \frac{dP_2}{dP}\bigg|_{\mathcal{F}_T} = \frac{e^{-\int_0^T r_u(x)du}}{B(0,T)}.$$

Then

$$V_0 = B(0,T^*)P_1\left(r_T(x)<r^*\right) - KB(0,T)P_2\left(r_T(x)<r^*\right).$$

Write

$$K_1 = \frac{\delta^2}{2}\cdot\frac{e^{\gamma T}-1}{\gamma(e^{\gamma T}+1)+(\sigma^2\psi_{0,1}(T^*-T)+b)(e^{\gamma T}-1)},$$

$$K_2 = \frac{\sigma^2}{2}\cdot\frac{e^{\gamma T}-1}{\gamma(e^{\gamma T}+1)+b(e^{\gamma T}-1)}.$$

Then it can be shown that the law of $\frac{r_T(x)}{K_1}$ under P_1 (resp. the law of $\frac{r_T(x)}{K_2}$ under P_2) is a decentral chi-square with $\frac{4a}{\sigma^2}$ degrees of freedom and decentral parameter ξ_1 (resp. ξ_2), where

$$\xi_1 = \frac{8r_0(x)\gamma^2 e^{\gamma T}}{\sigma^2(e^{\gamma T}-1)(\gamma(e^{\gamma T}+1))+(\sigma^2\psi_{0,1}(T^*-T)+b)(e^{\gamma T}-1)},$$

$$\xi_2 = \frac{8r_0(x)\gamma^2 e^{\gamma T}}{\sigma^2 \left(e^{\gamma T} - 1\right)\left(\gamma\left(e^{\gamma T} + 1\right) + b\left(e^{\gamma T} - 1\right)\right)}.$$

Consequently, if $F_{\delta,z}$ is the probability distribution function for a chi-square random variable with δ degrees of freedom and decentral parameter z, then

$$V_0 = B(0, T^*)F_{\frac{4a}{\sigma^2}, \xi_1}\left(\frac{r^*}{K_1}\right) - KB(0, T)F_{\frac{4a}{\sigma^2}, \xi_2}\left(\frac{r^*}{K_2}\right).$$

9.7 The Heath-Jarrow-Morton Model

Forward Rate Agreement

Suppose $0 \le t \le T < T + \varepsilon \le T^*$. 'Today' is time t. We wish to enter a contract to borrow \$1 at the future time T and repay it (with interest) at the time $T + \varepsilon$. The rate of interest to be paid between T and $T + \varepsilon$ is to be agreed today and so must be $\mathcal{F}_t$-measurable.

We could approximate this transaction by buying today a T-maturity zero for $B(t, T)$ and shorting an amount $\frac{B(t,T)}{B(t,T+\varepsilon)}$ of $(T+\varepsilon)$-maturity zeros.

The cost of this portfolio at time t is

$$B(t, T) - \frac{B(t, T)}{B(t, T + \varepsilon)} \cdot B(t, T + \varepsilon) = 0.$$

Now at the future time T we receive \$1 for the T-maturity zero. Then, at the time $(T + \varepsilon)$, we must pay $\frac{B(t,T)}{B(t,T+\varepsilon)}$ for the $(T + \varepsilon)$-maturity zeros.

In effect, we are looking at borrowing \$1 at the future time T and paying \$$\frac{B(t,T)}{B(t,T+\varepsilon)}$ at time $T + \varepsilon$. Consequently, the interest rate we are paying on the dollar received at time T is $R(t, T, T + \varepsilon)$, where

$$\frac{B(t, T)}{B(t, T + \varepsilon)} = \exp\left\{\varepsilon R(t, T, T + \varepsilon)\right\}$$

so

$$R(t, T, T + \varepsilon) = -\frac{1}{\varepsilon}[\log B(t, T + \varepsilon) - \log B(t, T)].$$

Definition 9.7.1. The instantaneous interest rate for money borrowed at time T, agreed upon at time $t \le T$, is the *forward rate* $f(t, T)$.

In fact,

$$f(t, T) = \lim_{\varepsilon \downarrow 0} R(t, T, T + \varepsilon) = \frac{-\partial}{\partial T} \log B(t, T).$$

Then, as $B(t, t) = 1$,

$$\log B(t, T) = \int_t^T \frac{\partial}{\partial T} \log B(t, u)\,du = -\int_t^T f(t, u)\,du.$$

Therefore, $B(t,T) = \exp(-\int_t^T f(t,u)du)$.

We note that this is an alternative representation for $B(t,T)$, in contrast to its expression in terms of the short-rate process r:

$$B(t,T) = E\left(\exp\left\{-\int_t^T r_u du\right\} | \mathcal{F}_t\right).$$

Agreeing at time t on the forward rate $f(t,u)$ means one agrees, at time t, that the instantaneous interest rate at time $u \in [t,T]$ will be $f(t,u)$.

Thus one agrees that investing \$1 at time t will give \$$\exp\int_t^T f(t,u)du$ at time T; investing \$$B(t,T)$ at time t will give

$$\$B(t,T) \cdot \exp\left\{\int_t^T f(t,u)du\right\} = \$1$$

at time T.

Lemma 9.7.2. *We have* $r_t = f(t,t)$.

Proof. We have two representations:

$$B(t,T) = E\left(\exp\left\{-\int_t^T r_u du\right\} | \mathcal{F}_t\right), \qquad (9.20)$$

$$B(t,T) = \exp\left\{-\int_t^T f(r,u)du\right\}. \qquad (9.21)$$

From (9.20),

$$\frac{\partial B(t,T)}{\partial T} = E\left(-r(T)\exp(-\int_t^T r_u du)|\mathcal{F}_t\right).$$

Evaluating at $T = t$, we have

$$\left.\frac{\partial B(t,T)}{\partial T}\right|_{T=t} = -r_t.$$

From (9.21),

$$\frac{\partial B(t,T)}{\partial T} = -f(t,T)\exp\left(-\int_t^T f(t,u)du\right)$$

and

$$\left.\frac{\partial B(t,T)}{\partial T}\right|_{T=t} = -f(t,t).$$

$\square$

The Heath-Jarrow-Morton model

The Heath-Jarrow-Morton (HJM) model for term structure describes a system of stochastic differential equations for the evolution of the forward rate $f(t,T)$. For each $T \in (0, T^*]$, suppose the dynamics of f are given by

$$df(t,T) = \alpha(t,T)dt + \sigma(t,T)dW_t. \tag{9.22}$$

Here the coefficients $\alpha(u,T)$ and $\sigma(u,T)$, for $0 \le u \le T$, are measurable (in (u, ω)) and adapted. The integral form of (9.22) is

$$f(t,T) = f(0,T) + \int_0^t \alpha(u,T)du + \int_0^t \sigma(u,T)dW_u. \tag{9.23}$$

Note that we have two time parameters and recall

$$B(t,T) = \exp\left\{ -\int_t^T f(t,u)du \right\}.$$

With d denoting a differential in the t variable,

$$d\left(-\int_t^T f(t,u)du \right) = f(t,t)dt - \int_t^T \left(df(t,u) \right) du$$

$$= r_t dt - \int_t^T [\alpha(t,u)dt + \sigma(t,u)dW_t]du$$

$$= r_t dt - \alpha^*(t,T)dt - \sigma^*(t,T)dW_t, \tag{9.24}$$

where

$$\alpha^*(t,T) = \int_t^T \alpha(t,u)du, \qquad \sigma^*(t,T) = \int_t^T \sigma(t,u)du.$$

Recall, that by definition, $f(t,u)$ is an $(\mathcal{F}_t)$-adapted process. Therefore,

$$X_t = -\int_t^T f(t,u)du$$

is an $\mathcal{F}_t$-adapted process. In fact, it is an Itô process with, as in (9.24),

$$dX_t = (r_t - \alpha^*(t,T))\, dt - \sigma^*(t,T)dW_t.$$

Also, $B(t,T) = e^{X_t}$, so

$$dB(t,T) = e^{X_t}\left(r_t - \alpha^*(t,T) + \frac{1}{2}\sigma^*(t,T)^2 \right) dt - e^{X_t}\sigma^*(t,T)dW_t$$

$$= B(t,T)\left(\left(r_t - \alpha^*(t,T) + \frac{1}{2}\sigma^*(t,T)^2 \right) dt - \sigma^*(t,T)dW_t \right).$$

Now, the discounted $B(t,T)$ will be a martingale under P (so P is a risk-neutral measure) if

$$\alpha^*(t,T) = \frac{1}{2}\left(\sigma^*(t,T)\right)^2 \text{ for } 0 \le t \le T \le T^*.$$

From the definitions of α^* and σ^*, this means that

$$\int_t^T \alpha(t,u)du = \frac{1}{2}\left(\int_t^T \sigma(t,u)du\right)^2.$$

This is equivalent to

$$\alpha(t,T) = \sigma(t,T)\int_0^T \sigma(t,u)du.$$

If P itself is not a risk-neutral measure, there may be a probability P^θ under which $\left(\frac{B(t,T)}{S_t^0}\right)$ is a martingale. This is the content of the following result due to Heath, Jarrow, and Morton [153].

Theorem 9.7.3. *For each $T \in (0,T]$, suppose $\alpha(u,T)$ and $\sigma(u,T)$ are adapted processes. We assume $\sigma(u,T) > 0$ for all u,T, and $f(0,T)$ is a deterministic function of T. The instantaneous forward rate $f(t,T)$ is defined by*

$$f(t,T) = f(0,T) + \int_0^t \alpha(u,T)du + \int_0^t \sigma(u,t)dW_u.$$

Then the term structure model determined by the processes $f(t,T)$ does not allow arbitrage if and only if there is an adapted process (θ_t) such that

$$\alpha(t,T) = \sigma(t,T)\int_t^T \sigma(t,u)du + \sigma(t,T)\theta_t \text{ for all } 0 \le t \le T \le T^*,$$

and the process

$$\Lambda_t^\theta = \exp\left\{-\int_0^t \theta_u dW_u - \frac{1}{2}\int_0^t \theta_u^2 du\right\}$$

is an $(\mathcal{F}_t, P)$-martingale.

Proof. Suppose θ is an adapted process such that $\Lambda^\theta(t)$ is an $(\mathcal{F}_t, P)$-martingale, and define a new probability measure P^θ by setting

$$\left.\frac{dP^\theta}{dP}\right|_{\mathcal{F}_{T^*}} = \Lambda^\theta(T^*).$$

By Girsanov's theorem, W^θ is a Brownian motion under P^θ, where

$$W_t^\theta = \int_0^t \theta_u du + W_t,$$

and

$$dB(t,T) = B(t,T)\left(\left(r_t - \alpha^*(t,T) + \frac{1}{2}\sigma^*(t,T)^2 + \sigma^*(t,T)\theta_t\right)dt\right.$$
$$\left. -\sigma^*(t,T)dW_t^\theta\right).$$

where, as before, $\alpha^*(t,T) = \int_t^T \alpha(t,u)du$ and $\sigma^*(t,T) = \int_t^T \sigma(t,u)du$.
    For $B(t,T)$ to have a rate of return r_t under P^θ, θ must satisfy

$$\alpha^*(t,T) = \frac{1}{2}\sigma^*(t,T)^2 + \sigma^*(t,T)\theta_t.$$

This must hold for all maturities T. Differentiating with respect to T, that
is

$$\alpha(t,T) = \sigma(t,T)\sigma^*(t,T) + \sigma(t,T)\theta_t \text{ for } 0 \le t \le T \le T^*.$$

$\square$

Remark 9.7.4. The point to note is that, if there is such a process θ_t, it is
independent of the time T maturity of the bond $B(t,T)$, and

$$\theta_t = -\left(\frac{-\alpha^*(t,T) + \frac{1}{2}\sigma^*(t,T)^2}{\sigma^*(t,T)}\right).$$

    Now, under the 'market' probability P, the rate of return of the bond
is

$$r_t - \alpha^*(t,T) + \frac{1}{2}\sigma^*(t,T)^2.$$

The rate of return above the interest rate r_t is therefore

$$-\alpha^*(t,T) + \frac{1}{2}\sigma^*(t,T)^2,$$

and the market price of risk is just

$$\frac{-\alpha^*(t,T) + \frac{1}{2}\sigma^*(t,T)^2}{\sigma^*(t,T)} = -\theta_t.$$

    The requirement of the theorem therefore is that the market price of
the risk be independent of the maturity times T. Substituting for θ, we
have that, under P^θ,

$$dB(t,T) = B(t,T)\left(r_t dt - \sigma^*(t,T)dW_t^\theta\right),$$
$$df(t,T) = \sigma(t,T)\sigma^*(t,T)dt + \sigma(t,T)dW_t^\theta.$$

9.8   A Markov Chain Model

Elliott, Hunter, and Jamieson ([117]) introduced a self-calibrating model for the short-term rate. It is supposed that the short-term rate r_t is a finite state space Markov chain defined on a probability space $(\Omega, \mathcal{F}, P)$ taking (positive) values $r_1, r_2, \ldots, r_N$. Each of these values can be identified with one of the canonical unit vectors e_i in $\mathbb{R}^N$, $e_i = (0, \ldots, 0, 1, 0, \ldots, 0)$. (In effect, we are considering an indicator function $\mathbf{1}_{\{r_i\}}(r)$ on the set $\{r_1, r_2, \ldots, r_N\}$). Without loss of generality, we can take the state space of our Markov chain $(X_t)_{t \geq 0}$ to be the set $S = \{e_1, e_2, \ldots, e_N\}$. Writing $r = (r_1, r_2, \ldots, r_N) \in \mathbb{R}^N$, we then have

$$r_t = r \cdot X_t = r(X_t),$$

where the central dot denotes the scalar product in $\mathbb{R}^N$. Considering the Markov chain to have state space S simplifies the notation.

The unconditional distribution of X_t is $E(X_t) = p_t = (p_t^1, \ldots, p^N(t))$, where

$$p_t^i = P(X_t = e_i) = E(e_i \cdot X_t) = P(r_t = r_i).$$

Suppose this distribution evolves according to the Kolmogorov equation

$$\frac{dp_t}{dt} = A p_t.$$

Here A is a 'Q-matrix'; that is, $A = (a_{ji})_{1 \leq i,j \leq N}$ with $\sum_{j=1}^N a_{ij} = 0$ and $a_{ji} \geq 0$ if $i \neq j$. The components a_{ji} could be taken to be time-varying, though this would complicate their estimation.

The price of a zero coupon bond at time t, with maturity T, in this model is

$$B(t,T) = E\left(\exp\left\{ -\int_t^T r(X_s)ds \right\} | \mathcal{F}_t \right),$$

where $(\mathcal{F}_t)$ is the filtration generated by X (or, equivalently, by r).

Because of the Markov property, this is

$$E\left(\exp\left\{ -\int_t^T r(X_s)ds \right\} | X_t \right) = B(t, T, X_t),$$

say, and so is a function of $X_t \in S$. Any (real) function of $X_t \in S$ is given as the scalar product of some function $\phi_t = (\phi_t^1, \phi_t^2, \ldots, \phi^N(t))' \in \mathbb{R}^N$ with X_t. That is, we can write

$$B(t, T, X_t) = \phi_t \cdot X_t,$$

where $\phi_t^i = B(t, T, e_i)$.

Now

$$e^{-\int_0^t r(X_s)ds} B(t, T, X_t) = e^{-\int_0^t r(X_s)ds} \phi_t \cdot X_t$$

$$= E\left(e^{-\int_0^t r(X_s)ds} \,|\, \mathcal{F}_t\right)$$

and so is a martingale.

Lemma 9.8.1. *Define the $\mathbb{R}^N$-valued process M by*

$$M_t = X_t - X_0 - \int_0^t AX_s ds.$$

Then M is an $(\mathcal{F}_t, P)$-martingale.

Proof. Consider the matrix exponential $e^{A(t-s)}$. By the Markov property,

$$E(X_t | X_s) = e^{A(t-s)} X_s$$

for $t \geq s$. (In effect, one solves the Kolmogorov equation with initial condition X_s.) Now, for $t \geq s$, if I is the $N \times N$ identity matrix,

$$
\begin{aligned}
E\left(M_t - M_s | \mathcal{F}_s\right) &= E\left(X_t - X_s | \mathcal{F}_s\right) - E\left(\int_s^t AX_u du | \mathcal{F}_s\right) \\
&= e^{A(t-s)} X_s - X_s - \int_s^t Ae^{A(u-s)} X_s du \\
&= \left(e^{A(t-s)} - I - \int_s^t Ae^{A(u-s)} du\right) X_s \\
&= \left(e^{A(t-s)} - I - [e^{A(u-s)}]_s^t\right) X_s = 0.
\end{aligned}
$$

$\square$

Corollary 9.8.2. *The semimartingale representation of X_t is therefore*

$$X_t = X_0 + \int_0^t AX_s ds + M_t.$$

Theorem 9.8.3. *The process $\phi_t \in \mathbb{R}^N$ has dynamics*

$$\frac{d\phi_t}{dt} = (\text{diag } r - A^*)\phi_t$$

with terminal condition $\phi_T = \mathbf{1} = (1, 1, \ldots, 1)' \in \mathbb{R}^N$.

Proof. We have seen that

$$e^{-\int_0^t r(X_s)ds} B(t, T, X_t) = e^{-\int_0^t r(X_s)ds} \phi_t \cdot X_t$$

is an $(\mathcal{F}_t, P)$-martingale. Consequently, the dt term in its Itô process (or semimartingale) representation must be identically zero. Now

$$e^{-\int_0^t r(X_s)ds}\phi_t \cdot X_t = B(0,T,X_0) - \int_0^t r(X_s)e^{-\int_0^s r(X_u)du}\phi_s \cdot X_s ds$$

$$+ \int_0^t e^{-\int_0^s r(X_u)du}\left(\frac{d\phi_s}{ds} \cdot X_s + \phi_s \cdot (AX_s)\right) ds + \int_0^t e^{-\int_0^s r(X_u)du}\phi_s \cdot dM_s.$$

Consequently,

$$e^{-\int_0^s r(X_u)du}\left(-r(X_s)\phi_s \cdot X_s + \frac{d\phi_s}{ds} \cdot X_s + \phi_s \cdot (AX_s)\right) = 0.$$

Now $r(X_s) = r \cdot X_s$, where $r = (r_1, r_s, \ldots, r_s)'$, and $r(X_s)\phi_s \cdot X_s = (\text{diag } r)\phi_s \cdot X_s$, where diag r is the matrix with r on its diagonal. Therefore

$$\frac{d\phi_s}{ds} \cdot X_s + A^*\phi_s \cdot X_s - ((\text{diag } r)\phi_s) \cdot X_s = 0 \text{ for all } X_s.$$

Consequently, ϕ is given by the vector equation

$$\frac{d\phi_t}{dt} = (\text{diag } r - A^*)\phi_t$$

with terminal condition $\phi_T = (1, \ldots, 1)' = \mathbf{1}$. $\square$

Corollary 9.8.4. *Write* $B = diag\ r - A^*$*. Then* $\phi_t = e^{-B(T-t)}\mathbf{1}$*, and the price at time t of a zero coupon bond is*

$$B(t,T,X_t) = \phi_t \cdot X_t = (e^{-B(T-t)} \cdot X_t)\mathbf{1}.$$

The yield for such a bond is

$$y_{t,T} = -\frac{1}{T-t}\log B(t,T,X_t).$$

Yield values are quoted in the market.

In [117] it is supposed that yield values give noisy information about such a Markov chain term structure model. The techniques of filtering from Hidden Markov models (see [110]), are then applied to estimate the state of X and the model parameters.

Chapter 10

Consumption-Investment Strategies

10.1 Utility Functions

The results of this chapter are a presentation of the comprehensive, fundamental, and elegant contributions of Karatzas, Lehoczky, Sethi, Shreve and Xu. See, for example, the papers [186] through [190].

We first review in the multi-asset situation concepts relating to trading strategies, consumption processes, and utility functions.

On a probability space $(\Omega, \mathcal{F}, P)$, consider a market that includes a bond S^0 and n risky assets $S^1, \ldots, S^N$. Their dynamics are given by the equations

$$dS_t^0 = S_t^0 r_t dt, \qquad\qquad S_0^0 = 1, \qquad\qquad (10.1)$$

$$dS_t^i = S_t^i \left(\mu_i(t)dt + \sum_{j=1}^n \sigma_{ij}(t)dW_j(t) \right), \quad S_0^i = s_i \text{ for } i = 1, 2, \ldots, n.$$

$$(10.2)$$

Here $W(t) = (W_1(t), \ldots, W_d(t))$ is an n-dimensional Brownian motion defined on $(\Omega, \mathcal{F}, P)$ and $(\mathcal{F}_t)$ will denote the completion of the filtration

$$\sigma \{W(u) : 0 \le u \le t\} = \sigma \{S_u : 0 \le u \le t\}.$$

The interest rate r_t, mean rate of return $\mu(t) = (\mu_1(t), \ldots, \mu_n(t))'$, and the volatility $\sigma(t) = (\sigma_{ij}(t))_{1 \le i,j \le d}$ are taken to be measurable, adapted, and bounded processes. Note that we have taken the dimension, n, of the Brownian motion equal to the number of risky assets.

Write $a(t) = \sigma(t)\sigma^*(t)$. We assume there is an $\varepsilon > 0$ such that, with $|.|$ denoting the Euclidean norm in $\mathbb{R}^n$,

$$\xi^* a(t)\xi \ge \varepsilon |\xi|^2 \text{ for all } \xi \in \mathbb{R}^n \text{ and } (t,\omega) \in [0,\infty) \times \Omega.$$

Consequently, the inverses of σ and σ^* exist and are bounded:

$$\left|\sigma(t,\omega)^{-1}\xi\right| \le \varepsilon^{-\frac{1}{2}} |\xi|, \quad \left|\sigma^*(t,\omega)^{-1}\xi\right| \le \varepsilon^{-\frac{1}{2}} |\xi| \quad \text{for all } \xi \in \mathbb{R}^n. \qquad (10.3)$$

The filtration is then equivalently given as the completion of the filtration generated by the process S.

Therefore in this situation the market price of risk defined by equation (7.34) has the unique solution

$$\theta_t = \sigma(t)^{-1}\left(b(t) - r_t\mathbf{1}\right);$$

furthermore, θ is bounded and progressively measurable.

As in Chapter 8, introduce

$$\Lambda_t = \exp\left\{-\int_0^t \theta_s' dW(s) - \frac{1}{2}\int_0^t |\theta_s'|^2 ds\right\}$$

and define a new probability measure P^θ by setting

$$\left.\frac{dP^\theta}{dP}\right|_{\mathcal{F}_t} = \Lambda_t.$$

We know from Girsanov's theorem that W^θ is a Brownian motion under P^θ, where

$$W_t^\theta = W(t) + \int_0^t \theta_s ds.$$

Furthermore, under P^θ,

$$dS_t^i = S_t^i\left(r_t dt + \sum_{j=1}^n \sigma_{ij}(t)dW_j^\theta(t)\right) \quad \text{for } i = 1, 2, \ldots, n.$$

That is, in this situation, P^θ is the unique risk-neutral, or martingale, measure.

Definition 10.1.1. A utility function $U : [0, \infty) \times (0, \infty) \to \mathbb{R}$ is a $C^{0,1}$ function such that:

a) $U(t, \cdot)$ is strictly increasing and strictly concave.

b) The derivative $U'(t, c) = \frac{\partial}{\partial c}U(t, c)$ is such that, for every $t > 0$,

$$\lim_{c \to \infty} U'(t, c) = 0, \qquad \lim_{c \downarrow 0} U'(t, c) = U'(t, 0+) = \infty.$$

These conditions have natural economic interpretations. The increasing property of U represents the fact that the investor prefers higher levels of consumption or wealth. The strict concavity of $U(t, c)$ in c implies $U'(t, c)$ is decreasing in c; this models the concept that the investor is risk averse.

The condition that $U'(t, 0+) = \infty$ is not strictly necessary, but it simplifies some of the proofs.

$U'(t, c)$ is strictly decreasing in c; therefore, there is an inverse map $I(t, c)$ so that

$$I(t, U'(t, c)) = c = U'(t, I(t, c))$$

for $c \in (0, \infty)$. The concavity of U implies that

$$U(t, I(t, y)) \geq U(t, c) + y(I(t, y) - c) \text{ for all } c, y. \qquad (10.4)$$

For some later results, we require that $U(t, c)$ is C^2 in $c \in (0, \infty)$ for all $t \in [0, T]$ and $U''(t, c) = \frac{\partial^2 U}{\partial c^2}$ is non-decreasing in c for all $t \in [0, T]$. These two conditions imply that $I(t, c)$ is convex and of class C^1 in $c \in (0, \infty)$, and

$$\frac{\partial}{\partial y} U(t, I(t, y)) = y \frac{\partial}{\partial y} I(t, y).$$

10.2 Admissible Strategies

The definitions in this section are the natural counterparts of the discrete-time notions introduced and discussed briefly in Section 5.6.

We recall that in the continuous-time setting a portfolio process or trading strategy $H = (H^1, \ldots, H^n)'$ is a measurable $\mathbb{R}^n$-valued process that is adapted $(\mathcal{F}_t)$ and is such that

$$\int_0^T |H_s|^2 \, ds < \infty \text{ a.s.}$$

A consumption process $(c_t)_{0 \leq t \leq T}$ is a non-negative, measurable, adapted process (with respect to $(\mathcal{F}_t)$) such that

$$\int_0^T c_t dt < \infty \text{ a.s.}$$

The adapted condition means the investor cannot anticipate the future, so 'insider trading' is not allowed.

The wealth of the investor at time t is then

$$X_t = \sum_{i=0}^n H_t^i S_t^i - \int_0^t c_s ds.$$

Here $H_t^i S_t^i$ represents the amount invested in asset $i = 0, 1, \ldots, n$, and $\int_0^t c_s ds$ represents the total amount consumed up to time t.

If the strategy H is self-financing, changes in the wealth derive only from changes in the asset prices, interest on the bond, and from consumption, and then

$$dX_t = \sum_{i=1}^d H_t^i dS_t^i + \left(1 - \sum_{i=1}^n H_t^i\right) dS_t^0 - c_t dt.$$

From (10.1) and (10.2), this is

$$(r_tX_t - c_t)\,dt + H_t'\,(\mu(t) - r_t\mathbf{1})\,dt + H_t'\sigma(t)dW(t)$$
$$= (r_tX_t - c_t)\,dt + H_t'\sigma(t)dW_t^\theta.$$

Writing $\beta_t = \left(S_t^0\right)^{-1} = \exp\left\{-\int_0^t r_s ds\right\}$, we see that

$$\beta_t X_t = x - \int_0^t \beta_s c_s ds + \int_0^t \beta_s H_s'\sigma(s)dW^\theta(s), \qquad (10.5)$$

where $x = X_0$ is the initial wealth of the investor.

Consequently,

$$D_t = \beta_t X_t + \int_0^t \beta_s c_s ds = x + \int_0^t \beta_s H_s'\sigma(s)dW^\theta(s),$$

which is the present discounted wealth plus the total discounted consumption so far, is a continuous local martingale under P^θ.

Definition 10.2.1. The *deflator* for the market is the process ξ defined by

$$\xi_t = \beta_t \Lambda_t.$$

This equals the discount factor β modified by the Girsanov density Λ to take account of the financial market.

Now

$$\Lambda_t D_t = \Lambda_t \left(\beta_t X_t + \int_0^t \beta_s c_s ds\right)$$
$$= \Lambda_t \left(x + \int_0^t \beta_s H_s'\sigma(s)dW^\theta(s)\right)$$
$$= \xi_t X_t + \int_0^t \xi_s c_s ds - \int_0^t C_s \Lambda_s \theta_s' dW(s),$$

where $C_s = \int_0^s \beta_u c_u du$. For any $\mathcal{F}$-measurable, P^θ-integrable random variable Ψ, Bayes' rule states that

$$E^\theta \left(\Psi \,|\, \mathcal{F}_s\right) = \frac{E\left(\Lambda_t \Psi \,|\, \mathcal{F}_s\right)}{\Lambda_s}.$$

Therefore, ΛD is a continuous local martingale under P. Moreover, so is $\left(\int_0^t C_s \Lambda_s \theta_s' dW(s)\right)$. Consequently,

$$N_t = \xi_t X_t + \int_0^t \xi_s c_s ds \qquad (10.6)$$

is a continuous local martingale under P. Furthermore, from Bayes' rule, we see that N is a P-supermartingale if and only if D is a P^θ-supermartingale.

Definition 10.2.2. Similarly to the set of trading strategies $SF(\xi)$ of Chapter 8, we introduce the set $SF(K, x)$.

A portfolio process $H = (H_t^1, \ldots, H_t^n)'$ and a consumption process c belong to $SF(K, x)$ if, for initial capital $x \geq 0$ and some non-negative, P-integrable random variable $K = K(H, c)$, the corresponding wealth process satisfies

$$X_T \geq 0 \text{ a.s.}, \qquad \xi_t X_t \geq -K(\omega) \text{ for all } 0 \leq t \leq T.$$

Here ξ_t is the deflator process of Definition 10.2.1.

Consequently, for every $(H, c) \in SF(K, x)$, the P-local martingale N of (10.6) is bounded from below. Using Fatou's lemma as in Chapter 8, we deduce that N is a P-supermartingale; therefore, D is a P^θ supermartingale.

Write $\mathcal{T}_{u,v}$ for the set of stopping times with values in $[u, v]$. Using the Optional Stopping Theorem on N (or D), for any $\tau \in \mathcal{T}_{0,T}$, for $(H, c) \in SF(K, x)$,

$$E\left(\xi_\tau X_\tau + \int_0^\tau \xi_s c_s ds\right) \leq x$$

or, equivalently,

$$E^\theta\left(\beta_\tau X_\tau + \int_0^\tau \beta_s c_s ds\right) \leq x. \tag{10.7}$$

These inequalities state that the expected value of current wealth at any time τ, and consumption up to time τ, deflated to time 0, should not exceed the initial capital x.

We now introduce consumption rate processes and final claims whose (deflated) expected value is bounded by the initial investment $x \geq 0$.

Definition 10.2.3.   a) Write $\mathcal{C}(x)$ for the consumption rate processes c that satisfy

$$E^\theta\left(\int_0^T c_s e^{-\int_0^s r_u du} ds\right) \leq x.$$

  b) Write $\mathcal{L}(x)$ for the non-negative $\mathcal{F}_T$-measurable random variables B that satisfy

$$E^\theta\left(Be^{-\int_0^T r_u du}\right) \leq x.$$

From the inequality (10.6), we see that $(H, c) \in SF(0, x)$ implies $c \in \mathcal{C}(x)$ and $X_T \in \mathcal{L}(x)$.

We now investigate to what extent we can deduce the opposite implications.

Theorem 10.2.4. *For every $c \in \mathcal{C}(x)$ there is a portfolio H such that $(H, c) \in SF(0, x)$. Furthermore, if c belongs to the class*

$$\mathcal{D}(x) = \left\{c \in \mathcal{C}(x) : E^\theta\left(\int_0^T \beta_s c_s ds\right) = x\right\},$$

then the corresponding wealth process X satisfies $X_T = 0$ and the process M is a martingale.

Proof. For $c \in \mathcal{C}(x)$, write

$$C = C_T = \int_0^T \beta_s c_s ds$$

and define the martingale

$$m_t = E^\theta \left(C \,|\, \mathcal{F}_t \right) - E^\theta \left(C \right).$$

Then, from the martingale representation result, m can be expressed as

$$m_t = \int_0^t \phi_s' dW^\theta(s), 0 \le t \le T,$$

for some $(\mathcal{F}_t)$-adapted, measurable $\mathbb{R}^d$-valued process ϕ, with

$$\int_0^T |\phi_s|^2 \, ds \; < \; \infty \text{ a.s.}$$

Now the process

$$X_t = \left(E^\theta \left(\int_0^T e^{-\int_0^s r_u du} c_s ds \,|\, \mathcal{F}_t \right) + (x - E^\theta \left(C \right)) \right) \beta_t^{-1} \qquad (10.8)$$

is non-negative because $c \in \mathcal{C}(x)$. As $\beta_t = (S_t^0)^{-1} = \exp\left\{ -\int_0^t r_u du \right\}$,

$$X_t \beta_t = x + m_t - \int_0^t \beta_s c_s ds = x + \int_0^t \phi_s' dW^\theta(s) - \int_0^t \beta_s c_s ds.$$

Write $H_t = (H_t^1, \ldots, H_t^n) = e^{\int_0^t r_u du} \left(\sigma'(t) \right)^{-1} \phi_t$. From (10.3), this is a portfolio process, so

$$X_t \beta_t = x + \int_0^t \beta_s H_s' \sigma(s) dW^\theta(s) - \int_0^t \beta_s c_s ds,$$

and we see from (10.4) that X is a wealth process corresponding to $(H, c) \in SF(0, x)$.

Now if, furthermore, $c \in \mathcal{D}(x)$, then $X_T = 0$ from (10.8), so $D_T = \int_0^T \beta_s c_s ds$. We have seen that the process D is a P^θ-supermartingale and, in this situation, it has a constant expectation

$$x = E \left(D_T \right) = E \left(\int_0^T \xi_s c_s ds \right) = E \left(D_0 \right).$$

Therefore, D is a P-martingale. $\qquad \square$

The next result describes the levels of terminal wealth attainable from an initial endowment x.

Theorem 10.2.5. *a) If $B \in \mathcal{L}(x)$, there is a pair $(H, c) \in SF(0, x)$ such that the corresponding wealth process X satisfies $X_T = B$ a.s.*

b) Write $\mathcal{M}(x) = \{ B \in \mathcal{L}(x) : E^\theta (\beta_T B) = x \}$. Then, if $B \in \mathcal{M}(x)$, we can take $c \equiv 0$ and the process $(X_t \beta_t)_{0 \le t \le T}$ is a P^θ-martingale.

Proof. For $B \in \mathcal{L}(x)$, we define the non-negative process Y_t by

$$Y_t \beta_t = E^\theta \left(\overline{B} | \mathcal{F}_t \right) + \left(x - E^\theta \left(\overline{B} \right) \right) \left(1 - \frac{t}{T} \right) = x + v_t - \rho t,$$

where

$$\overline{B} = \beta_T B, \quad \rho = T^{-1} \left(x - E^\theta \left(\overline{B} \right) \right), \quad v_t = E^\theta \left(\overline{B} | \mathcal{F}_t \right) - E^\theta \left(\overline{B} \right).$$

Take the consumption rate process to be

$$c_t = \rho \beta_t^{-1},$$

and represent v_t as

$$\int_0^t \psi_s' dW^\theta(s) = \int_0^t \beta_s \widehat{H}_s' \sigma(s) dW^\theta(s),$$

where $\widehat{H}_s' = e^{\int_0^s r_u du} (\sigma'(s))^{-1} \psi_s$. The result follows as in Theorem 10.2.4. $\square$

Remark 10.2.6. Minor modifications show that Theorem 10.2.5 still holds when T is replaced by a stopping time $\tau \in \mathcal{T}_{0,T}$.

10.3 Maximising Utility of Consumption

We consider an investor with initial wealth $x > 0$. The problem discussed in this section is how the investor should choose his trading strategy $H_1(t)$ and consumption rate $c_1(t)$ in order to remain solvent and also to maximise his utility over $[0, T]$, with $(H_1, c_1) \in SF(0, x)$.

As above, prices will be discounted by $\beta_t = (S_t^0)^{-1} = \exp \left\{ - \int_0^t r_u du \right\}$. Consider a utility function U_1.

The problem then is to maximise the expected discounted utility from consumption,

$$J_1(x, H_1, c_1) = E \left(\int_0^T U_1(c_1(s)) ds \right),$$

over all strategies $(H_1, c_1) \in SF(0, x)$ satisfying

$$E \left(\int_0^T U_1^-(c_1(s)) ds \right) < \infty.$$

Write $SF_B(x)$ for the set of such strategies. Following Definition 10.2.3, we have seen that $(H_1, c_1) \in SF(0, x)$ implies $c_1 \in \mathcal{C}(x)$. Therefore,

$$E^\theta \left(\int_0^T \beta_s c_1(s) ds \right) \leq x.$$

In this situation, utility is coming only from consumption, so it is easily seen that one should increase consumption up to the limit imposed by the bound. Consequently, we should consider only consumption rate processes for which

$$E^\theta \left(\int_0^T \beta_s c_1(s) ds \right) = E \left(\int_0^T \Lambda_s \beta_s c_1(s) ds \right) = x.$$

That is, we consider $c_1 \in \mathcal{D}(x)$. In other words, if we define the value function

$$V_1(x) = \sup_{(H_1, c_1) \in SF_B(x)} J_1(x, H_1, c_1),$$

then

$$V_1(x) = \sup_{\substack{(H_1, c_1) \in SF_B(x) \\ c_1 \in \mathcal{D}(x)}} J_1(x, H_1, c_1). \tag{10.9}$$

For this constrained maximisation problem, we consider the Lagrangian

$$\Gamma(c_1, y) = E \left(\int_0^T U_1(c_1(s)) ds \right) - y \left(E \left(\int_0^T \Lambda_s \beta_s c_1(s) ds \right) - x \right).$$

The first-order conditions imply that the optimal consumption rate $c_1^*(s)$ should satisfy

$$U_1'(c_1^*(s)) = y \Lambda_s \beta_s, \qquad E \left(\int_0^T \Lambda_s \beta_s c_1^*(s) ds \right) = x. \tag{10.10}$$

Therefore, with I_1 the inverse function of the strictly decreasing map U_1',

$$c_1^*(s) = I_1(s, y \Lambda_s \beta_s),$$

and y is determined by the condition (10.10). In fact, write

$$L_1(y) = E \left(\int_0^T \Lambda_s \beta_s I_1(s, y \Lambda_s \beta_s) ds \right) \quad \text{for } 0 < y < \infty.$$

Assume that $L_1(y) < \infty$ for $0 < y < \infty$. Then, from the corresponding properties of I_1, L_1 is continuous and strictly decreasing, with

$$L_1(0+) = \infty, \qquad\qquad L_1(\infty) = 0.$$

Consequently, there is an inverse map for L_1, which we denote by G_1, so that

$$L_1(G_1(y)) = y.$$

That is, for any $x > 0$, there is a unique y such that

$$y = G_1(x).$$

Differentiating, we also see that $L_1'(G_1(y))G_1'(y) = 1$. The corresponding optimal consumption process is therefore

$$c_1^*(s) = I\left(s, G_1(x)\Lambda_s\beta_s\right) \text{ for } 0 \leq t \leq T. \tag{10.11}$$

By construction, $c_1^* \in \mathcal{D}(x)$. From Theorem 10.2.4, there is a unique portfolio process H_1^* (up to equivalence) such that $(H_1^*, c_1^*) \in SF(0, x)$. The corresponding wealth process is then X_1, where

$$\beta_t X_1(t) = E^\theta\left(\int_t^T \beta_s c^*(s)ds \,|\mathcal{F}_t\right)$$

$$= x - \int_0^t \beta_s c^*(s)ds + \int_0^t \beta_s H^*(s)'\sigma(s)dW^\theta(s).$$

Note that $X_1(t) > 0$ on $[0, T)$ and $X_1(T) = 0$ a.s.

Theorem 10.3.1. *Assume $L_1(y) < \infty$ for $0 < y < \infty$. Then, for any $x > 0$, with c_1^* given by (10.11), the pair (H_1^*, c_1^*) belongs to $SF_B(x)$ and is optimal for the problem (10.9). That is,*

$$V_1(x) = J_1(x, H_1^*, c_1^*).$$

Proof. Consider any other $c \in \mathcal{C}(x)$. From the concavity of U_1, inequality (10.4) implies that

$$U_1\left(t, c_1^*(t)\right) \geq U_1(t, c_t) + G_1(x)\Lambda_t\beta_t\left(I(t, G_1(x)\Lambda_t\beta_t) - c_t\right). \tag{10.12}$$

Write $\widehat{c}_t = x\left(E\left(\int_0^T \Lambda_u\beta_u du\right)\right)^{-1}$. Then $\widehat{c}$ is a constant rate of consumption and

$$E^\theta\left(\int_0^T \beta_u\widehat{c}_u du\right) = E\left(\Lambda_T\int_0^T \beta_u\widehat{c}_u du\right) = x,$$

so that $\widehat{c} \in \mathcal{D}(x)$. Also, substituting $\widehat{c}$ in the right-hand side of (10.12) and integrating, we obtain

$$E\left(\int_0^T U_1\left(t, \widehat{c}_t\right)dt\right) + G_1(x)\left(L_1(G_1(x)) - x\right) = E\left(\int_0^T U_1\left(t, \widehat{c}_t\right)dt\right).$$

Thus, integrating both sides of (10.12) yields $E\left(\int_0^T U_1^-\left(c^*(s)\right)ds\right) < \infty$. Finally, consider $c \in \mathcal{C}(x)$. Integrating both sides of (10.12), we have

$$
E\left(\int_0^T U_1\left(t, c_1^*(t)\right)dt\right) \geq
$$
$$
E\left(\int_0^T U_1(t, c_t)dt\right) + G_1(x)\left(x - E\left(\int_0^T \Lambda_t \beta_t c_t dt\right)\right).
$$

The final bracket equals

$$
E\left(\Lambda_T \int_0^T \beta_t c_t dt\right) = E^\theta\left(\int_0^T \beta_t c_t dt\right)
$$

and so is non-negative. Therefore, c_1^* is optimal. $\qquad\square$

Remark 10.3.2. From the optimality conditions we have seen that the optimal consumption rate $c_1^*(t)$ is of the form

$$
c_1^*(t) = I_1\left(t, y\xi_t\right) \text{ for some } y > 0.
$$

Here $\xi_t = \beta_t \Lambda_t$ is the market deflator of Definition 10.2.1. Let us consider the expected utility function associated with a consumption rate process of this form:

$$
K_1(y) = E\left(\int_0^T U_1\left(t, I_1(t, y\xi_t)\right)dt\right) \text{ for } 0 < y < \infty. \tag{10.13}
$$

We require

$$
E\left(\int_0^T |U_1\left(t, I(t, y\xi_t)\right)|\, dt\right) < \infty \text{ for all } y \in (0, \infty). \tag{10.14}
$$

Then K_1 is continuous and strictly decreasing in y. We have proved in Theorem 10.3.1 that
$$
V_1(x) = K_1\left(G_1(x)\right).
$$

Under the assumption, for example, that $U_1(t, y)$ is C^2 in $y > 0$ and $\frac{\partial^2 U_1(t,y)}{\partial y^2}$ is non-decreasing in y for all $t \in [0, T]$, we can perform the differentiations of $L_1(y)$ and $K_1(y)$ to obtain

$$
L_1'(y) = E\left(\int_0^T \xi_t^2 \frac{\partial}{\partial z} I_1\left(t, y\xi_t\right)dt\right).
$$

Recalling that

$$
\frac{\partial U_1}{\partial z}\left(t, I_1(t, z)\right) = z\frac{\partial}{\partial z}I_1(t, z),
$$

we have, with $z = y\xi_t$, that

$$K_1'(y) = E\left(\int_0^T \xi_t \frac{\partial U_1}{\partial z}(t, I_1(t, y\xi_t))\, dt\right)$$

$$= E\left(\int_0^T y\xi_t^2 \frac{\partial}{\partial z} I_1(t, y\xi_t)\, dt\right)$$

$$= yL_1'(y).$$

We can therefore state the following result.

Theorem 10.3.3. *Under the integrability conditions that $L_1(y) < \infty$ and (10.4) holds, the value function is given by*

$$V_1(x) = K_1(G_1(x)). \tag{10.15}$$

Also, if the utility function $U_1(t, y)$ is C^2 in y and $\frac{\partial^2 U}{\partial y^2}(t, y)$ is non-decreasing in y, then the strictly decreasing functions L_1 and K_1 are continuously differentiable and

$$K_1'(y) = yL_1'(y).$$

Furthermore, from (10.15),

$$V_1'(x) = K_1'(G_1(x))\, G_1'(x) = G_1(x)L_1'(G_1(x))\, G_1'(x) = G_1(x).$$

In addition, note that V_1 is strictly increasing and concave.

Example 10.3.4. Suppose $U_1(t, c) = \exp\left\{-\int_0^t \rho(u)du\right\}\log c$, where $\rho : [0, T] \to \mathbb{R}$ is measurable and bounded. Then

$$U_1'(t, c) = \exp\left\{-\int_0^t \rho(u)du\right\}c^{-1}, \quad I_1(t, c) = \exp\left\{-\int_0^t \rho(u)du\right\}c^{-1},$$

$$L_1(y) = \frac{a_1}{y}, \qquad\qquad K_1(y) = -a_1\log y + b_1,$$

so

$$V_1(x) = a_1\log\left(\frac{x}{a_1}\right) + b_1,$$

where

$$a_1 = \int_0^T \exp\left\{-\int_0^t \rho(u)du\right\}dt$$

and

$$b_1 = E\left(\int_0^T \exp\left\{-\int_0^t \rho(u)du\right\}\left(\int_0^t \left(r_u + \frac{1}{2}|\theta_u|^2 - \rho(u)\right)du\right)dt\right).$$

Example 10.3.5. Suppose $U_1(t, c) = -\exp\left\{-\int_0^t \rho(u)du\right\} c^{-1}$. Then

$$L_1(y) = d_1 y^{-\frac{1}{2}}, \qquad\qquad G_1(y) = -d_1 y^{\frac{1}{2}},$$

so

$$V_1(x) = -\frac{d_1^2}{x},$$

where

$$d_1 = E\left(\int_0^T \exp\left\{-\frac{1}{2}\int_0^t (\rho(u) + r_u)du\right\}\Lambda_t^{\frac{1}{2}}\,dt\right).$$

Note that conditions $L_1(y) < \infty$ and (10.14) are both satisfied in these examples.

10.4   Maximisation of Terminal Utility

The previous section discussed maximisation of consumption. This section considers the dual problem of maximization of terminal wealth. That is, for any $(H_2, c_2) \in SF(0, x)$, we consider

$$J_2(x, H_2, c_2) = E\left(U_2\left(X_T\right)\right)$$

for a utility function U_2.

We restrict ourselves to the subset $SF_C(0, x)$ consisting of those (H, c) such that

$$E\left(U_2^-\left(X_T\right)\right) < \infty.$$

Define the value function

$$V_2(x) = \sup_{(H_2, c_2) \in SF_C(0, x)} J_2(x, H_2, c_2). \tag{10.16}$$

The expected terminal wealth discounted to time 0 should not exceed the initial investment x; that is,

$$E^\theta\left(\beta_T X_T\right) = E\left(\xi_t X_T\right) \le x.$$

The methods are similar to those of Theorem 10.3.1, so we sketch the ideas and proofs. Define

$$L_2(y) = E\left(\xi_T I_2\left(T, y\xi_T\right)\right) \text{ for } y > 0.$$

We assume $L_2(y) < \infty$ for $y \in (0, \infty)$. Again L_2 is continuous and strictly decreasing with $L_2(0+) = \infty$ and $L_2(\infty) = 0$.

Write G_2 for the inverse function of L_2. For an initial capital x_2, consider

$$X_2(T) = I_2\left(T, G_2(x_2)\xi_T\right). \tag{10.17}$$

This belongs to the class $\mathcal{M}(x_2)$ of Theorem 10.2.5 because

$$E^\theta \left(X_2(T)\beta_T \right) = E\left(\xi_T X_2(T) \right) = E\left(\xi_T I_2\left(T, G_2(x_2)\xi_T \right) \right) = x_2.$$

Hence, by Theorem 10.3.1, there is a trading strategy $(H_2, c_2) \in SF(0, x_2)$ that attains the terminal wealth $X_2(T)$. This strategy is unique up to equivalence, and for this pair $c_2 \equiv 0$. Consequently, the corresponding wealth process is given by

$$\beta_t X_2(t) = E^\theta \left(\beta_T X_2(T) \, | \, \mathcal{F}_t \right)$$

$$= x_2 + \int_0^t \beta_s H_2'(s)\sigma(s)dW^\theta(s) \text{ for } 0 \leq t \leq T. \qquad (10.18)$$

Using again the inequality (10.4) for utility functions, we can parallel the proof of Theorem 10.3.1 to show that $X_2(T)$, defined by (10.17), satisfies

$$E\left(U_2^-\left(X_2(T) \right) \right) < \infty, \qquad E\left(U_2\left(X_2(T) \right) \right) \geq E\left(U_2\left(X_T \right) \right), \qquad (10.19)$$

where X_T is any other random variable satisfying (10.19).

Consequently, we have proved the following result.

Theorem 10.4.1. *If $L_2(y) < \infty$ for all $y \in (0, \infty)$, consider any $x_2 > 0$ and the random variable*

$$X_2(T) = I_2\left(T, G_2(x_2)\xi_T \right).$$

Then the trading strategy $(H_2, 0)$ belongs to $SF_C(0, x_2)$ and

$$V_2(x_2) = E\left(U_2\left(T, X_2(T) \right) \right).$$

That is, $(H_2, 0)$ achieves the maximum in (10.16).

Similarly to Theorem 10.3.3, we can also establish the following.

Theorem 10.4.2. *If $L_2(y) < \infty$ and if*

$$E\left(|U_2\left(I_2(T, y\xi_T) \right)| \right) < \infty$$

for all $y \in (0, \infty)$, then the value function V_2 is given by

$$V_2(x) = K_2\left(G_2(x) \right),$$

where

$$K_2(y) = E\left(U_2\left(T, I_2(T, y\xi_T) \right) \right). \qquad (10.20)$$

Note that K_2 is continuous and strictly decreasing for $0 < y < \infty$.

Also, if $U_2(t, y)$ belongs to $C^2(0, \infty)$ and $\frac{\partial^2 U(t,y)}{\partial y^2}$ is non-decreasing in y, then the functions L_2, K_2 are also in $C^2(0, \infty)$ and $K_2'(y) = yL_2'(y)$ for $0 < y < \infty$.

Furthermore,

$$V_2' = G_2,$$

implying that V_2 is strictly increasing and strictly concave.

Example 10.4.3. Again consider the utility function

$$U(T, c) = \exp\left(-\int_0^T \rho(u)du\right) \log c,$$

where ρ is bounded, real, and measurable. In this case,

$$L_2(y) = \frac{a_2}{y}, \quad G_2(y) = -a_2 \log y + d_2, \quad V_2(x) = a_2 \log\left(\frac{x}{a_2}\right) + d_2,$$

with

$$a_2 = \exp\left\{\int_0^T \rho(u)du\right\}$$

and

$$d_2 = E\left(\exp\left\{-\int_0^T \rho(u)du\right\}\left(\int_0^T \left(r_u + \frac{1}{2}|\theta_u|^2 - \rho(u)\right)du\right)\right).$$

With $\rho(u) \equiv 0$, we have

$$I_2(T, y) = L_2(y) = y^{-1}.$$

Consequently, from (10.17),

$$X_2(T) = (G_2(x_2)\xi_T)^{-1}.$$

In this example, $G_2(x_2) = x_2^{-1}$ and

$$\xi_T = \Lambda_T \beta_T, \quad \Lambda_T = \exp\left\{-\int_0^T \theta_u dW(u) - \frac{1}{2}\int_0^T |\theta_u|^2 du\right\}.$$

Then

$$\beta_T X_2(T) = x_2 \exp\left\{\int_0^T \theta_u dW(u) + \frac{1}{2}\int_0^T |\theta_u|^2 du\right\}.$$

Recalling $dW(t) = dW_t^\theta - \theta_t dt$, we have

$$\beta_T X_2(T) = x_2 \exp\left\{\int_0^T \theta_u dW_u^\theta - \frac{1}{2}\int_0^T |\theta_u|^2 du\right\}$$

and, since the right-hand side is the final value of a P^θ-martingale, (10.18) yields

$$\begin{aligned}
\beta_t X_2(t) &= E^\theta\left(\beta_T X_2(T)\,|\,\mathcal{F}_t\right) \\
&= x_2 \exp\left\{\int_0^t \theta_u dW_u^\theta - \frac{1}{2}\int_0^t |\theta_u|^2 du\right\}
\end{aligned}$$

$$= x_2 + \int_0^t \beta_u X_2(u) dW_u^\theta.$$

Comparing this with (10.18), we see

$$H_2(t) = X_2(t)\sigma'(t)^{-1}\theta_t.$$

Example 10.4.4. For the utility function

$$U_2(T, c) = -\exp\left\{-\int_0^T \rho(u) du\right\} c^{-1},$$

we can show that

$$L_2(y) = a_2 y^{-\frac{1}{2}}, \qquad G_2(y) = -a_2 y^{\frac{1}{2}}, \qquad V_2(x) = -\frac{a_2^2}{x},$$

with

$$a_2 = E\left(\exp\left\{-\frac{1}{2}\int_0^T (\rho(u) + r_u) \, du\right\} \Lambda_T^{\frac{1}{2}}\right).$$

10.5 Consumption and Terminal Wealth

We consider now an investor who wishes to both live well (consume) and also acquire terminal wealth at time $T > 0$. These two objectives conflict, so we determine the investor's best policy.

Consider two utility functions U_1 and U_2. As in Section 10.3, the investor's utility from consumption is given by

$$J_1(x, H, c) = E\left(\int_0^T U_1(c_u) \, du\right).$$

The investor's terminal utility, as in Section 10.4, is

$$J_2(x, H, c) = E\left(U_2(T, X_t)\right).$$

Write $SF_D(0, x) = SF_B(0, x) \cap SF_C(0, x)$ for the set of admissible trading and consumption strategies. Then, with

$$J(x, H, c) = J_1(x, H, c) + J_2(x, H, c),$$

the investor aims to maximise $J(x, H, c)$ over all strategies

$$(H, c) \in SF_D(0, x).$$

It turns out that the optimal policy for the investor is to split his initial endowment x into two parts, x_1 and x_2, with $x_1 + x_2 = x$, and then to use the optimal consumption strategy (H_1, c_1) of Section 10.3 with initial investment x_1 and the optimal investment strategy $(H_2, 0)$ of Section 10.4 with initial investment x_2.

Thus, consider an initial endowment x and a pair $(H, c) \in SF_D(0, x)$. Write

$$x_1 = E^\theta \left(\int_0^T \beta_u c_u du \right), \qquad\qquad x_2 = x - x_1.$$

If X_t is the wealth process for (H, c), then

$$X_t = \beta_t^{-1} \left(x - \int_0^t \beta_u c_u du + \int_0^t \beta_u H'(u) \sigma(u) dW_u^\theta \right),$$

$$J(x, H, c) = E \left(\int_0^T U_1(s, c_s) \, dt + U_2(T, X_T) \right).$$

By definition, $c \in \mathcal{D}(x_1)$ and $X_T \in \mathcal{L}(x_2)$.

Now, from Theorem 10.3.1 there is an optimal strategy $(H_1, c_1) \in SF_B(0, x_1)$ that attains the value

$$V_1(x_1) = \sup_{(H,c) \in SF_B(0,x_1)} J_1(x_1, H, c).$$

Also, from Theorem 10.4.1 there is an optimal strategy $(H_2, 0) \in SF_C(0, x_2)$ that attains the value

$$V_2(x_2) = \sup_{(H,c) \in SF_C(0,x_2)} J_2(x_2, H, c).$$

Now suppose $X_1(t)$ is the wealth process corresponding to (H_1, c_1) and $X_2(t)$ is the wealth process corresponding to $(H_2, 0)$. Then

$$X_1(t) = \beta_t^{-1} \left(x_1 - \int_0^t \beta_u c_1(u) du + \int_0^t \beta_u H_1'(u) \sigma(u) dW_u^\theta \right),$$

with $X_1(T) = 0$ and

$$X_2(t) = \beta_t^{-1} \left(x_2 + \int_0^t \beta_u H_2'(u) \sigma(u) dW_u^\theta \right).$$

Consider, therefore, the wealth process $\overline{X}$, which is the sum of X_1 and X_2 and corresponds to an investment strategy $\overline{H} = H_1 + H_2$ and consumption process $\overline{c} = c_1$. Then, with $x = x_1 + x_2$,

$$\overline{X}_t = X_1(t) + X_2(t) = \beta_t^{-1} \left(x - \int_0^t \beta_u \overline{c}_u du + \int_0^t \beta_u \overline{H}_u \sigma(u) dW_u^\theta \right).$$

However, for any initial endowment x, any decomposition of x into $x = x_1 + x_2$, and any strategy $(H, c) \in SF_D(0, x)$, we must have, because of the optimality of $V_1(x_1)$ and $V_2(x_2)$, that

$$J(x, H, c) \leq V_1(x_1) + V_2(x_2).$$

Consequently,

$$V(x) = \sup_{(H,c) \in SF_D(0,x)} J(x, H, c) \leq V^*(x) = \max_{\substack{x_1+x_2=x \\ x_1 \geq 0, x_2 \geq 0}} [V_1(x_1) + V_2(x_2)].$$

We shall show that the maximum on the right-hand side can be achieved by an appropriate choice of x_1 and x_2. For such x_1 and x_2, there are optimal strategies (H_1, c_1) and $(H_2, 0)$, so the strategy $(\overline{H}, \overline{c})$ is then optimal for the combined consumption and investment problem. In fact, the maximum on the right-hand side is found by considering

$$\gamma(x_1) = V_1(x_1) + V_2(x - x_1).$$

The critical point of γ arises when $\gamma'(x_1) = 0$; i.e., when $V_1'(x_1) = V_2'(x - x_1)$. This means we are looking for the values $x_1, x_2, x_1 + x_2 = x$ such that the marginal expected utilities from the consumption problem and terminal wealth problem are equal. From Theorems 10.3.3 and 10.4.2, $V_i' = G_i$, so this is when

$$G_1(x_1) = G_2(x_2).$$

Write z for this common value. The inverse function of G_i is $L_i, i = 1, 2$, so

$$x_1 = L_1(z), \qquad\qquad x_2 = L_2(z).$$

For any $y \in (0, \infty)$, consider the function

$$L(y) = L_1(y) + L_2(y) = E\left(\int_0^T \xi_t I_1(t, y\xi_t)dt + \xi_T I_2(T, y\xi_T)\right).$$

Here ξ is the 'deflator' of Definition 10.2.1.

Then L is continuous, strictly decreasing, and $L(0+) = \infty, L(\infty) = 0$. Write G for the inverse function of L. Then, for the optimal decomposition,

$$x = x_1 + x_2 = L_1(z) + L_2(z) = L(z), \qquad z = G(x).$$

Consequently, the optimal decomposition of the initial endowment x is given by

$$x_1 = L_1(G(x)), x_2 = L_2(G(x)).$$

Consider the function

$$K(y) = K_1(y) + K_2(y) = E\left(\int_0^T U_1(t, I_1(t, y\xi_t))\, dt + U_2(T, I_2(T, y\xi_T))\right).$$

K is continuous and decreasing on $(0, \infty)$. From (10.15) and (10.20)

$$V(x) = V^*(x) = K(G(x)).$$

Summarizing the above discussion we state the following theorem.

Theorem 10.5.1. *For an initial endowment $x > 0$, the optimal consumption rate is*

$$\overline{c} = I_1\left(t, G(x)\xi_t\right) \text{ for } 0 \leq t \leq T,$$

and the optimal terminal wealth level is

$$\overline{X}_T = I_2\left(T, G(x)\xi_T\right).$$

There is an optimal portfolio process $\overline{H}$ such that $(\overline{H}, \overline{c}) \in SF_D(0, x)$, and the corresponding wealth process $\overline{X}$ is

$$\overline{X}_t = \beta_t^{-1} E^\theta \left(\int_t^T \beta_u I_1\left(u, G(x)\xi(u)\right) du + \beta_T I_2\left(T, G(x)\xi_T\right) \mid \mathcal{F}_t \right)$$

for $0 \leq t \leq T$. Furthermore, the value function of the problem is given by

$$V(x) = K\left(G(x)\right).$$

Example 10.5.2. Suppose $U_1(t, c) = U_2(t, c) = \exp\left\{ -\int_0^t \rho(u)du \right\} \log c$. Then

$$L(y) = \frac{a}{y}, \quad K(y) = -a \log y + b, \quad V(x) = a \log\left(\frac{x}{a}\right) + b \text{ for } 0 < x < \infty.$$

Here $a = a_1 + a_2$, $b = b_1 + b_2$, where a_1, b_1 (resp., a_2, b_2) are given in Example 10.3.4 (resp. Example 10.4.3).

Example 10.5.3. Suppose $U_1(t, c) = U_2(t, c) = -\frac{1}{c} \exp\left\{ -\int_0^t \rho(u)du \right\}$. Then

$$L(y) = ay^{-\frac{1}{2}}, \qquad K(y) = -ay^{-\frac{1}{2}}, \qquad V(x) = -\frac{a^2}{x},$$

where $a = a_1 + a_2$ with a_1 as in Example 10.3.5 and a_2 as in Example 10.4.4.

Remark 10.5.4. In the case when the coefficients r, μ_i, and $\sigma = (\sigma_{ij})$ in the dynamics (10.1), (10.2) are constant, more explicit closed form solutions for the optimal strategies, in terms of feedback strategies as functions of the current level of wealth, can be obtained.

The solution of the dynamic programming equation can be obtained in terms of a function that is the value function of a European put option. Details can be found in [186] through [189].

Chapter 11

Measures of Risk

Trading in assets whose future outcomes are uncertain necessarily involves risk for the investor. The management of such risk is of fundamental concern for the operation of financial markets. For example:

- *Financial regulators* seek to minimise the occurrence and impact of the collapse of financial institutions by placing restrictions on the types and sizes of permitted trades, such as limits on short sales;

- *Risk managers* in investment firms place restrictions on the activities of individual traders, seeking to avoid levels of exposure that the firm may not be able to meet in extreme circumstances;

- *Individual investors* seek to diversity their holdings, so as to avoid undue exposure to sudden moves in particular stocks or sectors of the market.

The mathematical analysis of measures of risk has also been a principal concern of the actuarial and insurance professions since their inception. Equally, it plays a fundamental role in the theory of portfolio selection (which is not covered in this book - see, for example, [217],[36]).

At its simplest, the standard deviation σ_K of the return K on a risky investment provides a measure of the deviation of the values of K from their mean $E(K)$. We saw in Chapters 1 and 7 that in the binomial and Black-Scholes pricing models, a European call option C on a stock S satisfies $\sigma_C \geq \sigma_S$ for the standard deviations of the return on the option and stock, respectively, and the same inequality holds for their excess mean returns. We interpreted this as indicating that the option is inherently riskier than the stock, although potentially more profitable.

In portfolio selection, the objective is to find a portfolio that maximises expected return while minimising risk; i.e., given portfolios V_1 and V_2 with mean returns μ_1, μ_2 and standard deviations σ_1, σ_2, respectively, it is assumed that investors will prefer V_1 to V_2 provided that $\mu_1 \geq \mu_2$ and

$\sigma_1 \le \sigma_2$. V_1 is said to *dominate* V_2 in this event. An *efficient* portfolio is one that is not dominated by any other, and the set of these (among all attainable portfolios) is the *efficient frontier*. Elementary properties of the variance show that, in the absence of short sales, when (positive) fractions of the investor's wealth are placed in a portfolio comprising two stocks, the variance of the return on this portfolio will be no greater than the larger of the variances of the return on the individual stocks. This simple result is easily generalised to general portfolios and underlies the claim that 'diversification reduces risk', which lies at the heart of the Capital Asset Pricing Model (CAPM) - see [36] for an elementary account. It is reasonable to expect more sophisticated measures of risk to retain this property, and this informs many of the more recent developments that seek to provide an axiomatic basis for measures of risk.

Variance is symmetric, while in risk management one is primarily concerned with containing the *downside risk* (i.e., to place bounds on the amount of potential loss, or the amount by which the final position may fall short of an expected return). This leads to the definition of measures of risk that focus on the lower tail of the distribution of the random variable representing the final position. Currently the most widely used measure of exposure in risk management is Value at Risk, usually abbreviated to VaR. Value at Risk was developed and adopted in response to the financial disasters, such as those at Baring's Bank, Orange County, and Metallgesellschaft, of the early 1990s.

We shall give a precise definition of VaR and show that there are possible problems with this measure of risk. Continuing to work in a single-period framework, we then introduce the definition of coherent risk measure proposed by Artzner et al. [9], which leads to possible refinements of VaR.

11.1   Value at Risk

A standard treatment of VaR can be found in the book by Jorion [180]. It is noted that risk management has undergone a revolution since the mid-1990s, generated largely by the use of VaR. In fact, VaR has become the standard benchmark for measuring financial risks. JP Morgan has developed *Risk Metrics*^{TM} based on VaR.

In practice, given sufficient data, VaR is easy to apply. The idea is to determine the level of exposure in a position (portfolio) that we can be 'reasonably sure' will not be exceeded. For example, suppose one knows the monthly returns on US Treasury notes over a certain time period - some returns will be positive, others negative. A confidence level of (say) 95% is chosen. One then wishes to determine the loss that will not be exceeded in 95% of the cases, or, put another way, so that only 5% of the returns are lower than this level.

That level of return can be determined from the data. Suppose, for example, it is a return of -2.25%. If an investor holds \$100 million of such

Treasury notes, based on previous data he or she can be 95% sure that the portfolio will not fall by more than 2.25% of its holdings (i.e., by more than $2.25 million) over the next month.

Clearly, the confidence level of 95% could be changed, as could the time period of one month.

The idea behind VaR is therefore that some threshold probability level α (say 5%) is given. If the random variable representing some position, which may suffer a possible loss, is denoted by X, then there is a smallest x such that $P(X > x) < \alpha$. Here x represents an 'acceptable' level of loss. To make this more precise, we first have the following definition.

Definition 11.1.1. Suppose X is a real random variable defined on a probability space $(\Omega, \mathcal{F}, P)$ and $\alpha \in [0, 1]$. The number q is an α-*quantile* if

$$P(X < q) \leq \alpha \leq P(X \leq q).$$

The largest α-quantile is

$$q^{\alpha}(X) = \inf\{x : P(X \leq x) > \alpha\}. \tag{11.1}$$

The smallest α-quantile is

$$q_{\alpha}(X) = \inf\{x : P(X \leq x) \geq \alpha\}. \tag{11.2}$$

Note that $q_{\alpha} \leq q^{\alpha}$. Moreover, q is an α-quantile if and only if $q_{\alpha} \leq q \leq q^{\alpha}$.

It is helpful to describe $q^{\alpha}(X)$ in terms of the distribution $F_X(x) = P(X \leq x)$ of X. As a function of α, $q^{\alpha}(X)$ is the right-continuous inverse of F_X; i.e.,

$$q^{\alpha}(X) = \inf\{x \in \mathbb{R} : F_X(x) > \alpha\}. \tag{11.3}$$

The function $q(\alpha) = q^{\alpha}(X)$ is then increasing and right-continuous in the variable α on $(0, 1)$ and satisfies the inequalities

$$F_X(q(\alpha)-) \leq \alpha \leq F_X(q(\alpha)), \qquad q(F_X(x)-) \leq x \leq q(F_X(x)), \tag{11.4}$$

where $g(s+) = \lim_{t \downarrow s} g(t)$ and $g(s-) = \lim_{t \uparrow s} g(t)$ for any real function g. We also have

$$F_X(x) = \inf\{\alpha \in (0, 1) : q(\alpha) > x\}. \tag{11.5}$$

These results are elementary, and the proofs are left to the reader. (See, e.g. , [132, Lemma 2.72].)

Note that Figure 11.1 illustrates clearly that $q_{\alpha} = q^{\alpha}$ unless the distribution function F_X has a 'flat' piece, and then the set $J_{\alpha} = \{x : F_X(x) = \alpha\}$ is a non-trivial left-closed interval with endpoints q_{α} and q^{α}. In that case,

$$J_{\alpha} = [q_{\alpha}, q^{\alpha}] \text{ if } P(X = q^{\alpha}) = 0, \qquad J_{\alpha} = [q_{\alpha}, q^{\alpha}) \text{ if } P(X = q^{\alpha}) > 0.$$

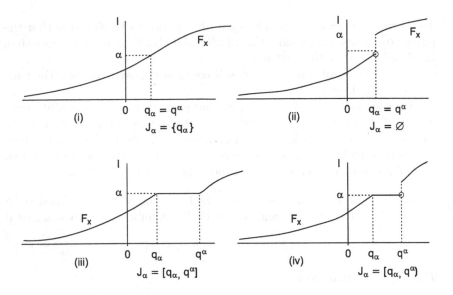

Figure 11.1

It is easily seen from Figure 11.1 that $q^\alpha(X) = \sup\{x : P(X < x) \leq \alpha\}$. It follows that for any X

$$
\begin{aligned}
q_{1-\alpha}(-X) &= \inf\{x : P(-X \leq x) \geq 1 - \alpha\} \\
&= \inf\{x : 1 - P(X < -x) \geq 1 - \alpha\} \\
&= \inf\{x : P(X < -x) \leq \alpha\} \\
&= -\sup\{y : P(X < y) \leq \alpha\} \\
&= -q^\alpha(X).
\end{aligned}
\tag{11.6}
$$

We are now ready to define VaR as follows.

Definition 11.1.2. Given a position described by the random variable X and a number $\alpha \in [0, 1]$, define

$$VaR_\alpha(X) = -q^\alpha(X) = q_{1-\alpha}(-X).$$

X is then said to be VaR_α-*acceptable* if

$$q^\alpha(X) \geq 0 \text{ or, equivalently, } VaR_\alpha(X) \leq 0.$$

The choice of q^α instead of q_α is somewhat arbitrary, and the discussion above shows that it only yields different results when the distribution F_X is 'flat' at α, so that J_α is a non-trivial interval. However, this occurs frequently in practical situations: for example, with discrete probability distributions. The significance of our choice will become clearer when we discuss 'expected shortfall', which is also known as 'conditional value at risk' and is prominent among the candidate risk measures proposed in recent years as potential replacements for VaR.

VaR can be considered as the amount of extra capital a firm needs to reduce to α the probability of bankruptcy, or the extra capital needing to be added (as a risk-free investment) to a given position to make an investing agency's financial exposure acceptable to an external regulator. A negative VaR implies that the firm could return some of its capital to shareholders or that it (or the investing agency) could accept more risk. Writing m instead of x in the third line of equations (11.6), we can express this by

$$VaR_\alpha(X) = \inf\left\{m \in \mathbb{R} : P(X + m < 0) \leq \alpha\right\}. \qquad (11.7)$$

This formulation provides an immediate proof of the following result.

Lemma 11.1.3. *VaR has the following properties:*

  (i) if $X \geq 0$, then $VaR_\alpha(X) \leq 0$;

  (ii) if $X \geq Y$, then $VaR_\alpha(X) \leq VaR_\alpha(Y)$;

  (iii) if $\lambda \geq 0$, $VaR_\alpha(\lambda X) = \lambda VaR_\alpha(X)$;

  (iv) $VaR_\alpha(X + k) = VaR_\alpha(X) - k$ for any real number k.

Note that (iv) implies that

$$VaR_\alpha(X + VaR_\alpha(X)) = 0. \qquad (11.8)$$

Thus we can interpret VaR as the minimum amount that will ensure that the probability that the absolute loss that could be suffered will be no more than this amount is at least $1 - \alpha$.

Remark 11.1.4. We observe that the properties (ii) and (iv), which are similar to those considered in an axiomatic context below, already suffice to make VaR Lipschitz-continuous with respect to the L^∞-norm. To see this, let X and Y be bounded random variables, and note that $X = Y + (X - Y) \leq Y + \|X - Y\|_\infty$ a.s. By properties (ii) and (iv), this yields, for any α, that

$$VaR_\alpha(X) \geq VaR_\alpha(Y + \|X - Y\|_\infty) = VaR_\alpha(Y) - \|X - Y\|_\infty,$$

so that

$$VaR_\alpha(Y) - VaR_\alpha(X) \leq \|X - Y\|_\infty.$$

Reversing the roles of X and Y, we also obtain

$$VaR_\alpha(X) - VaR_\alpha(Y) \leq \|Y - X\|_\infty = \|X - Y\|_\infty.$$

Therefore

$$|VaR_\alpha(Y) - VaR_\alpha(X)| \leq \|X - Y\|_\infty. \qquad (11.9)$$

However, a serious problem with VaR is that it is not subadditive, as the following simple example shows.

Example 11.1.5. Suppose a bank loans $100,000 to a company that will default on the loan with probability 0.008 (i.e., 0.8%). We are supposing the company either defaults on the whole amount or not at all. Writing X for the default amount, we have that $X = -\$100,000$ with probability 0.8%, and otherwise $X = \$0$. Therefore, with $\alpha = 0.01$ we see that $VaR_\alpha(X) \leq 0$. Suppose now that the bank makes two loans each of $50,000 to two different companies, each of which may default with probability 0.8%. Suppose the probabilities of default are independent. Then, with $\alpha = 0.01$, the VaR_α for the bank's diversified position is $50,000. While the probability of both companies defaulting remains below $\alpha = 0.01$, the probability of at least one default of $50000 is $0.016 > \alpha$.

Diversification is usually thought to reduce risk. However in this example it increases VaR. Moreover, as the next example, taken from [9], shows, VaR is also ineffective in recognising the dangers of concentrating credit risk.

Example 11.1.6. Consider the issue of corporate bonds in a market with zero base rate, all corporate bond spreads equal to 2%, and default by any company set at 1%. At a 5% quantile, VaR for a loan of $1,000,000 invested in bonds with a single company is $-\$20,000$; thus this measure indicates that there is no risk. On the other hand, suppose instead that the loan is placed in bonds issued independently by 100 companies at $10,000 each. The probability that two companies will default is $\binom{100}{2}(.01)^2(.99)^{98}$, which is approximately 0.184865, so the probability of at least 2 defaults is certainly greater than 0.18. Hence a positive VaR results at the 5% level; i.e., again diversification has increased risk as measured by VaR.

Finally, VaR does not give us any indication of the *severity* of the economic consequences of exposure to the rare events that it excludes from consideration. Consequently, in spite of its widespread use and its adoption by the Basel committee (see [9]), there are good reasons for rejecting VaR as an adequate measure of risk.

11.2   Coherent Risk Measures

The examples above show that, although it is widely used in practice as a management tool, there are problems with VaR: the VaR of a diversified position can be greater than the VaR of the original position; if a large loss occurs, VaR does not measure the actual size of the loss; and, because VaR is a single number, VaR does not indicate which item in a portfolio is responsible for the largest risk exposure.

In this section, we shall define and discuss *coherent* measures of risk. These have been introduced by Artzner et al. [9]. This paper discusses why such measures should have the properties stated in the definition given below. Here we concentrate on their mathematical properties. Our discussion is largely based on the notes by Delbaen [74] and the paper of Nakano [235].

We work on a probability space $(\Omega, \mathcal{F}, P)$. Our time parameter t takes values 0 (now) and 1, which may represent tomorrow or some date next month or next year. We thus restrict attention to a single-period model, where Ω represents the possible states at time $t = 1$ of our economic model.

As before, write $L^\infty = L^\infty(\Omega)$ for the space of essentially bounded real-valued functions on Ω, and $L^1 = L^1(\Omega)$. We again denote by L^1_+ the cone of non-negative functions in L^1. Although risk measures can be defined more generally on the space L^0 of all real-valued random variables on Ω, we choose to restrict attention to L^1, which is large enough for interesting applications and remains more tractable mathematically.

Definition 11.2.1. A coherent risk measure is a function $\rho : L^1 \to \mathbb{R}$ such that
    (i) if $X \geq 0$, then $\rho(X) \leq 0$;
    (ii) if $k \in \mathbb{R}$, then $\rho(X + k) = \rho(X) - k$;
    (iii) if $\lambda \geq 0$ in $\mathbb{R}$, then $\rho(\lambda X) = \lambda \rho(X)$;
    (iv) $\rho(X + Y) \leq \rho(X) + (Y)$.

Remark 11.2.2. In [9], the above definition is stated in terms of the actual final value of the position X at time 1, whereas our definition follows the more recent literature in assuming that X represents the discounted value of the position, or, alternatively, sets the discount rate to 0. This simplifies the formulation without loss of generality: working with a discount rate β and final position X', so that $X = \beta X'$ is the discounted value, one can express a risk measure ρ' in terms of X' by modifying (ii) to $\rho'(X' + \beta^{-1}m) = \rho'(X') - m$. The remaining axioms remain unchanged. Conversely, given such a risk measure ρ' defined on the set of undiscounted positions X', a coherent risk measure defined on discounted values is given by $\rho(X) = \rho'(\beta^{-1}X) = \rho'(X')$. Thus we shall assume throughout that X represents the discounted values.

Note that, while VaR satisfies properties (i)-(iii) (see Lemma 11.1.3), it fails to have the subadditivity property (iv), as the earlier examples illustrate.

It is easy to see that, in the presence of (iii), the subadditivity property (iv) is equivalent to convexity: let X, Y and $0 \leq \lambda \leq 1$ be given and note that, if a risk measure ρ satisfies (iii) and (iv), then

$$\rho(\lambda X + (1 - \lambda)Y) \leq \rho(\lambda X) + \rho((1 - \lambda)Y) = \lambda \rho(X) + (1 - \lambda)\rho(Y)$$

so that ρ is convex. Conversely, still assuming that (iii) holds, if ρ is convex, then for any X, Y

$$\rho(X + Y) = 2\rho\left(\frac{1}{2}(X + Y)\right) \leq 2\left(\frac{1}{2}\rho(X) + \frac{1}{2}\rho(Y)\right) = \rho(X) + \rho(Y),$$

so that ρ has the subadditivity property (iv).

Convexity provides a more general statement that diversification of the investor's portfolio does not increase risk, while the subadditivity property is important for risk managers in banks, as it ensures that setting risk limits independently for different trading desks (i.e., risk allocation) will not lead to a greater overall risk for the bank.

Convex risk measures (for which the property (iii) is typically *not* assumed, thus allowing risk to grow *non-linearly* as the position increases) were introduced by Foellmer and Schied and are studied extensively in [132]. However, we shall not pursue this and restrict our analysis to coherent risk measures.

Following Nakano, [235] we consider coherent risk measures that are lower semi-continuous in the L^1-norm; i.e., given $X \in L^1$ and $\varepsilon > 0$, we have

$$\rho(Y) > \rho(X) - \varepsilon \text{ when } \|X - Y\|_1 < \varepsilon.$$

Equivalently,

$$\lim_{Y \to X} \inf \rho(Y) \geq \rho(X). \tag{11.10}$$

In particular, (11.10) holds if the sequence (X_n) converges to X in L^1-norm.

Remark 11.2.3. In [74], coherent risk measures are initially defined on L^∞. Lower semi-continuity with respect to the topology of convergence in probability is assumed in this context and is then referred to as the *Fatou property*.

Lemma 11.2.4. *Let ρ be a coherent risk measure. Then*

(i) *if $a \leq X \leq b$, then $-b \leq \rho(X) \leq -a$;*

(i) *$\rho(X + \rho(X)) = 0$.*

Proof. As the random variable $X - a \geq 0$, $\rho(X-a) \leq 0$ by (i) and $\rho(X-a) = \rho(X) + a$ by (ii). Hence $\rho(X) \leq -a$. Taking $X = 0$ and $\lambda = 0$ in (iii) yields $\rho(0) = 0$. Taking $X = 0$ in (ii), we obtain $\rho(k) = -k$. As $X \leq b$, $b - X \geq 0$, so $\rho(-X + b) = \rho(-X) - b \leq 0$ using (ii) and (i). Therefore, $\rho(-X) \leq b$. Now $\rho(X - X) = \rho(0) = 0 \leq \rho(X) + \rho(-X)$ by (iv), so that $-\rho(X) \leq \rho(-X) \leq b$, giving $\rho(X) \geq -b$.

Taking $k = \rho(X)$, the second assertion follows immediately from (ii). $\square$

Example 11.2.5. Suppose that the probability space $(\Omega, \mathcal{F}, P)$ is equipped with a family $\mathcal{P}$ of probability measures, each of which is absolutely continuous with respect to P. Write

$$\rho_{\mathcal{P}}(X) = \sup \{E_Q(-X) : Q \in \mathcal{P}\}. \tag{11.11}$$

Then $\rho_{\mathcal{P}}$ is a coherent risk measure.

Exercise 11.2.6. Prove this assertion.

This example is fundamental. We shall show in Theorem 11.2.19, that under quite mild assumptions every coherent risk measure has this form. We give two examples of such risk measures for extreme choices of the family $\mathcal{P}$ that show that the choice of $\mathcal{P}$ needs to be made with some care in order to obtain 'sensible' risk measures: the family $\mathcal{P}$ should be neither too big nor too small.

Example 11.2.7. Suppose that $\mathcal{P} = \{P\}$. Then $\rho_{\mathcal{P}}(X) = E_P(-X)$. Thus a portfolio or position X is acceptable under this risk measure if and only if $E_P(X) \geq 0$.

This risk measure is too tolerant. It makes insufficient demand on the probability that the position X is positive.

Example 11.2.8. Suppose now that $\mathcal{P}$ is the set of *all* probability measures on $(\Omega, \mathcal{F})$ that are absolutely continuous with respect to P. In this case, we simply have $\sup\{E_Q(X) : Q \in \mathcal{P}\} = ess\sup X$, so that $\rho_{\mathcal{P}}(X) \leq 0$ if and only if $X \geq 0$ a.s. (P).

For this choice of $\mathcal{P}$, a position is acceptable if and only if it is almost surely non-negative. This risk measure is too strict. We thus seek families $\mathcal{P}$ that avoid these two extremes. Restrictions on the Radon-Nikodym derivatives $\frac{dQ}{dP}$ will ensure this.

Notation 11.2.9. Given the probability space $(\Omega, \mathcal{F}, P)$ and $k \in \mathbb{N}$, write

$$\mathcal{P}_k = \left\{ Q : Q \text{ is a probability measure, } Q \ll P \text{ and } \frac{dQ}{dP} \leq k \right\}. \quad (11.12)$$

Note that as Q is a probability measure, we must have $\frac{dQ}{dP} \geq 0$ a.s. (P). Moreover, we have the following.

Exercise 11.2.10. Show that if Q is a probability measure and $\frac{dQ}{dP} \leq 1$ a.s., then $Q = P$.

Consequently, we shall assume that $k > 1$. The following important result shows that when the distribution of the integrable random variable X does not have a jump at $q^{\alpha}(X)$, the family $\mathcal{P}_k$ provides us with a coherent risk measure that dominates VaR.

Theorem 11.2.11. *Suppose $X \in L^1$ and X has a continuous distribution function F_X. For $k > 1$, write $\alpha = \frac{1}{k}$. Then*

$$\rho_{\mathcal{P}_k}(X) = E_P(-X \mid X \leq q^{\alpha}(X)) \geq -q^{\alpha}(X) = VaR_{\alpha}(X).$$

Proof. As F_X is continuous, (11.4) shows that

$$P[X \leq q^{\alpha}(X)] = F_X(q^{\alpha}(X)) = \alpha = \frac{1}{k}.$$

Write $A = \{X \leq q^{\alpha}(X)\}$ and consider the measure Q_{α} defined by $\frac{dQ_{\alpha}}{dP} = k1_A$. Then $Q_{\alpha} \in \mathcal{P}_k$ and

$$E_{Q_{\alpha}}(-X) = E_P\left(-X\left(\frac{1}{\alpha}1_A\right)\right)$$

$$= \frac{1}{P(A)} E_P\left(-X \mathbf{1}_A\right)$$

$$= E_P\left(-X \mid A\right) \geq -q^\alpha(X) = VaR_\alpha(X).$$

Consider an arbitrary $Q \in \mathcal{P}_k$. Since $\frac{dQ}{dP} \leq k$, $A = \{X \leq q^\alpha(X)\}$ and $k = \frac{1}{P(A)}$, we obtain

$$E_Q\left(-X\right) = \int_A (-X)\frac{dQ}{dP}dP + \int_{A^c}(-X)\frac{dQ}{dP}dP$$

$$= k\int_A (-X)dP + \int_A (-X)\left(\frac{dQ}{dP} - k\right)dP + \int_{A^c}(-X)\frac{dQ}{dP}dP$$

$$\leq k\int_A (-X)dP - q^\alpha(X)\int_A \left(\frac{dQ}{dP} - k\right)dP + (-q^\alpha(X))\int_{A^c}\frac{dQ}{dP}dP$$

$$= k\int_A (-X)dP - q^\alpha(X)[Q(A) - kP(A) + Q(A^c)]$$

$$= k\int_A (-X)dP = E_{Q_\alpha}\left(-X\right).$$

□

Remark 11.2.12. However, it was shown in [4] that for general distributions the quantity $E_P(-X \mid A)$ does not define a subadditive function of X. Hence the risk measure so defined, which is known as the *tail conditional expectation at level* α and is sometimes written as $TCE^\alpha(X)$, can in particular circumstances suffer the same shortcomings as VaR.

Nonetheless, $TCE^\alpha(X)$ has been proposed in the literature as a possible improvement upon VaR. To illustrate some of its advantages, we have the following example, which is taken from [74].

Example 11.2.13. A bank has 150 clients, labelled $C_1, C_2, \ldots, C_{150}$. Write D_i for the random variable, which equals 1 if client i defaults on a loan and equals 0 if client i does not default. Suppose the bank lends \$1000 to each client $C_1, C_2, \ldots, C_{150}$. Initially we suppose that all the defaults are independent and that $P(D_i = 1) = 1.2\%$. The number $\Sigma_{i=1}^{150} D_i$ thus represents the total number of defaults, and the bank's total loss is therefore $1000(\Sigma_{i=1}^{150} D_i)$ dollars.

Now $D = \Sigma_{i=1}^{150} D_i$ has a binomial distribution and

$$P(D = k) = \frac{150!}{k!(150 - k)!}(0.012)^k (0.988)^{150-k}.$$

If we take $\alpha = 1\%$, it can be shown that $VaR_\alpha(D) = 5$ and $E\left(D \mid D \geq 5\right) = 6.287$.

Suppose, however, that the defaults are dependent. This can be modelled by introducing a probability Q, where $\frac{dQ}{dP} ce^{\varepsilon D^2}$. Here D and P are as above, $\varepsilon > 0$, and c is a normalising constant chosen so that Q is a probability measure. Then $Q[D_i = 1]$ increases as ε increases. Choosing $\varepsilon = 0.03029314$, $\alpha = 1\%$, and $p = 0.01$ (recalling that D is binomial), we obtain $Q[D_i = 1] = 1.2\%$ and $VaR_\alpha(D) = 6$, but $E(D|D \geq 6) = 14.5$.

Consequently, VaR does not distinguish between the two cases, while the tail conditional expectation $E(D|D \geq VaR_\alpha(D))$ distinguishes clearly between them.

The probability Q can model the situation where, if a number of clients default, there is a higher conditional probability that other clients will also default. We note that VaR is only a quantile and thus does not provide information about the size of the potential losses, whereas the tail conditional expectation is an average of all the worse cases and so provides better information about the tail distribution of the losses. It is, however, more difficult to calculate in many practical examples. The amendment required to rescue the proof of Theorem 11.2.11 is as follows (compare the definition of $CVaR$ Section 11.3).

Corollary 11.2.14. *If the distribution of X has a discontinuity at q_α, the proof of Theorem 11.2.11 applies with the modification*

$$\frac{dQ_\alpha}{dP} = k\mathbf{1}_{\{\mathbf{X}<\mathbf{q}_\alpha\}} + \beta\mathbf{1}_{\{\mathbf{X}=\mathbf{q}_\alpha\}},$$

where $k = \frac{1}{P(X<q_\alpha)}$ and $\beta = \frac{1}{P(X=q_\alpha)}$.

Remark 11.2.15. The complications introduced by the presence of jumps in the distribution function F_X have led to a proliferation of proposals for risk measures that dominate VaR. We shall not examine them further but refer the reader to [4] for a clear account of their main features. For our purposes, it suffices to note that if F_X is continuous, then Theorem 11.2.11 shows that the tail conditional expectation $TCE^\alpha(X)$ coincides with the so-called *worst conditional expectation* $WCE_\alpha(X)$, which can equivalently be defined as $-\inf\{E_Q(-X) : P(A) > \alpha\}$. Moreover, in this case these measures are the same as the so-called *conditional value at risk*, $CVaR_\alpha(X)$, although this description is really a misnomer since in the general case this quantity, which has also become known as *expected shortfall*, cannot be expressed as a conditional expectation of a quantity defined solely in terms of X. We shall return briefly to a consideration of the properties of expected shortfall in the next section.

Polar Sets and the Bipolar Theorem

In this brief subsection, we introduce definitions and results from functional analysis, which we state without proof. More details and proofs of the quoted results can be found in standard texts such as [264], [97].

Recall that the dual E^* of a real Banach space E is the vector space of all continuous linear functionals $f : E \to \mathbb{R}$ on E, and that E^* is a Banach space under the norm $|f| = \sup\{|f(x)| : |x| \leq 1\}$. We shall need to consider various topologies on E and E^*. The weak* topology on E^* is the locally convex topology induced by the family of seminorms $\mathcal{S} = \{p_x : x \in E\}$, where $p_x(f) = |f(x)|$ for $f \in E^*$. Thus the sets

$$\{\{f : p_x(f - f_0) < \varepsilon\} : p_x \in \mathcal{S}, f_0 \in E^*, \varepsilon > 0\}$$

form a subbase for the weak* topology $\sigma(E^*, E)$ on E^*. It is traditional to write x^* for elements of E^*, and we do so below.

Our first result is commonly known as the Krein-Smulian theorem.

Theorem 11.2.16. *Suppose E is a Banach space with dual space E^*. A convex set $S \subset E^*$ is weak*-closed if and only if for each $n \in \mathbb{N}$ its intersection with the closed ball $B_n = \{e^* : \|e^*\| \le n\}$ is weak*-closed; i.e., each set $S_n = S \cap B_n$ is weak*-closed.*

We shall also need the Bipolar theorem, which describes the closed convex balanced hull (see below) of a set $A \subset E$ in terms of the dual E^* of E. First, define the *polar* of $A \subset E$ by $A^\circ = \{x^* \in E^* : |x^*(a)| \le 1 \forall a \in A\}$. This set is convex (i.e., closed under convex combinations) and balanced (i.e., if $x^* \in A^\circ$, $|\lambda| \le 1$ then $\lambda x \in A^\circ$).

Note in particular that when A is itself closed under multiplication by positive scalars (e.g., when A is a cone), then the polar cone A° may equivalently be defined as $\{x^* \in E^* : x^*(a) \ge 0 \forall a \in A\}$. The operation may equally be applied to A° to define the bipolar $A^{\circ\circ} = (A^\circ)^\circ$. The *Bipolar theorem* then states the following.

Theorem 11.2.17. *In any locally convex space E, the bipolar of a set $A \subset E$ is its closed convex balanced hull (i.e., the smallest set with these properties containing A).*

This is a consequence of the Hahn-Banach theorem. For the dual pair (L^1, L^∞), we note in particular that if $A \subset L^1$ is a closed convex cone and $Z \in L^1 \setminus A$, then we can find $Y \in L^\infty$ such that $E(ZY) < 0$ and $E(XY) \ge 0$ for all X in A. But then the polar A° of A is the set $\{Y \in L^\infty : E(XY) \ge 0$ for $X \in A\}$ so that Z cannot be in the polar of A°. Since trivially $A \subset A^{\circ\circ}$, it follows that $A = A^{\circ\circ}$.

Definition 11.2.18. Let $(\Omega, \mathcal{F}, P)$ be a probability space and $\rho : L^1(\Omega) \to \mathbb{R}$ a coherent risk measure. Write $\mathcal{A} = \{X \in L^1(\Omega) : \rho(X) \le 0\}$. We call $\mathcal{A}$ the set of *acceptable* positions, or the *acceptance set* for ρ.

Note that because ρ is subadditive and positive homogeneous, $\mathcal{A}$ is a convex cone.

Representation of Coherent Risk Measures

We now have the following result, specifying conditions under which we can represent every coherent risk measure in the form given in Example 11.2.5. Write $\mathcal{Q}$ for the set of all probability measures on $(\Omega, \mathcal{F})$ that are absolutely continuous with respect to P. Write $Z_Q = \frac{dQ}{dP}$ for $Q \in \mathcal{Q}$.

Theorem 11.2.19. *Suppose $\rho : L^1 \to \mathbb{R}$. The following are equivalent.*

(i) The function ρ is a lower semi-continuous coherent risk measure.

(ii) There is a subset $\tilde{Q}$ of Q such that $\left\{ Z_Q : Q \in \tilde{Q} \right\}$ is a weak*-closed convex subset of L^∞ and for $X \in L^1$

$$\rho(X) = \sup_{Q \in \tilde{Q}} E_Q(-X). \tag{11.13}$$

Proof. That the second statement implies the first is immediate. For the converse, write $\phi(X) = -\rho(X)$ and recall that $\mathcal{A} = \left\{ X \in L^1 : \rho(X) \le 0 \right\} = \left\{ X \in L^1 : \phi(X) \ge 0 \right\}$ is the set of acceptable positions. Then $\mathcal{A}$ is clearly a convex cone. As ϕ is upper semi-continuous, the set $\mathcal{A}$ is also closed in the L^1-norm. To see this, let (X_n) be a sequence in L^1 with $\|X_n - X\|_1 \to 0$. By lower semi-continuity, $\rho(X) = \rho(\lim_n X_n) \le \liminf_n \rho(X_n) \le 0$, so that $X \in \mathcal{A}$. Applying the comments following the Krein-Smulian theorem to the cone $\mathcal{A}$, we see that

$$\mathcal{A}^\circ = \{Y \in L^\infty : E(XY) \ge 0 \text{ for all } X \in \mathcal{A}\}.$$

Thus $\mathcal{A}^\circ$ is a weak*-closed convex cone in L^∞, and, writing

$$\mathcal{C} = \{Y \in \mathcal{A}^\circ : E(Y) = 1\},$$

it follows that $\mathcal{A}^\circ = \cup_{\lambda \ge 0} \lambda \mathcal{C}$. In fact, if $A \in \mathcal{A}^\circ$ and $E(Y) > 0$, then $Y = \lambda \tilde{Y}$ for $\tilde{Y} = \frac{Y}{E(Y)} \in \mathcal{C}$, and $\lambda = E(Y)$. Further, we have $L^1_+ \subset \mathcal{A}$ since all indicator functions 1_A $(A \in \mathcal{F})$ belong to L^1_+, so that if $E(Y 1_A) \ge 0$ for all $A \in \mathcal{F}$, then $Y \ge 0$ a.s. Hence, if $Y \in \mathcal{A}^\circ$ and $E(Y) = 0$, then $Y = 0$ a.s.

The bipolar theorem now implies that

$$\mathcal{A} = \left\{ X \in L^1 : E(XY) \ge 0 \text{ for all } Y \in \mathcal{C} \right\}.$$

Consequently, $\phi(X) \ge 0$ if and only if $E(XY) \ge 0$ for all $Y \in \mathcal{C}$.

Now $\phi(X - \phi(X)) = 0$, so $E(X - \phi(X)Y) \ge 0$ for all $Y \in \mathcal{C}$. This implies that $\inf_{Y \in \mathcal{C}} E(XY) \ge \phi(X)$.

For any $\varepsilon > 0$, we have $\phi(X - \phi(X)) - \varepsilon < 0$, so there is a Y in $\mathcal{C}$ such that $E(X - \phi(X) - \varepsilon) < 0$, or $E(XY) \le \phi(X) + \varepsilon$. But ε is arbitrary, so $\inf_{Y \in \mathcal{C}} E(XY) \le \phi(X)$. Hence they are equal.

If we write

$$\tilde{Q} = \{Q \in Q : Z_Q = Y \text{ for some } Y \in \mathcal{C}\},$$

then the identity

$$\phi(X) = \inf_{Y \in \mathcal{C}} E(XY)$$

implies that $\tilde{Q}$ is a weak*-closed subset of L^∞. But $\mathcal{C} = \left\{ Z_Q : Q \in \tilde{Q} \right\}$, so this is the required representation for ρ. $\qquad\square$

11.3   Deviation Measures

An alternative approach to risk measures has been proposed in [246], [293]. This is based on the concept of a *deviation measure* and is related to generalisations of standard deviation or variance. We give an axiomatic description and derive the most basic properties, while briefly relating deviation measures to coherent risk measures.

As we have seen, the minimisation of standard deviation or variance is a familiar objective in portfolio optimisation. Problems with this approach are that it penalises up and down deviations equally and that it does not take account of 'fat tails' in loss distributions.

A related criticism of coherent risk measures and VaR is that they measure a negative outcome of the position X. For practitioners, 'loss' often refers to the shortfall relative to expectation. That is, for practitioners, risk measures usually refer to $X - E(X)$.

Working on the probability space $(\Omega, \mathcal{F}, P)$ we shall define a deviation measure on the space $L^2(\Omega)$.

Definition 11.3.1. A deviation measure is a functional $D : L^2(\Omega) \to [0, \infty]$ satisfying:
  D1. $D(X + C) = D(X)$ for $X \in L^2(\Omega)$ and $C \in \mathbb{R}$;
  D2. $D(\lambda X) = \lambda D(X)$ for $\lambda > 0$;
  D3. $D(X + Y) \le D(X) + D(Y)$ for $X, Y \in L^2(\Omega)$;
  D4. $D(C) = 0$ for $C \in \mathbb{R}$, and $D(X) > 0$ if X is non-constant.

Note that $D(X - E(X)) = D(X)$ from D1. It follows from D4 that $D(X) = 0$ if and only if $X - E(X) = 0$ since $D(Y) = 0$ if and only if $Y = c$ is constant. But $X - E(X) = c$ implies $c = 0$ since $E(X - E(X)) = 0$. However, in general, D may not be symmetric; that is, it is possible that $D(-X) \ne D(X)$.

Note that if D is a deviation measure, then its reflection $\overline{C}$, given by $\overline{C}(X) = D(-X)$, is also a deviation measure, and its symmetrisation, $\widetilde{D}$, given by $\widetilde{D}(X) = \frac{1}{2}[D(X) + \overline{C}(X)]$, is a deviation measure.

Example 11.3.2. Standard deviation $\sigma(X) = (E((X - E(X)))^2)^{\frac{1}{2}}$ is a deviation measure, as are

$$\sigma_+(X) = \left(E\left((\max\{X - E(X), 0\})^2 \right) \right)^{\frac{1}{2}}$$

and

$$\sigma_-(X) = \left(E\left((\max\{E(X) - X, 0\})^2 \right) \right)^{\frac{1}{2}}.$$

To relate deviation measures and coherent risk, expectation-bounded risk measures are introduced in [246].

Definition 11.3.3. An *expectation-bounded* risk measure on $L^2(\Omega)$ is a functional $R : L^2(\Omega) \to (-\infty, \infty]$ satisfying:

R1. $R(X + C) = R(X) - C$ for $X \in L^2(\Omega)$ and $C \in \mathbb{R}$;

R2. $R(0) = 0$ and $R(\lambda X) = \lambda R(X)$ for $X \in L^2(\Omega)$ and $\lambda > 0$;

R3. $R(X + Y) \le R(X) + R(Y)$ for $X, Y \in L^2(\Omega)$;

R4. $R(X) > E(-X)$ for non-constant X and $R(X) = -X$ for constant X.

An expectation-bounded risk measure is *coherent* if, further,

R5. $R(X) \le R(Y)$ when $X \ge Y$.

From R1 and R2 it is clear that $R(C) = -C$.

Property R4 is described as *expectation-boundedness*.

Property R5 is again *monotonicity*. Although R5 is apparently stronger than condition (i) of Definition 11.2.1, we see that if $X \le Y$ a.s., then $Y = X + (Y - X)$ where $(Y - X) \ge 0$. Consequently, if ρ satisfies (i) and (iv) of Definition 11.2.1, then $\rho(Y) \le \rho(X) + \rho(Y - X) \le \rho(X)$. That is, a coherent risk measure satisfies condition R5.

Note that if R is a functional satisfying R1-R4, then, on $L^2(\Omega)$, it satisfies the conditions of Definition 11.2.1 and so is a coherent risk measure.

The next result relates deviation measures to expectation-bounded risk measures.

Theorem 11.3.4. *Suppose $D : L^2(\Omega) \to [0, \infty]$ is a deviation measure. Then $R(X) = D(X) - E(X)$ is an expectation-bounded risk measure. Conversely, if R is this expectation-bounded risk measure, then $D(X) = R(X - E(X))$.*

Proof. Suppose D is a deviation measure. The properties R2 and R3 follow from D2 and D3. Also,

$$R(X + C) = D(X + C) - E(X) - C$$
$$= D(X) - E(X) - C$$
$$= R(X) - C,$$

so R satisfies R1.

From D4, if X is non-constant,

$$D(X) = R(X) + E(X) > 0,$$

and R4 follows.

Conversely, if $D(X) = R(X - E(X))$, then

$$D(X + C) = R((X + C) - E(X) - C)$$
$$= R(X) + E(X)$$
$$= D(X),$$

so D1 is satisfied. Again, D2 and D3 follow from R2 and R3. Also, for non-constant X, R1 and R4 imply

$$R(X - E(X)) = R(X) + E(X) > 0.$$

Therefore D4 is satisfied. This completes the proof. $\square$

Example 11.3.5. For $X \in L^2(\Omega)$, write

$$D(X) = E(X) - ess\inf X = ess\sup\{E(X) - X\}.$$

This is a deviation measure describing the lower range of X. $R(X) = ess\sup(-X)$ is the corresponding risk measure. Both D and R are coherent, and R is expectation-bounded.

Conditional Value at Risk, or Expected Shortfall

A popular risk measure is conditional value at risk, $CVaR$. If we assume that there is a zero probability that $X = VaR_\alpha(X)$, we can define this as a true conditional expectation: for $\alpha \in (0,1)$ and $X \in L^2(\Omega)$,

$$CVaR_\alpha(X) = -E(X \mid X \leq VaR_\alpha(X)).$$

When X has a general distribution (i.e., possibly with jumps), this breaks down. Thus we define $CVaR$ as follows: let $U = \{X \leq q_\alpha(X)\}$ and write

$$CVaR_\alpha(X) = -\alpha^{-1}E(X\mathbf{1}_U) + q_\alpha(X)(\alpha - P(U)).$$

This quantity is also called the *expected shortfall* by some authors and has other attractive features, such as continuity in the quantile level α, which can be seen immediately from its representation in integral form; see [4] for a derivation:

$$CVaR_\alpha(X) = -\frac{1}{\alpha}\int_0^\alpha q_\beta(X)d\beta. \tag{11.14}$$

We introduce the following notation.

Notation 11.3.6. For $\alpha \in (0,1)$, write

$$\mathbf{1}_{\{X \leq x\}}^\alpha = \begin{cases} \mathbf{1}_{\{X \leq x\}} + \frac{\alpha - P(X \leq x)}{P(X = x)}\mathbf{1}_{\{X = x\}} & \text{if } P(X = x) > 0, \\ \mathbf{1}_{\{X \leq x\}} & \text{if } P(X = x) = 0. \end{cases}$$

Then

$$\mathbf{1}_{\{X \leq q_\alpha(X)\}}^\alpha \in [0,1], \tag{11.15}$$

$$E\left(\mathbf{1}_{\{X \leq q_\alpha(X)\}}^\alpha\right) = \alpha - \alpha^{-1}E\left(X\mathbf{1}_{X \leq q_\alpha(X)}^\alpha\right) = CVaR_\alpha(X). \tag{11.16}$$

We now show that $CVaR$ is a coherent risk measure.

Theorem 11.3.7. *Suppose $\alpha \in (0,1)$. Write $\rho : L^2(\Omega) \to \mathbb{R}$ for $\rho(X) = CVaR_\alpha(X)$. Then:*

(i) if $X \geq 0$, $\rho(X) \leq 0$;

(ii) if $\lambda \geq 0$, then $\rho(\lambda X) = \lambda\rho(X)$;

(iii) if $k \in \mathbb{R}$, then $\rho(X + k) = \rho(X) - k$;

(iv) if $X, Y \in L^2(\Omega)$, then $\rho(X + Y) \leq \rho(X) + \rho(Y)$.

Proof. (i) From the definition, if $X \geq 0$, then $\rho(X) = CVaR_\alpha(X) \leq 0$.
(ii) For $\lambda \geq 0$, $P(\lambda X \leq \lambda x) = P(X \leq x)$, so

$$q_\alpha(\lambda X) = \inf\{\lambda x : P(\lambda X \leq \lambda x) \geq \alpha\}$$
$$= \lambda \inf\{x : P(X \leq x) \geq \alpha\}$$
$$= \lambda q_\alpha(X).$$

Therefore, setting $D(U) = \{U \leq q_\alpha(U)\}$ for any random variable U, we have

$$\rho(\lambda X) = CVaR_\alpha(X)$$
$$= -\alpha^{-1}\left(E\left(\lambda X \mathbf{1}_{D(\lambda X)}\right) + q_\alpha(X)(\alpha - P(D(\lambda X)))\right)$$
$$= -\alpha^{-1}\lambda\left(E\left(X \mathbf{1}_{D(\lambda X)} + q_\alpha(X)(\alpha - P(D(\lambda X)))\right)\right)$$
$$= \lambda CVaR_\alpha(X) = \lambda\rho(X).$$

(iii) For $k \in \mathbb{R}$, $P(X + k \leq x + k) = P(X \leq x)$, so that

$$q_\alpha(X + k) = \inf_x\{x + k : P(X + k \leq x + k) \geq \alpha\}$$
$$= k + \inf_x\{x : P(X \leq x) \geq \alpha\}$$
$$= k + q_\alpha(X).$$

Therefore

$$\rho(X + k) = CVar_\alpha(X + k)$$
$$= -\alpha^{-1}\left(E\left((X + k)\mathbf{1}_{\{D(X+k)\}}\right) + q_\alpha(X + k)(\alpha - P(D(X + k)))\right)$$
$$= -\alpha^{-1}\left(E\left(X \mathbf{1}_{\{D(X)\}}\right) + q_\alpha(X)(\alpha - P(D(X)))\right)$$
$$\quad - \alpha^{-1}k\left(E\left(\mathbf{1}_{\{D(X+k)\}}\right) + \alpha - P(D(X + k))\right)$$
$$= \rho(X) - k.$$

(iv) Using the notation introduced above, we prove that ρ is subadditive. Suppose that $X, Y \in L^2(\Omega)$ and write $Z = X + Y$. Then, from (11.7),

$$\alpha(\rho(X) + \rho(Y) - \rho(Z))$$
$$= E\left(Z\mathbf{1}_{\{D(Z)\}}^\alpha - X\mathbf{1}_{\{D(X)\}}^\alpha - Y\mathbf{1}_{\{D(Y)\}}^\alpha\right)$$
$$= E\left(X(\mathbf{1}_{\{D(Z)\}}^\alpha - \mathbf{1}_{\{D(X)\}}^\alpha)\right) + E\left(Y\left(\mathbf{1}_{\{D(Z)\}}^\alpha - \mathbf{1}_{\{D(Y)\}}^\alpha\right)\right)$$
$$\geq q_\alpha(X)E\left(\mathbf{1}_{\{D(Z)\}}^\alpha - \mathbf{1}_{\{D(X)\}}^\alpha\right) + q_\alpha(Y)E\left(\mathbf{1}_{\{D(Z)\}}^\alpha - \mathbf{1}_{\{D(Y)\}}^\alpha\right)$$
$$= q_\alpha(X)(\alpha - \alpha) + q_\alpha(Y)(\alpha - \alpha) = 0.$$

We have used the facts that

$$1^\alpha_{\{Z \leq q_\alpha(Z)\}} - 1^\alpha_{\{X \leq q_\alpha(X)\}} \geq 0 \text{ if } X > q_\alpha(X)$$

and

$$1^\alpha_{\{Z \leq q_\alpha(Z)\}} - 1^\alpha_{\{\{X \leq q_\alpha(X)\}\}} \leq 0 \text{ if } X < \overline{q}_\alpha(X).$$

This follows from the definition of 1^α. $\square$

Remark 11.3.8. This brief review of various approaches to measuring risk, including VaR and deviation measures, has only skimmed the surface of recent work in this very active field of research. Importantly, this research has revealed deficiencies of VaR, which is still the dominant risk-management tool used in practice. The concept of coherent risk measure was created to address this situation, and to aid computation and the construction of concrete examples for particular needs, a representation result for such measures was established. In particular, conditional value at risk, $CVaR$, has been shown to be a coherent measure of risk. Deviation measures and the related bounded expectation measures were introduced with similar objectives in view, and we have shown how relationships with coherent risk measures can be established. Though this field is one of intense current research, it may take time for the newer concepts touched upon here to settle down and become common in financial practice.

An area of much current work is the extension of these ideas to a multi-period setting, where martingales and generalised Snell envelopes come to the fore. The interested reader is referred to the recent papers [10], [11] for this material, which is beyond the scope of this book.

11.4   Hedging Strategies with Shortfall Risk

This final section outlines how risk measures can be applied to the construction of hedging strategies for financial assets, which is one of the principal topics covered in this book. We have seen how, in a viable financial market model, derivative securities can be priced by arbitrage considerations alone, and that this price, as well as the replicating strategy, are uniquely determined when the market is complete. For incomplete markets, we were able to reproduce these results for attainable claims, but in the general case the buyer's and seller's prices represent the limits of an *arbitrage interval* of possible prices for the claim, and additional optimality criteria are needed to identify both the optimal price and optimal hedging strategy.

An investor can always play safe by employing a 'superhedging strategy' - an approach outlined in Chapters 2 and 7 for discrete and continuous-time pricing models, respectively (also see [184] for a fuller account). However, the initial capital required to eliminate all risk may be considered too high by the investor, who may be willing instead to accept the risk of loss at a specified level. The question then becomes: how much initial capital can be saved by accepting the risk of having to find additional capital at maturity in (say) 1% of all possible outcomes? A second question is then: by what criteria should the shortfall risk be measured, or what measure of risk should be employed?

In [128],[129] Föllmer and Leukert introduced these ideas and showed how the problem of such 'quantile hedging' against a given contingent claim H can be reduced to consideration of an optimisation problem for the modified claim ϕH, where ϕ ranges over the class of 'randomised tests' (i.e., $\mathcal{F}_T$-measurable random variables with values in the interval $[0,1]$). This allows an application of the Neyman-Pearson lemma from the theory of hypothesis testing to provide an optimal solution (see, e.g., [303] for a detailed treatment). Here we confine attention to integrable claims, and, in particular, adapt the treatment given in [235] using coherent measures of risk.

Quantile Hedging in a Complete Market

Assume that the price process $(S_t)_{t\in[0,T]}$ is given as a semimartingale defined on a probability space $(\Omega, \mathcal{F}, P)$ adapted to a filtration $\mathbb{F} = (\mathcal{F}_t)_{t\in[0,T]}$, where $\mathcal{F}_0$ is assumed to be trivial and $\mathcal{F}_T = \mathcal{F}$. We assume that this market model is viable, so that the set $\mathcal{P}$ of equivalent martingale measures is non-empty. In this market, a self-financing strategy (V_0, θ) is determined by the initial capital V_0 and a predictable process θ such that the resulting value process $V = (V_t)$ satisfies, P-a.s. for all t,

$$V_t = V_t(\theta) = V_0 + \int_0^t \theta_u dS_u, \qquad (11.17)$$

where we shall assume the usual integrability conditions without further mention (see Chapter 7). The strategy is *admissible* if also $V_t \geq 0$ P-a.s. for all t.

In a complete market, there is a unique measure $Q \sim P$ under which the (discounted) price process is a martingale. For simplicity, we shall assume that the discount rate is 0, so that S_t already represents the discounted asset price. Now let $H \in L^1_+(Q)$ be a contingent claim. There is a perfect *hedging strategy* θ^H such that for all t, P-a.s.,

$$E_Q(H|\mathcal{F}_t) = H_0 + \int_0^t \theta_u^H dS_u. \qquad (11.18)$$

Thus the claim H is replicated by the strategy (H_0, θ^H), provided the investor allocates initial capital $H_0 = E_Q(H)$ to the hedge. However, suppose the investor is willing to allocate initial capital at most V_0^* to hedge against the claim H. We may then seek the strategy that provides maximum probability that the hedge will be successful (i.e., will suffice to cover the liability of the claim at time T). In other words, we seek the strategy (V_0, θ) that maximises the probability of the set

$$A(H, \theta) = \{V_T \geq H\} = \left(\omega : V_0 + \int_0^T \theta_u(\omega) dS_u(\omega) \geq H(\omega)\right) \quad (11.19)$$

subject to the constraint

$$V_0 \leq V_0^*. \quad (11.20)$$

In [128], $A(H, \theta)$ is called the *success set* for the claim and the resulting strategy. For any measurable set B, we can consider the *knockout option* $H_B = H1_B$, which, at time T, pays out $H(\omega)$ if $\omega \in B$ and 0 otherwise. Note that with our assumptions $H_B \in L^1_+(Q)$. As the market model is complete, this contingent claim can be hedged perfectly by a unique admissible strategy. Now let A^* be a success set for H with maximal probability; i.e., such that

$$P(A^*) = \max P(A(H, \theta)) \quad (11.21)$$

subject to the constraint

$$E_Q(H1_{A(H, \theta)}) \leq V_0^*. \quad (11.22)$$

Denote the perfect hedging strategy for the knockout option $H_{A^*} = H1_{A^*}$ by θ^*. Thus we have P-a.s for all $t \leq T$,

$$E_Q(H1_{A^*}|\mathcal{F}_t) = E_Q(H1_{A^*}) + \int_0^t \theta_u^* dS_u. \quad (11.23)$$

This allows us to reduce the original optimisation problem to the question of constructing a success set of maximal probability.

Proposition 11.4.1. *Suppose that, as defined above, A^* is a success set with maximal probability under the constraint (11.22). Then the perfect hedging strategy (V_0^*, θ^*) for the knockout option H_{A^*} solves the optimisation problem defined by (11.19),(11.20), and its success set is P-a.s. equal to A^*.*

Proof. First consider any admissible strategy (V_0, θ) with $V_0 \leq V_0^*$. The process $V_t = V_0 + \int_0^t \theta_u dS_u$ is a non-negative local martingale and hence a supermartingale (see Lemma 7.5.3) under Q. Since $V_T \geq 0$ P-a.s., the success set $A = A(H, \theta)$ for this strategy satisfies $V_T \geq H1_A$ P-a.s., so that

$$V_0^* \geq V_0 \geq E_Q(V_T) \geq E_Q(H1_A).$$

This shows that A satisfies the constraint (11.22), and therefore, by the definition of A^*, we conclude that $P(A) \leq P(A^*)$.

We will show that any strategy (V_0, θ^*) satisfying $E_Q(H1_{A^*}) \leq V_0 \leq V_0^*$ is optimal. Such a strategy is admissible since, P-a.s., $H1^*_A \geq 0$, so that by (11.23),

$$V_0 + \int_0^t \theta^*_u dS_u \geq E_Q(H1_{A^*}) + \int_0^t \theta^*_u dS_u = E_Q(H1_{A^*}|\mathcal{F}_t) \geq 0. \quad (11.24)$$

Consider the success set $A(H, \theta^*)$ for the strategy (V_0, θ^*). We have

$$A^* \subset \{H1_{A^*} = H\} \subset A(H, \theta^*)$$

since $V_0 \geq E_Q(H1_{A^*})$ and (11.23) imply that $V_T(\theta^*) \geq H$ a.s. on A^*. On the other hand, A^* has maximal P-measure among success sets, so it follows that $A^* = A(H, \theta^*)$ P-a.s. Hence the strategy (V_0, θ^*) is an optimal solution of the original problem (11.19), (11.20), as required. □

Remark 11.4.2. Having reduced the problem to that of finding a maximal success set, we briefly recall the basic elements of the Neyman-Pearson theory of hypothesis testing: to discriminate between two given probability measures P and P^*, one may try to devise a *pure test* (i.e., a random variable $\phi : \Omega \to \{0, 1\}$), under which we reject P^* if the event $\{\phi = 1\}$ occurs. This allows for two kinds of erroneous conclusions: $P^*(\phi = 1)$ is the probability that we reject P^* in error, and $P(\phi = 0) = 1 - P(\phi = 1)$ is the probability that P^* is accepted in error. In general, it is not possible to minimise both probabilities simultaneously. However, one can accept a tolerance level α (e.g., $\alpha = .01$) for the first kind of error - much as is done for VaR - and seek instead to solve a constrained optimisation problem for the second kind, i.e., we seek to maximise $P(\phi = 1)$ subject to the constraint

$$P^*(\phi = 1) \leq \alpha. \quad (11.25)$$

A solution for this optimisation problem can be found by choosing a third probability measure Q such that P and P^* are both absolutely continuous with respect to Q, with densities Z_P and Z_{P^*}, respectively. The key quantity is then the *likelihood ratio* Z_P/Z_{P^*} : the optimal test is the function

$$\phi^* = 1_{\{a^* Z_{P^*} < Z_P\}}, \quad (11.26)$$

where $a^* \in (0, \infty)$ is chosen so that $P^*(a^* Z_{P^*} < Z_P) = \alpha$. Thus the test ϕ^* rejects P^* if and only if the likelihood ratio exceeds the level a^*.

To construct a maximal success set in the family $A(H, \theta)$, we therefore introduce the measure $P^* \ll Q$ with Radon-Nikodym derivative

$$\frac{dP^*}{dQ} = \frac{H}{E_Q(H)} = \frac{H}{H_0}. \quad (11.27)$$

The constraint $E_Q(H1_A) \leq V_0^*$ becomes

$$P^*(A) = E_{P^*}(1_{A^*}) = \frac{1}{H_0} E_Q(H1_A) \leq \frac{V_0^*}{H_0}. \quad (11.28)$$

Write $\alpha = \frac{V_0^*}{H_0}$ and define the set

$$\widetilde{A} = \left(\frac{dP}{dQ} > aH\right). \tag{11.29}$$

Define the level a^* by

$$a^* = \inf\{a : P^*[\widetilde{A}] \leq \alpha\}. \tag{11.30}$$

The Neyman-Pearson lemma now allows us to deduce that $\widetilde{A}$ is a success set of maximal measure as follows.

Theorem 11.4.3. *Suppose that $P^*(\widetilde{A}) = \alpha$. Then the optimal strategy solving (11.19), (11.20) is the unique replicating strategy (V_0^*, θ^*) for the knockout option $H1_{\widetilde{A}}$.*

Proof. Both P and P^* are absolutely continuous with respect to the unique EMM Q, and the set $\widetilde{A}$ consists precisely of the points $\omega \in \Omega$ at which $\frac{dP}{dQ}(\omega) > \widetilde{a}H_0\frac{dP^*}{dQ}(\omega)$, so that the likelihood ratio is bounded below by the constant $\widetilde{a}H_0$. Then the Neyman-Pearson lemma states that for any measurable set A, $P^*(A) \leq P^*(\widetilde{A})$ implies $P(A) \leq P(\widetilde{A})$. Hence the constraint (11.22) is satisfied and $\widetilde{A}$ is a success set of maximal measure, so that, by Proposition 11.4.1, the strategy (V_0^*, θ^*) solves the original optimisation problem. $\square$

Remark 11.4.4. These ideas are taken much further in [128], where explicit results are given for the Black-Scholes model and the theory is developed further for incomplete markets. We do not pursue this here but will instead sketch briefly how the same ideas may be used in the context of coherent risk measures.

However, in the more general situation, we need to extend the class of 'tests' that allows us to discriminate between alternative hypotheses since the 'level' a^* defined in (11.30) need not exist in general. To deal with this, we replace the $\{0, 1\}$-valued test function ϕ^* by a more general 'randomised' test ϕ with possible values ranging through the interval $[0, 1]$. The interpretation of these tests is that, in the event that the outcome $\omega \in \Omega$ is observed, then P^* is rejected with probability $\phi(\omega)$ and rejected with probability $1 - \phi(\omega)$. This means that $E_P(\phi)$ provides for us the probability of rejecting the hypothesis P^* when it is false (and thus defines the *power* of the test ϕ), while $E_{P^*}(\phi)$ gives the probability of error of the first kind (rejecting P^* when it is true). The optimisation problem to be solved is therefore to maximise $E_P(\phi)$ over all tests ϕ that satisfy the constraint $E_P^*(\phi) \leq \alpha$. This problem again has an explicit solution, as will be seen in the general situation outlined in the next subsection.

Efficient Hedging with Coherent Risk Measures

We outline the results obtained in [235]. As in the previous subsection, assume as given a viable market model $(\Omega, \mathcal{F}, P, (\mathcal{F}_t)_{t\in[0,T]}, (S_t)_{t\in[0,T]})$ and denote the non-empty set of equivalent martingale measures by $\mathcal{P}$. Assume further that the integrable contingent claim H satisfies $\sup_{Q \in \mathcal{P}} E_Q(H) < \infty$.

Now let $\rho : L^1 \to \mathbb{R}$ denote a coherent risk measure that is lower semi-continuous in the L^1-norm. We wish to minimise the shortfall risk when using admissible hedging strategies with given initial capital V_0^*, so that we seek the admissible strategy (V_0, θ) that minimises

$$\rho(\min[(V_T - H), 0]) = \rho \left(\min \left(\left(V_0 + \int_0^T \theta_u dS_u - H \right), 0 \right) \right) \quad (11.31)$$

subject to the constraint

$$V_0 \leq V_0^*. \quad (11.32)$$

Defining the set of 'randomised tests' (see [67] for an explanation of the terminology, which comes from the theory of hypothesis testing) by

$$\mathcal{R} = \{\phi : \Omega \to [0,1] : \phi \text{ is } \mathcal{F}\text{-measurable}\}$$

and the constrained set of tests

$$\mathcal{R}_0 = \{\phi \in \mathcal{R} : \sup_{Q \in \mathcal{P}} E_Q(\phi H) \leq V_0^*\}, \quad (11.33)$$

we can use the representation theorem for coherent risk measures to prove the following proposition.

Proposition 11.4.5. *There exists a randomised test ϕ^* in $\mathcal{R}_0$ such that*

$$\inf_{\phi \in \mathcal{R}_0} \rho(-(1 - \phi)H) = \rho(-(1 - \phi^*)H). \quad (11.34)$$

Proof. The set $\mathcal{R}$ is $\sigma(L^\infty, L^1)$-compact in L^∞, and the map

$$\phi \to \sup_{Q \in \mathcal{P}} E_Q(\phi H)$$

is lower semi-continuous in the weak* topology on L^∞. Hence the set $\mathcal{R}_0$ is weak*-closed and hence also weak*-compact.

We recall the essential features of the proof of Theorem 11.2.19: if $\mathcal{Q}$ denotes the set of all probability measures absolutely continuous with respect to P, and $\mathcal{C} = \{Y \in \mathcal{A}^\circ : E[Y] = 1\}$, where $\mathcal{A}$ denotes the set of acceptable positions for ρ, then the subset of $\mathcal{Q}$ given by

$$\tilde{\mathcal{Q}} = \{Q \in \mathcal{Q} : Z_Q = Y \text{ for some } Y \in \mathcal{C}\}$$

satisfies, for any $X \in L^1$,

$$\rho(X) = \sup_{Q \in \tilde{\mathcal{Q}}} E_Q(-X) \quad (11.35)$$

and the set $\{\frac{dQ}{dP} : Q \in \tilde{Q}\}$ is convex and weak*-closed in L^1.

But the L^∞-functional

$$\phi \to \sup_{Q \in \tilde{Q}} E_Q[(1 - \phi)H]$$

is also lower semi-continuous in the weak* topology, so its infimum over $\mathcal{R}_0$ is attained. □

This again reduces the original optimisation problem of finding an admissible strategy that solves (11.31), (11.32) to the question of finding an optimal randomised test ϕ^*. However, we first need to generalise the concept of 'success set', which applies when ϕ is an indicator function, to this more general context.

Definition 11.4.6. For any admissible strategy (V_0, θ), the *success ratio* is the function

$$\phi(V_0, \theta) = 1_{\{V_T \geq H\}} + \frac{V_T}{H} 1_{\{V_T < H\}}. \tag{11.36}$$

The role of the simple knockout option is now taken by $\phi^* H$. We denote by V^* the right-continuous version of the process

$$V_t^* = ess \sup_{Q \in \mathcal{P}} E_Q(\phi^* H | \mathcal{F}_t).$$

This is a supermartingale for every Q in $\mathcal{P}$, and thus the optional decomposition theorem (see [201], [129]) applies to provide an admissible strategy (V_0^*, θ^*) and an increasing optional process C^* with $C_0^* = 0$ such that

$$V_t^* = V_0^* + \int_0^t \theta_u^* dS_u - C_t^*.$$

Remark 11.4.7. The force of the optional decomposition theorem is to provide a characterisation of the wealth processes defined in Chapters 7 and 10. The collection of processes V defined by $V_t = V_0 + \int_0^t \theta_u dS_u - C_u$, where $C = (C_u)_{u \in [0,T]}$ is adapted and increasing, with $C_0 = 0$, is identical with the collection of $\mathcal{P}$- supermartingales (i.e., processes that are supermartingales for every EMM Q). The decomposition is non-unique, unlike the Doob-Meyer decomposition of the P-supermartingale $V_t = V_0 + M_t - A_t$, where A is increasing and predictable, with $A_0 = 0$. However, under the stronger condition that V is a supermartingale for each EMM Q the martingale M can be taken to be the stochastic integral (i.e., a gains process) generated by some admissible strategy θ at the cost of relaxing the requirements on the 'compensator' C.

The strategy (V_0^*, θ^*) provides the solution to our original optimisation problem whenever the randomised test has the minimisation property described in Proposition 11.4.5.

Theorem 11.4.8. *Suppose that ϕ^* solves the minimisation problem posed in Proposition 11.4.5 and (V_0^*, θ^*) is the admissible strategy for the claim $\phi^* H$ determined by its optional decomposition, then this strategy solves the optimisation problem* (11.31), (11.32), *and its success ratio is ϕ^* P-a.s.*

Proof. The proof follows the same pattern as for the quantile hedging case. Take an admissible strategy (V_0, θ) satisfying the constraint $V_0 \leq V_0^*$ and with success ratio ϕ. Since $\phi H = V_T \wedge H$, we have

$$(V_T - H) \wedge 0 = -(V_T - H)^+ = -(H - V_T \wedge H) = -(1 - \phi)H.$$

Also, the supermartingale property of V implies that

$$E_Q(\phi H) \leq E_Q(V_T) \leq V_0 \leq V_0^*.$$

Hence the success ratio ϕ is in $\mathcal{R}_0$ and so

$$\rho((V_T - H) \wedge 0) = \sup_{Q \in \widetilde{\mathcal{Q}}} E_Q((1 - \phi)H) \geq \sup_{Q \in \widetilde{\mathcal{Q}}} E((1 - \phi^*)H). \quad (11.37)$$

In particular, the success ratio $\phi(V_0^*, \theta^*)$ satisfies this inequality, while on the other hand

$$\phi(V_0^*, \theta^*)H = V_T^* \wedge H \geq \phi^* H,$$

so that, for all $Q \in \widetilde{\mathcal{Q}}$,

$$E_Q[(1 - \phi(V_0^*, \theta^*)H)] \leq E_Q[(1 - \phi^*)H].$$

This shows that the two quantities are equal, and so

$$\rho((V_T^* - H) \wedge 0) = \sup_{Q \in \widetilde{\mathcal{Q}}} E_Q((1 - \phi(V_0^*, \theta^*)H)) = \sup_{Q \in \widetilde{\mathcal{Q}}} E((1 - \phi^*)H).$$

$\square$

Remark 11.4.9. In [235] and [129], these general results are applied to particular examples of coherent risk measures. In the context of the Black-Scholes model, for example, that with ρ as the worst conditional expectation, the amount of capital 'saved' by accepting a given level of loss can be computed explicitly. For a European call option H, where in the present setting, with Φ denoting the cumulative normal distribution function, the cost of replication is

$$E_Q(H) = S_0 \Phi(d_+) - K\Phi(d_-),$$

let $V_0^* \leq E_Q(H)$ be a given level of initial capital, and assume further that the drift $\mu \geq 0$ and that

$$P(H > 0) = \Phi(\mu\sqrt{T + d_-}) \leq \alpha.$$

Then it is shown in [235] that the minimisation problem for ϕ is solved by the most powerful randomised test $\phi^* = \mathbf{1}_{\{S_T > c\}}$, so that

$$V_0^* = E_Q(H\mathbf{1}_{\{S_T > c\}}$$

and the constant c can be determined from the identity

$$E_Q(H\mathbf{1}_{\{S_T > c\}}$$
$$= S_0 \Phi \left(\frac{1}{\sigma\sqrt{T}} \log \left(\frac{S_0}{c} \right) + \frac{1}{2}\sigma\sqrt{T} \right) - K\Phi \left(\frac{1}{\sigma\sqrt{T}} \log \left(\frac{S_0}{c} \right) - \frac{1}{2}\sigma\sqrt{T} \right).$$

Bibliography

[1] K.K. Aase and B. Oksendal. Admissible investment strategies in continuous trading. *Stochastic Process. Appl.*, 30:291–301, 1988.

[2] C. Acerbi, C. Nordio, and C. Sirtori. Expected shortfall as a tool for financial risk management. Working paper, Abaxbank, 2001.

[3] C. Acerbi and D. Tasche. Expected shortfall: A natural coherent alternative to value at risk. Working paper, Abaxbank, 2001.

[4] C. Acerbi and D. Tasche. On the coherence of expected shortfall. Working paper, Abaxbank, 2002.

[5] W. Allegretto, G. Barone-Adesi, and R.J. Elliott. Numerical evaluation of the critical price and American options. *Eur. J. of Finance*, 1:69–78, 1995.

[6] J.-P. Ansel and C. Stricker. Lois de martingale, densités et décomposition de Föllmer-Schweizer. *Ann. Inst. H. Poincaré Probab. Statist.*, 28:375–392, 1992.

[7] P. Artzner and F. Delbaen. Term structure of interest rates: The martingale approach. *Adv. Appl. Math.*, 10:95–129, 1989.

[8] P. Artzner, F. Delbaen, J. Eber, and D. Heath. Thinking coherently. *Risk*, 10:68–71, 1997.

[9] P. Artzner, F. Delbaen, J. Eber, and D. Heath. Coherent measures of risk. *Math. Finance*, 9:203–228, 1999.

[10] P. Artzner, F. Delbaen, J. Eber, D. Heath, and H. Ku. Coherent multiperiod risk measurement. Preprint, ETH, 2002.

[11] P. Artzner, F. Delbaen, J. Eber, D. Heath, and H. Ku. Multiperiod risk and multiperiod risk measurement. Preprint, ETH, 2002.

[12] P. Artzner, F. Delbaen, J. Eber, D. Heath, and H. Ku. Coherent multiperiod risk-adjusted values and Bellman's principle. Preprint, ETH, 2003.

[13] P. Artzner and D. Heath. Approximate completeness with multiple martingale measures. *Math. Finance*, 5:1–11, 1995.

[14] L. Bachelier. Theory of speculation. In P.H. Cootner, editor, *The Random Character of Stock Market Prices*, volume 1018 (1900) of *Ann. Sci. École Norm. Sup.*, pages 17–78. MIT Press, Cambridge, Mass., 1964.

[15] G. Barone-Adesi and R.J. Elliott. Approximations for the values of American options. *Stochastic Anal. Appl.*, 9:115–131, 1991.

[16] G. Barone-Adesi and R.J. Elliott. Pricing the treasury bond futures contract as the minimum value of deliverable bond prices. *Rev. Futures Markets*, 8:438–444, 1991.

[17] G. Barone-Adesi and R. Whaley. The valuation of American call options and the expected ex-dividend stock price decline. *J. Finan. Econ.*, 17:91–111, 1986.

[18] G. Barone-Adesi and R. Whaley. Efficient analytic approximation of American option values. *J. Finance*, 42:301–320, 1987.

[19] E.M. Barron and R. Jensen. A stochastic control approach to the pricing of options. *Math. Oper. Res.*, 15:49–79, 1990.

[20] E. Barucci. *Financial Markets Theory: Equilibrium, Efficiency and Information*. Springer, Heidelberg, 2003.

[21] B. Bensaid, J.-P. Lesne, H. Pagès, and J. Scheinkman. Derivative asset pricing with transaction costs. *Math. Finance*, 2:63–68, 1992.

[22] A. Bensoussan. On the theory of option pricing. *Acta Appl. Math.*, 2:139–158, 1984.

[23] A. Bensoussan and R.J. Elliott. Attainable claims in a Markov model. *Math. Finance*, 5:121–132, 1995.

[24] A. Bensoussan and J.L. Lions. *Applications of Variational Inequalities in Stochastic Control*. North Holland, Amsterdam, New York, Oxford, 1982.

[25] J.M. Bismut. Martingales, the Malliavin calculus and hypoellipticity under general Hörmander's conditions. *Z. Wahrsch. Verw. Gebiete.*, 56:469–505, 1981.

[26] F. Black and M. Scholes. The valuation of option contracts and a test of market efficiency. *J. Finance*, 27:399–417, 1972.

[27] F. Black and M. Scholes. The pricing of options and corporate liabilities. *J. Polit. Econ.*, 81:637–659, 1973.

[28] N. Bouleau and D. Lamberton. Residual risks and hedging strategies in Markovian markets. *Stochastic Process. Appl.*, 33:131–150, 1989.

[29] P. Boyle and T. Vorst. Option replication in discrete time with transaction costs. *J. Finance*, 47:271–293, 1992.

[30] P.P. Boyle. Options: A Monte-Carlo approach. *J. Finan. Econ.*, 4:323–338, 1977.

[31] A. Brace and M. Musiela. A multifactor Gauss Markov implementation of Heath, Jarrow, and Morton. *Math. Finance*, 4:259–283, 1994.

[32] M. Brennan, G. Courtadon, and M. Subrahmanyan. Options on the spot and options on futures. *J. Finance*, 40:1303–1317, 1985.

[33] M. Brennan and E. Schwartz. The valuation of American put options. *J. Finance*, 32:449–462, 1976.

[34] M. Brennan and E. Schwartz. A continuous-time approach to the pricing of bonds. *J. Bank Finance*, 3:135–155, 1979.

[35] M. Capinski and E. Kopp. *Measure, Integral and Probability*. Springer, London, 2004.

[36] M. Capinski and T. Zastawniak. *Mathematics for Finance*. Springer, London, 2003.

[37] P. Carr, R. Jarrow, and R. Myneni. Alternative characterizations of American put options. *Math. Finance*, 2:87–106, 1992.

[38] A.P. Carverhill. When is the short rate Markovian? *Math. Finance*, 4:305–312, 1994.

[39] P. Cheridito, F. Delbaen, and M. Kupper. Coherent and convex risk measures for cadlag processes. Preprint, ETH, 2003.

[40] M. Chesney, R. Elliott, and R. Gibson. Analytical solutions for the pricing of American bond and yield options. *Math. Finance*, 3:277–294, 1993.

[41] M. Chesney and R.J. Elliott. Estimating the instantaneous volatility and covariance of risky assests. *Appl. Stochastic Models Data Anal.*, 11:51–58, 1995.

[42] M. Chesney, R.J. Elliott, D. Madan, and H. Yang. Diffusion coefficient estimation and asset pricing when risk premia and sensitivities are time varying. *Math. Finance*, 3:85–100, 1993.

[43] M. Chesney and L. Scott. Pricing European currency options: A comparison of the modified Black-Scholes model and a random variance model. *J. Finan. Quant. Anal.*, 24:267–284, 1989.

[44] N. Christopeit and M. Musiela. On the existence and characterization of arbitrage-free measures in contingent claim valuation. *Stochastic Anal. Appl.*, 12:41–63, 1994.

[45] K-L. Chung. *A Course in Probability Theory*. Academic Press, Princeton, 2000.

[46] D.B. Colwell and R.J. Elliott. Discontinuous asset prices and nonattainable contingent claims. *Math. Finance*, 3:295–308, 1993.

[47] D.B. Colwell, R.J. Elliott, and P.E. Kopp. Martingale representation and hedging policies. *Stochastic Process. Appl.*, 38:335–345, 1991.

[48] Basle Committee. Credit risk modelling: Current practices and applications. Technical report, Basle Committee on Banking Supervision, 1999.

[49] A. Conze and R. Viswanathan. Path dependent options: The case of lookback options. *J. Finance*, 46:1893–1907, 1991.

[50] G. Courtadon. A more accurate finite difference approximation for the valuation of options. *J. Finan. Quant. Anal.*, 18:697–700, 1982.

[51] J.C. Cox and C.-F. Huang. Optimal consumption and portfolio policies when asset prices follow a diffusion process. *J. Econ. Theory*, 49:33–83, 1989.

[52] J.C. Cox, J.E. Ingersoll, and S.A. Ross. Duration and the measurement of basic risk. *J. Business*, 52:51–61, 1979.

[53] J.C. Cox, J.E. Ingersoll, and S.A. Ross. The relation between forward prices and futures prices. *J. Finan. Econ.*, 9:321–346, 1981.

[54] J.C. Cox, J.E. Ingersoll, and S.A. Ross. An intertemporal general equilibrium model of asset prices. *Econometrica*, 53:363–384, 1985.

[55] J.C. Cox, J.E. Ingersoll, and S.A. Ross. A theory of the term structure of interest rates. *Econometrica*, 53:385–407, 1985.

[56] J.C. Cox and S.A. Ross. The pricing of options for jump processes. Rodney L. White Center Working Paper 2-75, University of Pennsylvania, 1975.

[57] J.C. Cox and S.A. Ross. A survey of some new results in financial options pricing theory. *J. Finance*, 31:382–402, 1976.

[58] J.C. Cox and S.A. Ross. The valuation of options for alternative stochastic processes. *J. Finan. Econ.*, 3:145–166, 1976.

[59] J.C. Cox, S.A. Ross, and M. Rubinstein. Option pricing: A simplified approach. *J. Finan. Econ.*, 7:229–263, 1979.

[60] J.C. Cox and M. Rubinstein. A survey of alternative option-pricing models. In M. Brenner, editor, *Option Pricing, Theory and Applications*, pages 3–33. 1983.

[61] J.C. Cox and M. Rubinstein. *Options Markets*. Prentice-Hall, Englewood Cliffs, N.J., 1985.

[62] N. Cutland, E. Kopp, and W. Willinger. A nonstandard approach to option pricing. *Math. Finance*, 1(4):1–38, 1991.

[63] N. Cutland, E. Kopp, and W. Willinger. From discrete to continuous financial models: New convergence results for option pricing. *Math. Finance*, 3:101–124, 1993.

[64] J. Cvitanić and I. Karatzas. Convex duality in constrained portfolio optimization. *Ann. Appl. Probab.*, 2:767–818, 1992.

[65] J. Cvitanić and I. Karatzas. Hedging contingent claims with constrained portfolios. *Ann. Appl. Probab.*, 3:652–681, 1993.

[66] J. Cvitanić and I. Karatzas. Hedging and portfolio optimization under transaction costs: A martingale approach. *Math. Finance*, 6, 1996.

[67] J. Cvitanić and I. Karatzas. Generalized Neyman-Pearson lemma via convex duality. *Bernoulli*, 7:79–97, 2001.

[68] R.C. Dalang, A. Morton, and W. Willinger. Equivalent martingale measures and no-arbitrage in stochastic securities market model. *Stochastics Stochastics Rep.*, 29:185–201, 1990.

[69] R-A. Dana and M. Jeanblanc. *Financial Markets in Continuous Time*. Springer, Heidelberg, 2003.

[70] M.H.A. Davis and A.R. Norman. Portfolio selection with transaction costs. *Math. Oper. Res.*, 15:676–713, 1990.

[71] M.H.A. Davis, V.P. Panas, and T. Zariphopoulou. European option pricing with transaction costs. *SIAM J. Control Optim.*, 31:470–493, 1993.

[72] F. Delbaen. Representing martingale measures when asset prices are continuous and bounded. *Math. Finance*, 2:107–130, 1992.

[73] F. Delbaen. Consols in CIR model. *Math. Finance*, 3:125–134, 1993.

[74] F. Delbaen. Coherent risk measures. Lecture notes, Scuola Normale Superiore di Pisa, 2000.

[75] F. Delbaen. Coherent risk measures on general probability spaces. Preprint, ETH, 2000.

[76] F. Delbaen and W. Schachermayer. A general version of the fundamental theorem of asset pricing. *Math. Ann.*, 300:463–520, 1994.

[77] F. Delbaen and W. Schachermayer. The no-arbitrage property under a change of numeraire. *Stochastics Stochastics Rep.*, 53:213–226, 1995.

[78] F. Delbaen and W. Schachermayer. The Banach space of workable contingent claims in arbitrage theory. *Ann. Inst. H. Poincaré Probab. Statist.*, 33:113–144, 1997.

[79] F. Delbaen and W. Schachermayer. A compactness principle for bounded sequences of martingales with applications. *Prog. Probab.*, 45:137–173, 1999.

[80] C. Dellacherie and P.-A. Meyer. *Probabilities and Potential A*. North-Holland, Amsterdam, 1975.

[81] C. Dellacherie and P.-A. Meyer. *Probabilities and Potential B*. North-Holland, Amsterdam, 1982.

[82] L.U. Dothan. On the term structure of interest rates. *J. Finan. Econ.*, 6:59–69, 1978.

[83] L.U. Dothan. *Prices in Financial Markets*. Oxford University Press, New York, 1990.

[84] L.U. Dothan and D. Feldman. Equilibrium interest rates and multi-period bonds in a partially observable economy. *J. Finance*, 41:369–382, 1986.

[85] R. Douady. Options á limite et options á limite double. Working paper, 1994.

[86] J.-C. Duan. The GARCH option pricing model. *Math. Finance*, 5:13–32, 1995.

[87] D. Duffie. An extension of the Black-Scholes model of security valuation. *J. Econ. Theory*, 46:194–204, 1988.

[88] D. Duffie. *Security Markets: Stochastic Models*. Academic Press, Boston, 1988.

[89] D. Duffie. *Futures Markets*. Prentice-Hall, Englewood Cliffs, N.J., 1989.

[90] D. Duffie. *Dynamic Asset Pricing Theory*. Princeton University Press, Princeton, N.J., 1992.

[91] D. Duffie and C. Huang. Multiperiod security markets with differential information. *J. Math. Econ.*, 15:283–303, 1986.

[92] D. Duffie and C.-F. Huang. Implementing Arrow-Debreu equilibria by continuous trading of few long-lived securities. *Econometrica*, 53:1337–1356, 1985.

[93] D. Duffie and R. Kan. Multi-factor term structure models. *Philos. Trans. R. Soc. London*, 347:577–586, 1994.

[94] D. Duffie and P. Protter. From discrete- to continuous-time finance: Weak convergence of the financial gain process. *Math. Finance*, 2:1–15, 1992.

[95] D. Duffie and H.P. Richardson. Mean-variance hedging in continuous time. *Ann. Appl. Probab.*, 1:1–15, 1991.

[96] D. Duffie, M. Schroder, and C. Skiadas. Recursive valuation of defaultable securities and the timing of resolution of uncertainty. *Ann.Appl. Prob.*, 6:1075–1090, 1996.

[97] N. Dunford and J.T. Schwartz. *Linear Operators, Part I*. Interscience, New York, 1956.

[98] E. Eberlein. On modeling questions in security valuation. *Math. Finance*, 2:17–32, 1992.

[99] N. El Karoui. Les aspects probabilistes du contrôle stochastique. In *Lecture Notes in Mathematics*, volume 876, pages 73–238. Springer, Berlin, 1981.

[100] N. El Karoui and H. Geman. A probabilistic approach to the valuation of floating rate notes with an application to interest rate swaps. *Adv. Options Futures Res.*, 7:47–63, 1994.

[101] N. El Karoui, H. Geman, and V. Lacoste. On the role of state variables in interest rate models. Working paper, 1995.

[102] N. El-Karoui, H. Geman, and J.C. Rochet. Changes of numeraire, arbitrage and option prices. *J. Appl. Probab.*, 32:443–458, 1995.

[103] N. El Karoui, M. Jeanblanc-Picqué, and S. Shreve. Robustness of the Black and Scholes formula. *Math. Finance*, 8, 1998.

[104] N. El Karoui and I. Karatzas. A new approach to the Skorohod problem and its applications. *Stochastics Stochastics Rep.*, 34:57–82, 1991.

[105] N. El Karoui, S. Peng, and M.C. Quenez. Backward stochastic differential equations in finance. Preprint 260, Université Paris VI, 1994.

[106] N. El Karoui and M.C. Quenez. Dynamic programming and pricing of contingent claims in an incomplete market. *SIAM J. Control Optim.*, 33:29–66, 1995.

[107] N. El Karoui and J.C. Rochet. A pricing formula for options on coupon bonds. In *Modeles Mathematiques de la finance*. INRIA, Paris, 1990.

[108] N. El Karoui and D. Saada. A review of the Ho and Lee model. International Conference in Finance, June 1992.

[109] R.J. Elliott. *Stochastic Calculus and Applications.* Springer-Verlag, Berlin, 1982.

[110] R.J. Elliott, L. Aggoun, and J.B. Moore. *Hidden Markov Models: Estimation and Control.* Applications of Mathematics 29. Springer-Verlag, New York, 1994.

[111] R.J. Elliott and M. Chesney. Estimating the volatility of an exchange rate. In J. Janssen and C. Skiadis, editors, *6th International Symposium on Applied Stochastic Models and Data Analysis*, pages 131–135. World Scientific, Singapore, 1993.

[112] R.J. Elliott and D.B. Colwell. Martingale representation and non-attainable contingent claims. In P. Kall, editor, *15th IFIP Conference*, Lecture Notes in Control & Information Sciences 180, pages 833–842. Springer, Berlin, 1992.

[113] R.J. Elliott and H. Föllmer. Orthogonal martingale representation. In *Liber Amicorum for M. Zakai*. Academic Press, New York.

[114] R.J. Elliott, H. Geman, and R. Korkie. Portfolio optimization and contingent claim pricing with differential information. *Stochastics Stochastic Rep.*, 60:185–203, 1997.

[115] R.J. Elliott, H. Geman, and D. Madan. Closed form formulae for valuing portolios of American options. Working paper.

[116] R.J. Elliott and W.C. Hunter. Filtering a discrete time price process. In *29th IEEE Asilomar Conference on Signals Systems and Computers, Asilomar, CA. November 1995*, pages 1305–1309. IEEE Computer Society Press, Los Alamos, 1996.

[117] R.J. Elliott, W.C. Hunter, and B.M. Jamieson. Drift and volatility estimation in discrete time. *J. Econ. Dynamics and Control*, 22, 1998), PAGES= 209-218.

[118] R.J. Elliott, W.C. Hunter, and B.M. Jamieson. Financial signal processing. *Int. J. Theor. Appl. Finance*, 4:561–584, 2001.

[119] R.J. Elliott, W.C. Hunter, P.E. Kopp, and D.B. Madan. Pricing via multiplicative price decomposition. *J. Finan. Eng.*, 4:247–262, 1995.

[120] R.J. Elliott and P.E. Kopp. Option pricing and hedge portfolios for Poisson processes. *J. Stochastic Anal. Appl.*, 8:157–167, 1990.

[121] R.J. Elliott and P.E. Kopp. Equivalent martingale measures for bridge processes. *J. Stochastic Anal. Appl.*, 9:429–444, 1991.

[122] R.J. Elliott and D.B. Madan. A discrete time equivalent martingale measure. *Math. Finance*, 8:127–152, 1998.

[123] R.J. Elliott, D.B. Madan, and C. Lahaie. Filtering derivative security evaluations from market prices. In M.H. Dempster and S.R Pliska, editors, *Mathematics of Derivative Securities*, Proceedings of the Newton Institute, pages 141–162. Cambridge University Press, Cambridge, 1997.

[124] R.J. Elliott and R.W. Rishell. Estimating the implicit interest rate of a risky asset. *Stochastic Process. Appl.*, 49:199–206, 1994.

[125] R.J. Elliott and J. van der Hoek. An application of hidden Markov models to asset allocation problems. *Finance Stochastics*, 3:229–238, 1997.

[126] R.J. Elliott and J. van der Hoek. *Binomial Models in Finance*. Springer, New York, to appear, 2005.

[127] D. Feldman. The term structure of interest rates in a partially observed economy. *J. Finance*, 44:789–811, 1989.

[128] H. Föllmer and P. Leukert. Quantile hedging. *Finance Stochastics*, 3:251–273, 1999.

[129] H. Föllmer and P. Leukert. Efficient hedging: Cost versus shortfall risk. *Finance Stochastics*, 4:117–146, 2000.

[130] H. Föllmer and A. Schied. Convex measures of risk and trading constraints. *Finance Stochastics*, 6:429–447, 2002.

[131] H. Föllmer and A. Schied. Robust preferences and measures of risk. In K. Sandmann and P. Schonbucher, editors, *Advances in Finance and Stochastics*. Springer Verlag, New York, 2002.

[132] H. Föllmer and A. Schied. *Stochastic Finance*. de Gruyter, Berlin, 2002.

[133] H. Föllmer and M. Schweizer. Hedging by sequential regression: An introduction to the mathematics of option trading. *ASTIN Bull.*, 18:147–160, 1989.

[134] H. Föllmer and M. Schweizer. Hedging of contingent claims under incomplete information. In M.H.A. Davis and R.J. Elliott, editors, *Applied Stochastic Analysis*, Stochastic Monographs 5, pages 389–414. Gordon and Breach, New York, 1991.

[135] H. Föllmer and M. Schweizer. A microeconomic approach to diffusion models for stock prices. *Math. Finance*, 3:1–23, 1993.

[136] H. Föllmer and D. Sondermann. Hedging of non-redundant contingent claims. In W. Hildebrandt and A. Mas-Colell, editors, *Contributions to Mathematical Economics*, pages 205–223. North-Holland, Amsterdam, 1986.

[137] A. Frachot and J.P. Lesne. Expectation hypothesis with stochastic volatility. Working paper, Banque de France, 1993.

[138] A. Frachot and J.P. Lesne. Modèle facoriel de la structure par terms des taux d'interet theorie et application econometrique. *Ann. Econ. Stat.*, 40:11–36, 1995.

[139] M. Garman and S. Kohlhagen. Foreign currency option values. *J. Int. Money Finance*, 2:231–237, 1983.

[140] H. Geman. L'importance de la probabilité "forward neutre" dans une approach stochastique des taux d'intérêt. Working paper, ESSEC, 1989.

[141] H. Geman and A. Eydeland. Domino effect. *Risk*, 8(4):65–67, 1995.

[142] H. Geman and M. Yor. Bessel processes, Asian options and perpetuities. *Math. Finance*, 4:345–371, 1993.

[143] H. Geman and M. Yor. The valuation of double-barrier options: A probabilistic approach. Working paper, 1995.

[144] R. Geske. The valuation of corporate liabilities as compound options. *J. Finan. Quant. Anal.*, 12:541–552, 1977.

[145] R. Geske. The pricing of options with stochastic dividend yield. *J. Finance*, 33:617–625, 1978.

[146] R. Geske and H.E. Johnson. The American put option valued analytically. *J. Finance*, 39:1511–1524, 1984.

[147] J.M. Harrison. *Brownian Motion and Stochastic Flow Systems*. Wiley, New York, 1985.

[148] J.M. Harrison and D.M. Kreps. Martingales and arbitrage in multi-period securities markets. *J. Econ. Theory*, 20:381–408, 1979.

[149] J.M. Harrison and S.R. Pliska. Martingales and stochastic integrals in the theory of continuous trading. *Stochastic Process. Appl.*, 11:215–260, 1981.

[150] J.M. Harrison and S.R. Pliska. A stochastic calculus model of continuous trading: Complete markets. *Stochastic Process. Appl.*, 15:313–316, 1983.

[151] H. He. Convergence from discrete-time to continuous-time contingent claims prices. *Rev. Finan. Stud.*, 3:523–546, 1990.

[152] D. Heath and R. Jarrow. Arbitrage, continuous trading, and margin requirement. *J. Finance*, 42:1129–1142, 1987.

[153] D. Heath, R. Jarrow, and A. Morton. Bond pricing and the term structure of interest rates: A discrete time approximation. *J. Finan. Quant. Anal.*, 25:419–440, 1990.

[154] D. Heath, R. Jarrow, and A. Morton. Bond pricing and the term structure of interest rates: A new methodology for contingent claim valuation. *Econometrica*, 60:77–105, 1992.

[155] T.S.Y. Ho and S.-B. Lee. Term structure movements and pricing interest rate contingent claims. *J. Finance*, 41:1011–1029, 1996.

[156] C.-F. Huang. Information structures and equilibrium asset prices. *J. Econ. Theory*, 35:33–71, 1985.

[157] C.-F. Huang and R.H. Litzenberger. *Foundations for Financial Economics*. North-Holland, New York, 1988.

[158] J. Hull. *Options, Futures and Other Derivative Securities*. Prentice-Hall, Englewood Cliffs, N.J., 1989.

[159] J. Hull. *Introduction to Futures and Options Markets*. Prentice-Hall, Englewood Cliffs, N.J., 1991.

[160] J. Hull and A. White. The pricing of options on assets with stochastic volatilities. *J. Finance*, 42:281–300, 1987.

[161] J. Hull and A. White. An analysis of the bias in option pricing caused by a stochastic volatility. *Adv. Futures Options Res.*, 3:29–61, 1988.

[162] J. Hull and A. White. Pricing interest-rate derivative securities. *Rev. Finan. Stud.*, 3:573–592, 1990.

[163] J. Hull and A. White. Valuing derivative securities using the explicit finite difference method. *J. Finan. Quant. Anal.*, 25:87–100, 1990.

[164] S.D. Jacka. Optimal stopping and the American put. *Math. Finance*, 1(2):1–14, 1991.

[165] S.D. Jacka. A martingale representation result and an application to incomplete financial markets. *Math. Finance*, 2:239–250, 1992.

[166] S.D. Jacka. Local times, optimal stopping and semimartingales. *Ann. Probab.*, 21:329–339, 1993.

[167] J. Jacod. *Calcul stochastique et problèmes de martingales*. Lecture Notes in Mathematics 714. Springer, Berlin, 1979.

[168] J. Jacod and A.N. Shiryayev. *Limit Theorems for Stochastic Processes*. Grundlehren der Mathematischen Wissenschaften 288. Springer-Verlag, Berlin, 1987.

[169] P. Jaillet, D. Lamberton, and B. Lapeyre. Variational inequalities and the pricing of American options. *Acta Appl. Math.*, 21:263–289, 1990.

[170] F. Jamshidian. An exact bond option pricing formula. *J. Finance*, 44:205–209, 1989.

[171] F. Jamshidian. An analysis of American options. Working paper, Merrill Lynch Capital Markets, 1990.

[172] F. Jamshidian. Bond and option evaluation in the Gaussian interest rate model. *Res. Finance*, 9:131–170, 1991.

[173] F. Jamshidian. Forward induction and construction of yield curve diffusion models. *J. Fixed Income*, pages 62–74, June 1991.

[174] R. Jarrow. *Finance Theory*. Prentice-Hall, Englewood Cliffs, N.J., 1988.

[175] R.A. Jarrow, D. Lando, and S. Turnbull. A markov model for the term structure of credit risk spreads. Working paper, Cornell University, 1993.

[176] R.A. Jarrow and D.B. Madan. A characterization of complete markets on a Brownian filtration. *Math. Finance*, 1:31–43, 1991.

[177] R.A. Jarrow and G.S. Oldfield. Forward contracts and futures contracts. *J. Finan. Econ.*, 9:373–382, 1981.

[178] R.A. Jarrow and S.M. Turnbull. Delta, gamma and bucket hedging of interest rate derivatives. *Appl. Math. Finance*, 1:21–48, 1994.

[179] H. Johnson. An analytic approximation for the American put price. *J. Finan. Quant. Anal.*, 18:141–148, 1983.

[180] P. Jorion. *Value at Risk: The New Benchmark for Managing Financial Risk*. McGraw-Hill, NewYork, 2000.

[181] Yu.M. Kabanov and Ch. Stricker. A teachers' note on no-arbitrage criteria. *Lecture Notes in Mathematics 1755*, pages 149–152, 2001.

[182] I. Karatzas. On the pricing of American options. *Appl. Math. Optim.*, 17:37–60, 1988.

[183] I. Karatzas. Optimization problems in the theory of continuous trading. *SIAM J. Control Optim.*, 27:1221–1259, 1989.

[184] I. Karatzas. *Lectures in Mathematical Finance*. American Mathematical Society, Providence, 1997.

[185] I. Karatzas and S.-G. Kou. On the pricing of contingent claims under constraints. *Finance Stochastics*, 3:215–258, 1998.

[186] I. Karatzas, J.P. Lehoczky, S.P. Sethi, and S.E. Shreve. Explicit solution of a general consumption/investment problem. *Math. Oper. Res.*, 11:261–294, 1986.

[187] I. Karatzas, J.P. Lehoczky, and S.E. Shreve. Optimal portfolio and consumption decisions for a "small investor" on a finite horizon. *SIAM J. Control Optim.*, 25:1557–1586, 1987.

[188] I. Karatzas, J.P. Lehoczky, and S.E. Shreve. Existence and uniqueness of multi-agent equilibrium in a stochastic, dynamic consumption/investment model. *Math. Oper. Res.*, 15:80–128, 1990.

[189] I. Karatzas, J.P. Lehoczky, and S.E. Shreve. Equilibrium models with singular asset prices. *Math. Finance*, 1:11–29, 1991.

[190] I. Karatzas, J.P. Lehoczky, S.E. Shreve, and G.-L. Xu. Martingale and duality methods for utility maximization in an incomplete market. *SIAM J. Control Optim.*, 29:702–730, 1991.

[191] I. Karatzas and D.L. Ocone. A generalized Clark representation formula with application to optimal portfolios. *Stochastics Stochastics Rep.*, 34:187–220, 1992.

[192] I. Karatzas, D.L. Ocone, and J. Li. An extension of Clark's formula. *Stochastics Stochastics Rep.*, 32:127–131, 1991.

[193] I. Karatzas and S. Shreve. *Methods of Mathematical Finance*. Springer Verlag, New York, 1998.

[194] I. Karatzas and S.E. Shreve. *Brownian Motion and Stochastic Calculus*. Springer, Berlin, 1988.

[195] I. Karatzas and X.-X. Xue. A note on utility maximization under partial observations. *Math. Finance*, 1:57–70, 1991.

[196] D.P. Kennedy. The term structure of interest rates as a Gaussian random field. *Math. Finance*, 4:247–258, 1994.

[197] D.P. Kennedy. Characterizing and filtering Gaussian models of the term structure of interest rates. Preprint, University of Cambridge, 1995.

[198] I.J. Kim. The analytic valuation of American options. *Rev. Finan. Stud.*, 3:547–572, 1990.

[199] P.E. Kopp. *Martingales and Stochastic Integrals.* Cambridge University Press, Cambridge, 1984.

[200] P.E. Kopp and V. Wellmann. Convergence in incomplete financial market models. *Electronic J. Probab.*, 5(15):1–26, 2000.

[201] D.O. Kramkov. Optional decomposition of supermartingales and hedging contingent claims in incomplete models. *Probab. Theory and Rel. Fields*, 105:459–749, 1996.

[202] D.M. Kreps. Multiperiod securities and the efficient allocation of risk: A comment on the Black-Scholes model. In J. McCall, editor, *The Economics of Uncertainty and Information*. University of Chicago Press, Chicago, 1982.

[203] N.V. Krylov. *Controlled Diffusion Processes.* Applications of Mathematics 14. Springer Verlag, New York, 1980.

[204] H. Kunita. *Stochastic Partial Differential Equations Connected with Nonlinear Filtering.* Springer, New York, 1981.

[205] S. Kusuoka. On law invariant coherent risk measures. *Adv. Math. Econ.*, 3:83–95, 2001.

[206] P. Lakner. Martingale measure for a class of right-continuous processes. *Math. Finance*, 3:43–53, 1993.

[207] D. Lamberton. Convergence of the critical price in the approximation of American options. *Math. Finance*, 3:179–190, 1993.

[208] D. Lamberton and B. Lapeyre. Hedging index options with few assets. *Math. Finance*, 3:25–42, 1993.

[209] D. Lamberton and B. Lapeyre. *Introduction to Stochastic Calculus Applied to Finance.* Chapman & Hall, London, 1995.

[210] Levy M. Levy, H. and S. Solomon. *Microscopic Simulation of Financial Markets.* Academic Press, New York, 2003.

[211] F.A. Longstaff. The valuation of options on coupon bonds. *J. Bank. Finance*, 17:27–42, 1993.

[212] F.A. Longstaff and E.S. Schwartz. Interest rate volatility and the term structure: A two-factor general equilibrium model. *J. Finance*, 47:1259–1282, 1992.

[213] D.B. Madan and F. Milne. Option pricing with V.G. martingale components. *Math. Finance*, 1:39–55, 1991.

[214] D.B. Madan, F. Milne, and H. Shefrin. The multinomial option pricing model and its Brownian and Poisson limits. *Rev. Finan. Stud.*, 2:251–265, 1989.

[215] D.B. Madan and E. Senata. The variance gamma (V.G.) model for share market returns. *J. Business*, 63:511–524, 1990.

[216] M.J.P. Magill and G.M. Constantinides. Portfolio selection with transactions costs. *J. Econ. Theory*, 13:245–263, 1976.

[217] H.M. Markowitz. Portfolio selection. *J. Finance*, 7(1):77–91, 1952.

[218] H.P. McKean. Appendix: A free boundary problem for the heat equation arising from a problem in mathematical economics. *Ind. Manage. Rev.*, 6:32–39, 1965.

[219] R.C. Merton. Lifetime portfolio selection under uncertainty: The continuous-time model. *Rev. Econ. Statist.*, 51:247–257, 1969.

[220] R.C. Merton. Optimum consumption and portfolio rules in a continuous-time model. *J. Econ. Theory*, 3:373–413, 1971.

[221] R.C. Merton. An intertemporal capital asset pricing model. *Econometrica*, 41:867–888, 1973.

[222] R.C. Merton. Theory of rational option pricing. *Bell J. Econ. Manage. Sci.*, 4:141–183, 1973.

[223] R.C. Merton. On the pricing of corporate debt: The risk structure of interest rates. *J. Finance*, 29:449–470, 1974.

[224] R.C. Merton. Option pricing when underlying stock returns are discontinuous. *J. Finan. Econ.*, 3:125–144, 1976.

[225] R.C. Merton. On estimating the expected return on the market: An exploratory investigation. *J. Finan. Econ.*, 8:323–361, 1980.

[226] R.C. Merton. *Continuous-Time Finance*. Basil Blackwell, Cambridge, 1990.

[227] P.A. Meyer. *Un cours sur les intégrales stochastiques*. Séminaire de Probabilités X. Lecture Notes in Mathematics, 511. Springer-Verlag, Berlin, 1976.

[228] F. Modigliani and M.H. Miller. The cost of capital, corporation finance and the theory of investment. *Am. Econ. Rev.*, 48:261–297, 1958.

[229] M. Musiela. Stochastic PDEs and term structure models. Technical report, La Baule, June 1993.

[230] M. Musiela. Nominal annual rates and lognormal volatility structure. Preprint, The University of New South Wales, 1994.

[231] M. Musiela. General framework for pricing derivative securities. *Stochastic Process. Appl.*, 55:227–251, 1995.

[232] M. Musiela and M. Rutkowski. *Martingale Methods in Financial Modelling*. Applications of Mathematics, 36. Springer-Verlag, New York, 1997.

[233] M. Musiela and D. Sondermann. Different dynamical specifications of the term structure of interest rates and their implications. Preprint, University of Bonn, 1993.

[234] R. Myneni. The pricing of the American option. *Ann. Appl. Probab.*, 2:1–23, 1992.

[235] Y. Nakano. Efficient hedging with coherent risk measure. Preprint, Hokkaido University, 2001.

[236] J. Neveu. *Discrete-Parameter Martingales*. North-Holland, Amsterdam, 1975.

[237] D.L. Ocone and I. Karatzas. A generalized Clark representation formula with application to optimal portfolios. *Stochastics Stochastics Rep.*, 34:187–220, 1991.

[238] N.D. Pearson and T.-S. Sun. Exploiting the conditional density in estimating the term structure: An application to the Cox, Ingersoll and Ross model. *J. Finance*, 49:1279–1304, 1994.

[239] M. Picquet and M. Pontier. Optimal portfolio for a small investor in a market with discontinuous prices. *Appl. Math. Optim.*, 22:287–310, 1990.

[240] S.R. Pliska. A stochastic calculus model of continuous trading: Optimal portfolios. *Math. Oper. Res.*, 11:371–382, 1986.

[241] S.R. Pliska. *Introduction to Mathematical Finance: Discrete Time Models*. Blackwell, Oxford, 1997.

[242] S.R. Pliska and C.T. Shalen. The effects of regulations on trading activity and return volatility in futures markets. *J. Futures Markets*, 11:135–151, 1991.

[243] S. Port and C. Stone. *Brownian Motion and Classical Potential Theory*. Academic Press, New York, 1978.

[244] D. Revuz and M. Yor. *Continuous Martingales and Brownian Motion*. Springer, New York, 1991.

[245] R.T. Rockafellar. *Convex Analysis*. Princeton University Press, Princeton, N.J., 1970.

[246] R.T. Rockafellar and S. Uryasev. Conditional value-at-risk for general loss distributions. Research Report 2001-5, University of Florida, 2001.

[247] C. Rogers and Z. Shi. The value of an Asian option. *J. Appl. Probab.*, 32, 1995.

[248] L.C.G. Rogers. Equivalent martingale measures and no-arbitrage. *Stochastics Stochastics Rep.*, 51:41–49, 1994.

[249] L.C.G. Rogers and S.E. Satchell. Estimating variance from high, low and closing prices. *Ann. Appl. Probab.*, 1:504–512, 1991.

[250] S.A. Ross. The arbitrage theory of capital asset pricing. *J. Econ. Theory*, 13:341–360, 1976.

[251] M. Rubinstein. The valuation of uncertain income streams and the pricing of options. *Bell J. Econ.*, 7:407–425, 1976.

[252] M. Rubinstein. A simple formula for the expected rate of return of an option over a finite holding period. *J. Finance*, 39:1503–1509, 1984.

[253] M. Rubinstein. Exotic options. Working paper, 1991.

[254] M. Rubinstein and H.E. Leland. Replicating options with positions in stock and cash. *Finan. Analysts J.*, 37:63–72, 1981.

[255] M. Rubinstein and E. Reiner. Breaking down the barriers. *Risk*, 4(8):28–35, 1991.

[256] P.A. Samuelson. Rational theory of warrant prices. *Ind. Manage. Rev.*, 6:13–31, 1965.

[257] P.A. Samuelson. Lifetime portfolio selection by dynamic stochastic programming. *Rev. Econ. Statist.*, 51:239–246, 1969.

[258] P.A. Samuelson. Mathematics of speculative prices. *SIAM Rev.*, 15:1–42, 1973.

[259] K. Sandmann. The pricing of options with an uncertain interest rate: A discrete-time approach. *Math. Finance*, 3:201–216, 1993.

[260] K. Sandmann and D. Sondermann. A term structure model and the pricing of interest rate options. Discussion Paper B-129, University of Bonn, 1989.

[261] K. Sandmann and D. Sondermann. A term structure model and the pricing of interest rate derivatives. Discussion paper B-180, University of Bonn, 1991.

[262] W. Schachermayer. A Hilbert space proof of the fundamental theorem of asset pricing in finite discrete time. *Insurance Math. Econ.*, 11:249–257, 1992.

[263] W. Schachermayer. A counterexample to several problems in the theory of asset pricing. *Math. Finance*, 3:217–230, 1993.

[264] H.H. Schaefer. *Topological Vector Spaces*. Springer, Heidelberg, 1966.

[265] S.M. Schaefer and E.S. Schwartz. A two-factor model of the term structure: An approximate analytical solution. *J. Finan. Quant. Anal.*, 4:413–424, 1984.

[266] S.M. Schaefer and E.S. Schwartz. Time-dependent variance and the pricing of bond options. *J. Finance*, 42:1113–1128, 1987.

[267] M. Scholes. Taxes and the pricing of options. *J. Finance*, 31:319–332, 1976.

[268] M. Schweizer. Risk-minimality and orthogonality of martingales. *Stochastics Stochastics Rep.*, 30:123–131, 1990.

[269] M. Schweizer. Option hedging for semimartingales. *Stochastic Process. Appl.*, 37:339–363, 1991.

[270] M. Schweizer. Martingale densities for general asset prices. *J. Math. Econ.*, 21:363–378, 1992.

[271] M. Schweizer. Mean-variance hedging for general claims. *Ann. Appl. Probab.*, 2:171–179, 1992.

[272] M. Schweizer. Approximation pricing and the variance-optimal martingale measure. *Ann. Probab.*, 96:206–236, 1993.

[273] M. Schweizer. Approximating random variables by stochastic integrals. *Ann. Probab.*, 22:1536–1575, 1994.

[274] M. Schweizer. A projection result for semimartingales. *Stochastics Stochastics Rep.*, 50:175–183, 1994.

[275] M. Schweizer. Risk-minimizing hedging strategies under restricted information. *Math. Finance*, 4:327–342, 1994.

[276] M. Schweizer. On the minimal martingale measure and the Föllmer-Schweizer decomposition. *Stochastic Anal. Appl.*, 13:573–599, 1995.

[277] M. Schweizer. Variance-optimal hedging in discrete time. *Math. Oper. Res.*, 20:1–32, 1995.

[278] L. Shepp and A.N. Shiryayev. The Russian option: Reduced regret. *Ann. Appl. Probab.*, 3:631–640, 1993.

[279] H. Shirakawa. Interest rate option pricing with Poisson-Gaussian forward rate curve processes. *Math. Finance*, 1:77–94, 1991.

[280] A. Shiryayev. *Essentials of Stochastic Finance: Facts, Models, Theory*. World Scientific, Singapore, 1999.

[281] A.N. Shiryayev. *Probability.* Graduate Texts in Mathematics 95. Springer-Verlag, Berlin, 1984.

[282] A.N. Shiryayev. On some basic concepts and some basic stochastic models used in finance. *Theory Probab. Appl.*, 39:1–13, 1994.

[283] A.N. Shiryayev, Y.M. Kabanov, O.D. Kramkov, and A.V. Melnikov. Toward the theory of pricing of options of both European and American types, I. Discrete time. *Theory Probab. Appl.*, 39:14–60, 1994.

[284] A.N. Shiryayev, Y.M. Kabanov, O.D. Kramkov, and A.V. Melnikov. Toward the theory of pricing of options of both European and American types, II. Continuous time. *Theory Probab. Appl.*, 39:61–102, 1994.

[285] S.E. Shreve. A control theorist's view of asset pricing. In M.H.A. Davis and R.J. Elliott, editors, *Applied Stochastic Analysis*, Stochastic Monographs, 5, pages 415–445. Gordon and Breach, New York, 1991.

[286] S.E. Shreve, H.M. Soner, and G.-L. Xu. Optimal investment and consumption with two bonds and transaction costs. *Math. Finance*, 1:53–84, 1991.

[287] C. Stricker. Integral representation in the theory of continuous trading. *Stochastics*, 13:249–257, 1984.

[288] C. Stricker. Arbitrage et lois de martingale. *Ann. Inst. H. Poincaré Probab. Statist.*, 26:451–460, 1990.

[289] M. Taksar, M.J. Klass, and D. Assaf. A diffusion model for optimal portfolio selection in the presence of brokerage fees. *Math. Oper. Res.*, 13:277–294, 1988.

[290] M.S. Taqqu and W. Willinger. The analysis of finite security markets using martingales. *Adv. Appl. Probab.*, 19:1–25, 1987.

[291] S.J. Taylor. Modeling stochastic volatility: A review and comparative study. *Math. Finance*, 4:183–204, 1994.

[292] S.M. Turnbull and F. Milne. A simple approach to the pricing of interest rate options. *Rev. Finan. Stud.*, 4:87–120, 1991.

[293] S. Uryasev. Conditional value-at-risk: Optimisation, algorithms and applications. *Finan. Eng. News*, 2(3):21–41, 2000.

[294] Van der Hoek, J. and E. Platen. Pricing contingent claims in the presence of transaction costs. Working paper, 1995.

[295] P. Van Moerbeke. On optimal stopping and free boundary problem. *Arch. Rational Mech. Anal.*, 60:101–148, 1976.

[296] O. Vasicek. An equilibrium characterisation of the term structure. *J. Finan. Econ.*, 5:177–188, 1977.

[297] R. Whaley. Valuation of American call options on dividend-paying stocks: Empirical tests. *J. Finan. Econ.*, 10:29–58, 1982.

[298] R. Whaley. Valuation of American futures options: Theory and empirical tests. *J. Finance*, 41:127–150, 1986.

[299] D. Williams. *Probability with Martingales*. Cambridge University Press, Cambridge, 1991.

[300] W. Willinger and M.S. Taqqu. Pathwise stochastic integration and applications to the theory of continuous trading. *Stochastic Process. Appl.*, 32:253–280, 1989.

[301] W. Willinger and M.S. Taqqu. Toward a convergence theory for continuous stochastic securities market models. *Math. Finance*, 1:55–99, 1991.

[302] P. Wilmot, J. Dewynne, and S. Howison. *Option Pricing: Mathematical Models and Computation*. Oxford University Press, Oxford, 1994.

[303] H. Witting. *Mathematische Statistik I*. B.G. Teubner, Stuttgart, 1985.

[304] J.A. Yan. Characterisation d'une classe d'ensembles convexes de l^1 ou h^1. *Lecture Notes in Mathematics*, 784:220–222, 1980.

[305] P.G. Zhang. *Exotic Options: A Guide to Second Generation Options*. World Scientific, Singapore, 1997.

Index

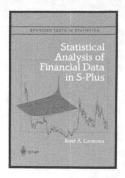
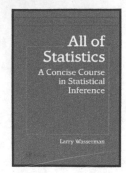